Water Resources Management

Water Resources Management

Edited by **Sarah Luck**

SYRAWOOD
PUBLISHING HOUSE
New York

Published by Syrawood Publishing House,
750 Third Avenue, 9th Floor,
New York, NY 10017, USA
www.syrawoodpublishinghouse.com

Water Resources Management
Edited by Sarah Luck

International Standard Book Number: 978-1-68286-223-0 (Hardback)

Printed in China.

Contents

Preface

The world is advancing at a fast pace like never before. Therefore, the need is to keep up with the latest developments. This book was an idea that came to fruition when the specialists in the area realized the need to coordinate together and document essential themes in the subject. That's when I was requested to be the editor. Editing this book has been an honour as it brings together diverse authors researching on different streams of the field. The book collates essential materials contributed by veterans in the area which can be utilized by students and researchers alike.

Water resource management deals with developing, planning and using water resources efficiently. The topics covered in this comprehensive book deal with some of the vital topics related to water resources management such as controlling water pollution, water conservation, and water cycle management, etc. The aim of this book is to present researches that have transformed this discipline and aided its advancement. Students, researchers, experts and all associated with water resources management will benefit alike from this book.

Each chapter is a sole-standing publication that reflects each author's interpretation. Thus, the book displays a multi-facetted picture of our current understanding of applications and diverse aspects of the field. I would like to thank the contributors of this book and my family for their endless support.

Editor

Performance assessment of some developed surface irrigation methods

N. B. Abdelmageed

Faculty of Engineering, Shobra, Benha University, Egypt.

Irrigation development is a gateway to increased agricultural, water and land productivity, increased household and national food security. However, irrigation development has been a major challenge in many developing countries, including Egypt. The overall objective of this study is to detect the influence of different irrigation systems on water-use efficiency, crop and soil salinity in highly soil salinity. Two techniques were applied in experiment, the first technique was siphon irrigation, and the second one was gated pipe. The monitored parameters were water table depth, water and soil salinity and crop yield. The study revealed that the intensive management is a very important aspect for the success of siphon technique and cotton crop yield was higher by 17% compared to gated pipe irrigation treatment. The total soil salinity increased in both treatments. The siphon method increases the average salinity by 2.7% while the gated pipe increased it by 12.9%; converging the value of the crop coefficient in all relations used in most stages of growth, except Penman relationship which gave the highest values.

Key words: Siphon irrigation method, gated pipe irrigation method, subsurface drainage, water-table management, water-table salinity, soil salinity, and crop yield.

INTRODUCTION

As water is becoming more and more a scarce resource all over the world, proper management of the available water is essential. For an optimal use of the available water resources, water management strategies have to be developed. Soil salinity problems generally occur in arid and semiarid regions and reduce crop production at different levels. Salinity is also a major limiting factor for crop yield in poorly drained soils (Mikati, 1997; Rogers, 2002; Patel et al., 2002). Nhundu et al. (2010) recommend that national governments should formulate and hold sound irrigation development strategies and encouraged to partner with public and private institutions in defining and implementing such comprehensive strategies for sustainable irrigation development. Landwirtschaftlichen et al. (2011) recommend a proper field preparation, including for instance a laser-guided land leveling is necessary before bed making to facilitate a uniform distribution of irrigation water; a suitable bed height, that is, 10 to 15 cm height, is needed for efficient salt leaching; adequate soil moisture content needs to be ensured during planting to obtain a proper plant stand; the use of appropriate herbicides for weed control is advantageous; the use of appropriate machinery to drill seed and fertilizer at the proper depth is compulsory; a reshaping of beds during planting, if necessary; the use of short-maturing crop varieties is advantageous. Abdel Ghaffer et al. (2006) studied the sub-irrigation method to manage the water table and the effect of method on wheat crop.

The beneficial and adverse effects of trace element levels in crop foods that make up major proportion of dietary intake on human health are particularly important (Grant et al., 2008). Agronomic and genetic practices can be important tools both to increase the concentration of desirable trace elements such as Zn and reduce that of potentially harmful trace elements such as Cd. While soil Cd concentrations are relatively low in most wheat growing areas in Iran, some agricultural lands are contaminated with Cd due to excessive application of low-quality phosphate-fertilizers containing Cd as impurities (Afyuni et al., 2007). In some cases the concentrations of Cd in wheat grains have increased to levels above the maximum permissible limit of 0.2 ppm (Codex, 2008). Bouksila et al. (2010) estimate the soil salinity over a shallow saline water table in semiarid Tunisia.

The objective of this research is studying the evolution of soil salinity after and during the use of gated pipe method and siphon method in irrigation, and study the effect of both methods on water-use efficiency, soil salinity, and their effect on the cotton.

METHODOLOGY

Experimental site

Field experiments were conducted in the irrigated area of western Delta, Egypt. The irrigated area covers 2.8 ha and the main crops are cotton. The climate is Mediterranean semiarid with little rainfall. Only 100 to 200 mm of rain falls on the delta area during an average year, and most of this falls in the winter months. The experimental area is divided into lines where each line 240 m in length and 0.75 m in width and has a sandy silt loam to clay loam texture. The field hydraulic conductivity was measured using the auger hole method and the average value is 2.0 m/day. The main source of the irrigation water is supplied from field canal, both gated pipe and siphon irrigation is used in the system. The site is served by a subsurface drainage system. The collector drains (PVC corrugated plastic pipe) have been installed at about 1.5 m depth and all laterals drains (PVC corrugated plastic pipe covered by synthetic envelope materials) have been installed at a depth of 1.2 m with an average space of 80 m. The lateral drains were sloped at 10% and exit directly to the main collector through a manhole. Figure 1 shows the experimental study. In total 77 sampling plots, spaced at about 50 by 50 m were investigated. In each plot, soil samples were collected at 0, 15, 30, 45, 60, 75, 90, 105, 120, 135, and 150 cm depth. The soil samples were analyzed to determine soil particle size and ECe. Soil particle size was measured in the laboratory using the sedimentation method (pipette and hydrometer). Five fractions were measured, clay (d<2 μm), fine silt (2<d<20 μm), coarse silt (20<d<50 μm), fine sand (50<d<200 μm), and coarse sand (200 μm<d<2 mm). Beside soil samples, the depth to groundwater table from the soil surface (Dgw) and electrical conductivity of the groundwater (ECgw) were measured at each of the 77 plots. Table 1 shows the average soil salinity along the soil depth. The table shows that the pattern of soil salinity started with low value of 1.7 dS/m at the upper layer and increases with depth to a value of 3.9 dS/m.

The gated pipe irrigation method

The pipes are 6 m in length, 150 mm in diameter and with distance holes 0.75 m which can communicate with each. The pipe holes can be changed to give the flow required by using the Equation 1. Pipe connected with the basin to secure the appropriate pressurized by counter discharge (Kemper et al., 1981).

$$Q = 0.0066d^2 * h^{0.5} \qquad (1)$$

Where Q is the discharge "m^3/s", d is opening diameter "m", h is the water head above the opening center "m".

The siphons system

The siphons with 0.037 m in diameter and 1.5 to2.0 m are used to irrigate area. The formula used to compute the discharge from siphons is based on Wigginton (2004) formula as shown in Equation 2.

$$Q = \frac{\pi D^2}{4}\left[\frac{2g\Delta g}{1.9 + f\frac{L}{D}}\right]^{0.5} \qquad (2)$$

Where Q is the discharge (m^3/s), g is acceleration due to gravity (9.8 m/s^2), H is the head (m). D is actual internal diameter (m), f is the friction loss coefficient of Darcy Weisbach (0.019 for example, for "small diameter pipes"), L is the length (m). The water velocity and the water slope were measured at the middle line at each 20 m and Parshall flume with 5 cm contraction at 3 to 5 m from the start line. The class A basin with 121.5 cm in diameter and 25 cm in height rested on wooden block used to measure the evaporation. Figure 1 shows the layout of the experimental site. The water drained by subsurface drainage 10 cm in diameter with spacing 80 m and depth 150 cm.

Measurements

Measurements included water table depth, irrigation and water table salinity, rainfall, temperature, evaporation, and soil salinity.

Irrigation water salinity

Saline soils are of increasing importance both in Egypt and worldwide. Richards (1954) define SAR as:

$$SAR = \frac{Na^+}{\sqrt{\frac{Ca^{++} + Mg^{++}}{2}}} \qquad (3)$$

Where SAR is sodium adsorption ration "%", Na^+ is the sodium "meg/L", Ca^{++} is the calcium ratio "meg/L", Mg^{++} is the Magnesium ratio "meg/L". It was measured before each irrigation gift by a handheld electrical conductivity meter in (dS/m). Irrigation water salinity varied from 0.83 to 2.74 with average of 1.78 dS/m and sodium adsorption ratio is 2.96% and salts total dissolved salts in water varied from 2331 to 2754 with rate of 2542 ppm.

Weather temperature

Weather has been observed daily during the study period and Figure 2 shows the temperature in the study area during the experiment.

Crop yield

The study area were planted in lines 75 cm in distance between

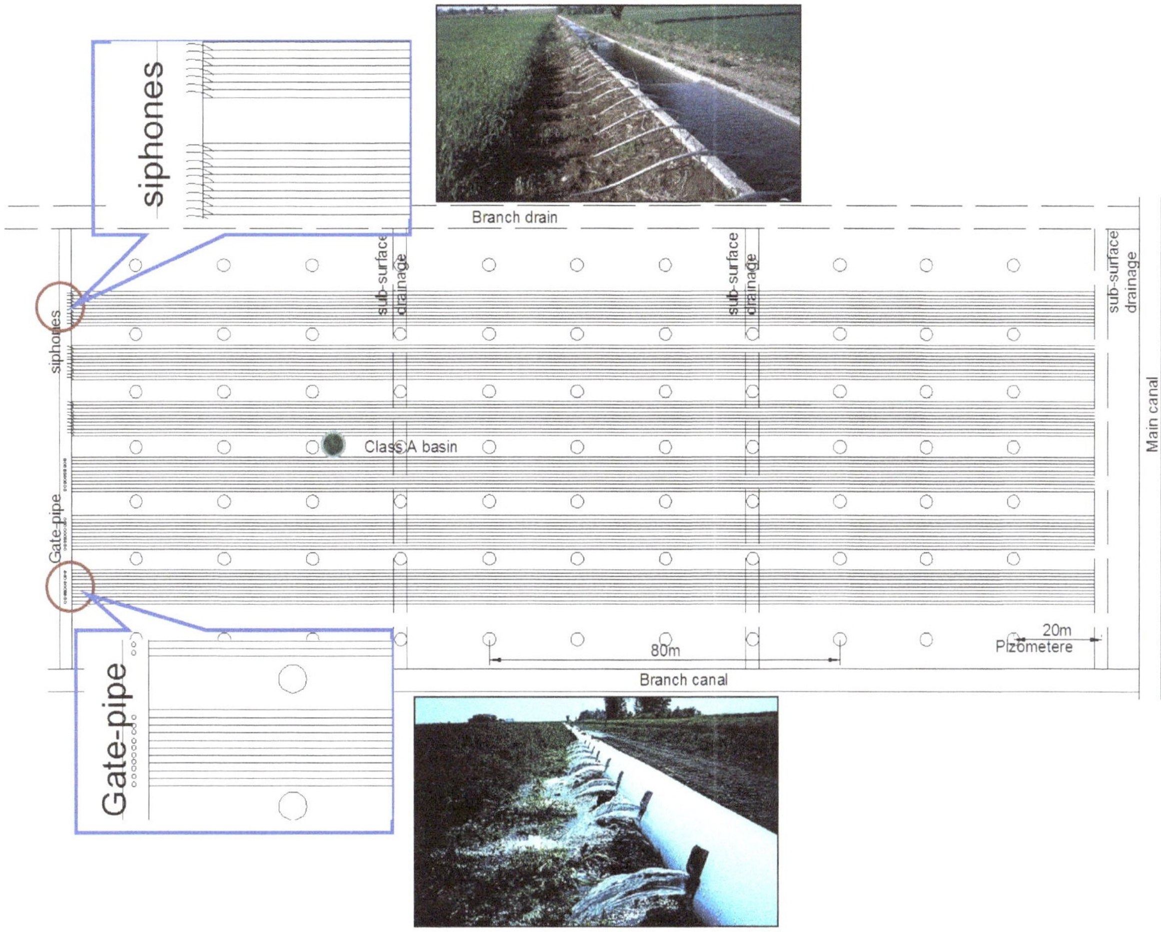

Figure 1. Experimental layout.

Table 1. Soil salinity.

Depth (cm)	E.C (dS/m)	$CaCO_3$ (meg/L)
0 - 35	1.7	24.25
35 - 47	2.95	28
47 - 105	3.3	22.5
105 - 125	3.9	22.75

them and 15 to 17 cm between the plants. Observation of cotton growth was followed and four crop samples were taken from each treatment at harvest time to determine the average cotton crop yield.

RESULTS AND DISCUSSION

Soil salinity

Samples were also obtained throughout the irrigation season to monitor changes in soil salinity. Figures 3 and 4 show the soil salinity profile at the before, mid and after the growing season for one sampling location. These data indicate that the soil salinity increased in both treatments at mid season then decreased after season in root zone depth due to the growth of cotton and increases in water consumption. At mid season cotton need more irrigation water for growth so the water table decreased and the soil salinity increased, after the season the crop did not need the irrigation water so the water table increased and the soil salinity decreased. In the gated pipe method, the electrical conductivity (E.C) was increased by average 13.5% at mid-season then decreased by 0.5% after the season. In siphon method, the average E.C was increased by about 6.9% at mid-season then decreased by 13.7% after the season as shown in Figure 4. This method increases the salinity at 0 to 15 cm depth at mid season then decreased it after season, and it decrease the soil salinity at 15 to 105 cm depth mid season and after season. The siphon method increases the average of total salinity by 2.7% while the gated pipe increased the salinity by 12.9%. This result is confirmed with the logical, where the irrigation water passes through the siphon system at depth lower than the ground level, so the seepage zone is lower than the ground level therefore the salinity is decreased in this zone but in gated pipe the water irrigation passes at the

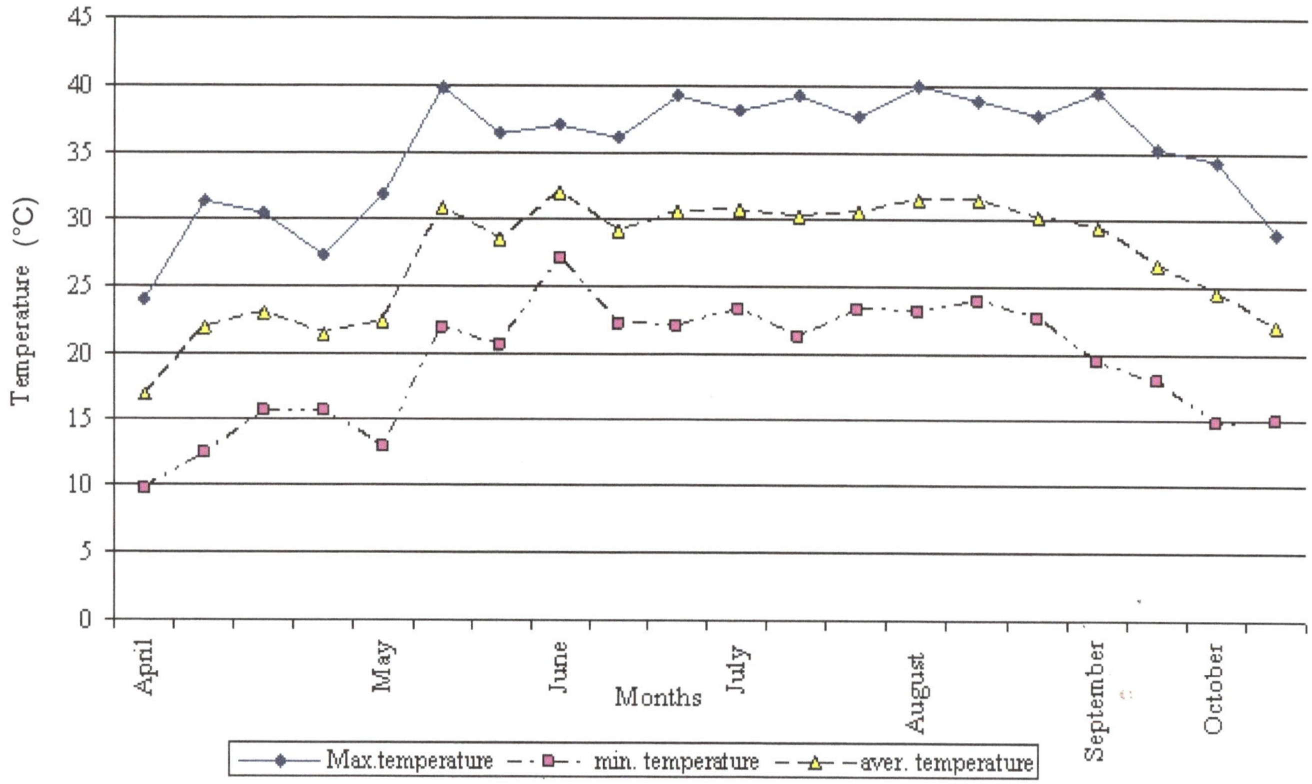

Figure 2. The weather temperature.

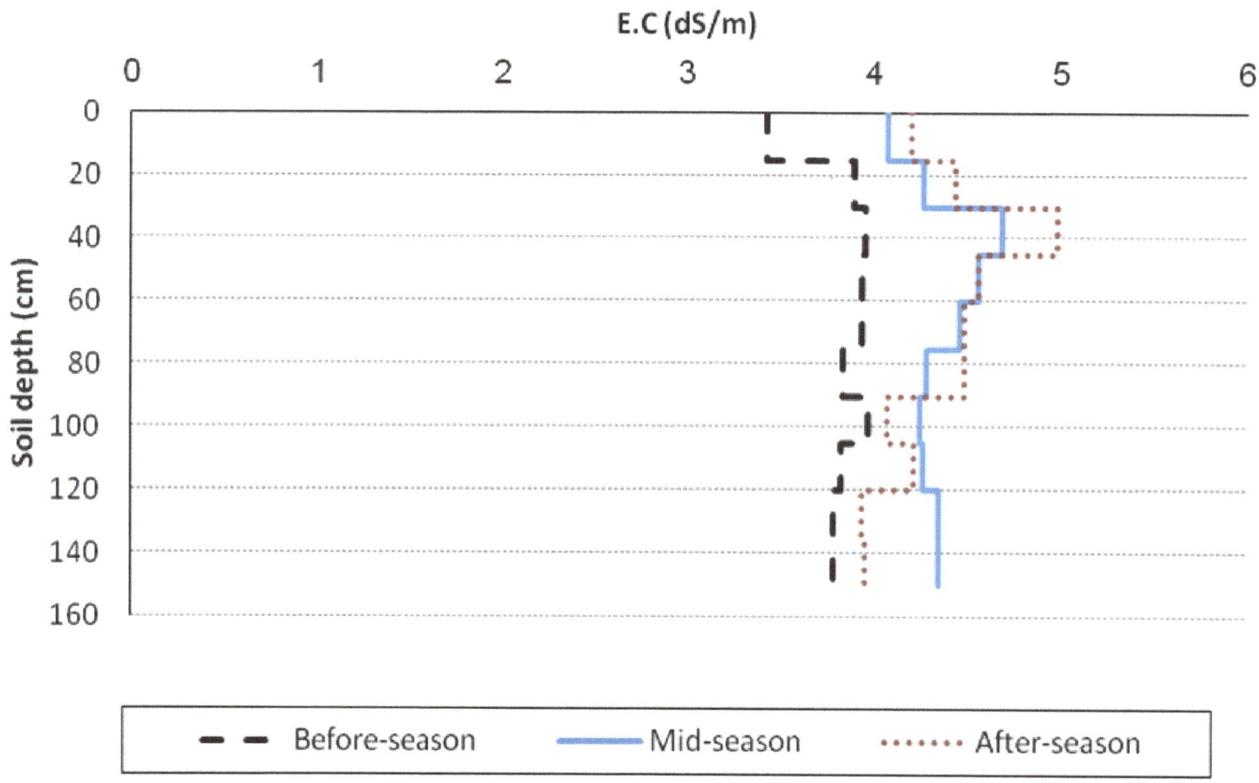

Figure 3. The electrical conductivity for gated pipe.

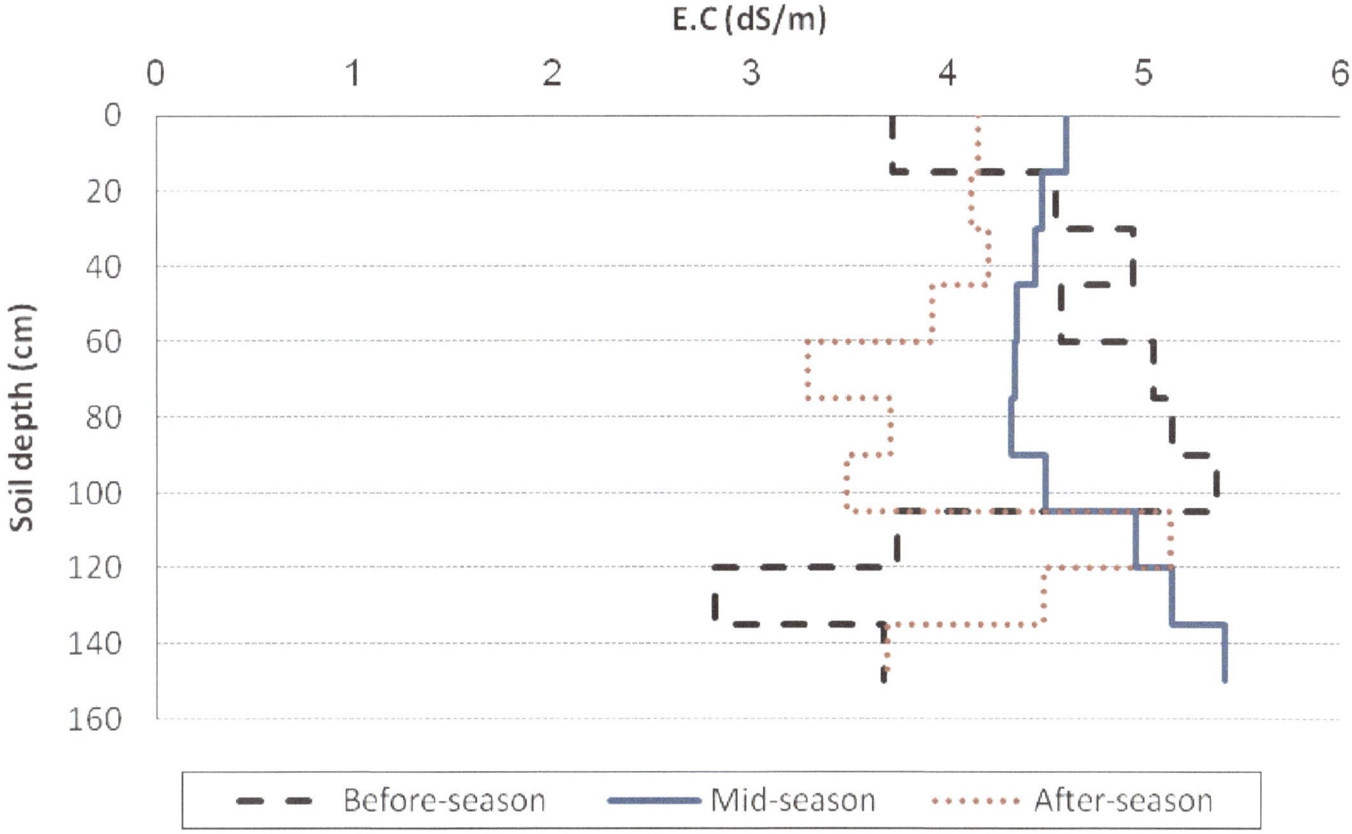

Figure 4. The electrical conductivity for siphon.

same level with the ground level. The sodium adsorption ratio (SAR) is a measure of the number of sodium ions attached to soil particles, relative to the number of calcium and magnesium ions. A large number of sodium ions (high SAR) will degrade soil particles and reduce the number of large pore spaces in the soil. Such a soil is not very permeable, and it is difficult to leach accumulated salts from such a soil by applying excess water. Sodium adsorption ratio (SAR) decreased for both methods. The siphon field average SARs in saturated soil extracts ranged from 4 to 15.83%, with an overall mean of 6.05% and a standard deviation of 1.1% before season and decrease with an overall mean 4.5% by decrease about 24% but in the gated pipe field average SARs in saturated soil extracts ranged from 4 to 12.83%, with an overall mean of 4.75% a standard deviation of 0.29% before season and decrease with an overall mean 4.05% by decrease about 11.4% as shown in Figure 5. The siphon method gives lower ECe and decrease SAR which increase the soil permeability.

The average toxic salts (sodium sulfate, sodium chloride and magnesium chloride) changed from 1.8058 mS/m before planting to 1.8376 mid agriculture then 1.9695 mS/m the end of the season for the treatment of gated pipe. And it decreased from 2.2323 mS/m before planting to 1.9113 mS/m at mid agriculture then increased to 1.9622 mS/m after season for siphon treatment. The gated pipe irrigation method increase the toxic salt by 1.8% at mid-season then it increase by 7.2% after season, but the siphon method decrease it by 14.4% at mid-season then increase it by 2.7% after season. In final the gated pipe increase the toxic salt by 9.1% where the siphon method decreases it by 12.1%. The average non-toxic salts (calcium bicarbonate and calcium Sulfate) changed from 1.0638 before planting to 1.4366 mS/m mid season then 0.9142 mS/m at the end of the season for the treatment of gated pipe. And it increased from 1.1675 mS/m before planting to 1.3758 mS/m at mid agriculture then increased to 0.8838 mS/m after season for siphon. The gated pipe irrigation method increases the non-toxic salt by 35.5% at mid-season then it decreases by 26.2% after season, but the siphon method increases it by 17.8% at mid-season then decreases it by 35.8% after season. In final the gated pipe decreases the non-toxic salt by 14.1% where the siphon method decreases it by 24.41%. The probability of producing alkaline soil in all transactions out of the question because the value of $\frac{K+N_a}{Ca+Mg}$ is less than 1 as shown in Table 2.

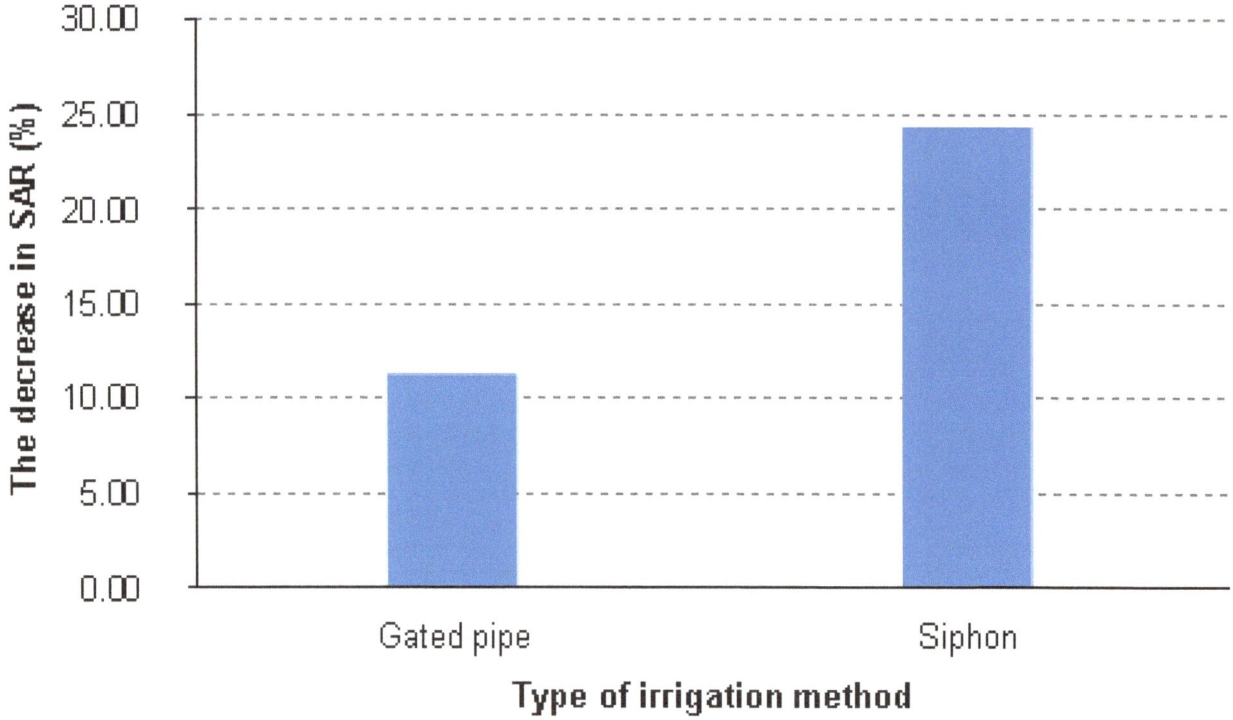

Figure 5. The sodium ratio for both treatments.

Table 2. Alkaline probability.

Season	Siphon	Gated pipe
Before season	0.761	0.61
Mid- season	0.432	0.48
After season	0.603	0.453

Water table salinity

The water table salinity for both treatments was represented in Figure 6. It ranges from 2.45 to 5.37 dS/m with an overall mean of 4 dS/m and a standard deviation of 1.17 dS/m for gated pipe treatment while the water table salinity for siphon treatment ranges from 2.3 to 5.25 dS/m with an average value of 3.9 dS/m and a standard deviation of 1.12 dS/m. It is obvious from these results that there is no difference between the water table salinity for both treatments except in flowers root phase in May due to the gated pipe method in this phase need more water as shown in Figure 8.

Water table level

The average water table levels for both treatments (Siphon and pipeline) along the cotton season are represented in Figure 7. The siphon field average water table level is ranged from 69 to 106 cm, with an overall mean of 92.5 cm and a standard deviation of 8.85 after season and in the gated pipe field it ranged from 39 to 107 cm, with an overall mean of 85.6 a standard deviation of 15.3. The gated pipe method decreases the water table level less than the siphon method by 7% so the soil salinity in gated pipe method increased than the siphon method. The soil salinity negatively correlated with the water table level.

Water consumption

For the gated pipe the water consumption during the flowering phase and form roots gives the largest amount which gives 4200 m^3/ha (42.4% from the total amount). The germination stage gives 2327 m^3/ha (23.5% from the total amount), While the water consumption during the floral buds gave the minimum amount. But for the siphon the water consumption during the flowering phase and form roots gives the largest amount which gives 3534 m^3/ha (36.61% from the total amount). The germination stage gives 2340 m^3/ha (24.2% from the total amount); while the water consumption during the floral buds gave the minimum amount equal 893 m^3/ha (9.2%) as shown in Figure 8. The total water consumption for the treatment of gated pipe is 9903 which exceed than siphon by 2.5%.

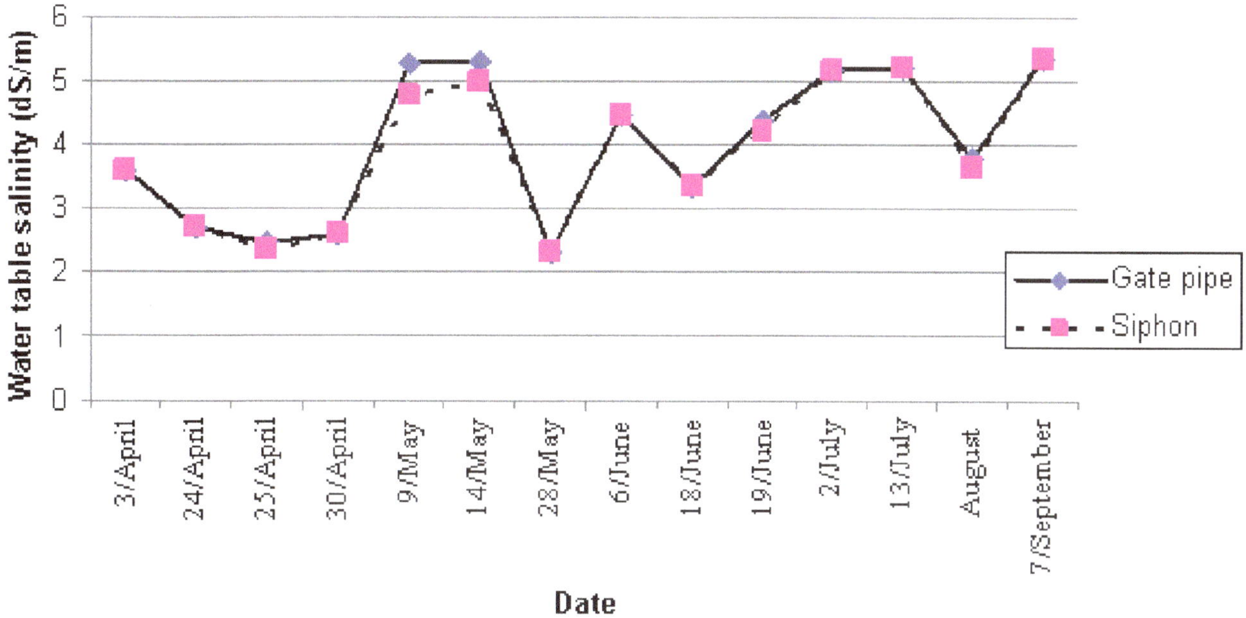

Figure 6. Water table salinity for both treatments.

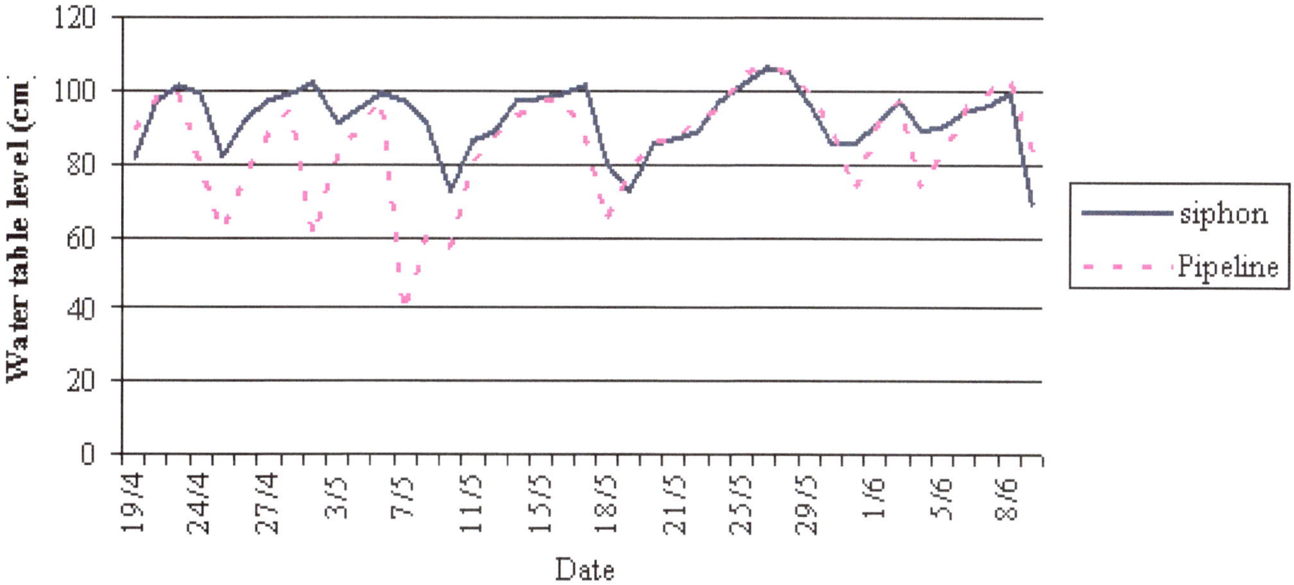

Figure 7. Water table level.

Water irrigation efficiency

The value of the additional water efficiency was measured by using Equation (4) which gives that working in siphons is higher than in the pipeline as shown in Table 3.

$$E_a = \frac{D_s}{D_A} \quad (4)$$

Where E_a is the additional water efficiency "%", D_s is the stored water depth in root zone "cm", D_A is the additional water depth "cm". The siphon method gives higher efficiency in all phases except in maturity of the plant phase; this result is due to the fact that the siphon method need more water in this phase. The irrigation efficiency is positively correlated with the soil salinity, and the permeability. The water distribution efficiency in

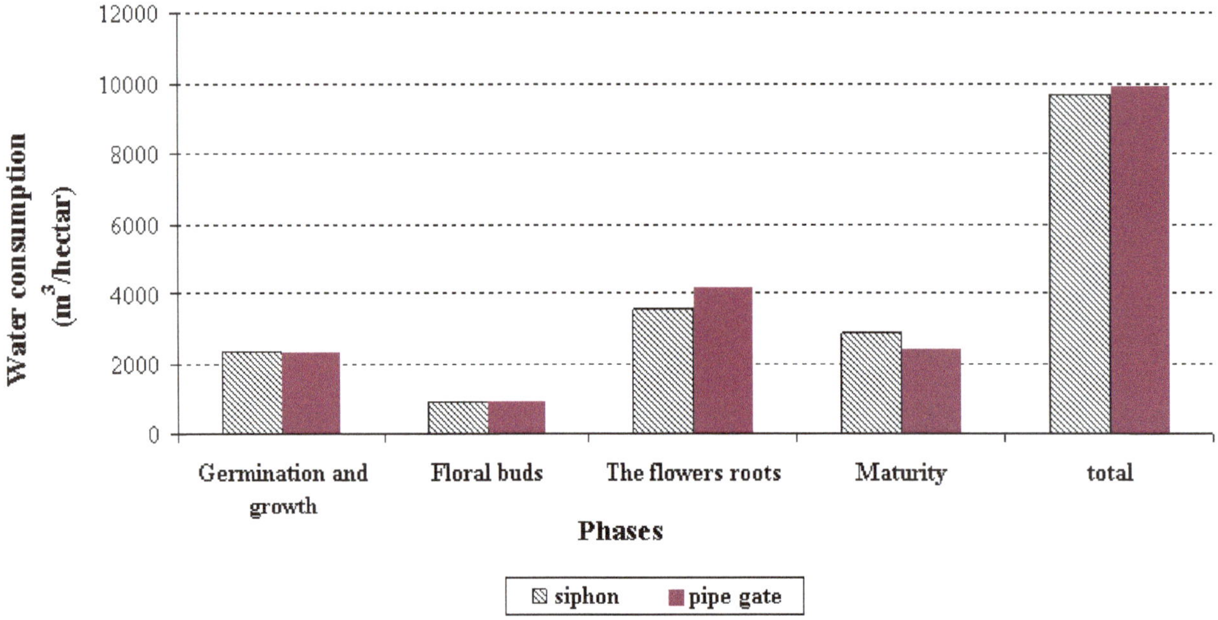

Figure 8. The water consumption for each treatment.

Table 3. The added Irrigation addition efficiency.

Phases	Pipeline	Siphon
Germination	56	66
Floral buds	59	67
Flowers	51	66
Maturity of the plant	55	52
Average	**55.25**	**62.75**

siphon is 80% where in gated pipe is 72%. The storage coefficient for pipeline is 100 and 99% for the siphon.

Time progress

Figure 9 shows the time progress inside the filed. The siphon treatment gives less time than the gated pipe by 14.3%. The applied time for the treatment of gated pipe is 223 min but in siphons is 191 min.

Crop production

The crop water need (ETP) is defined as the depth (or amount) of water needed to meet the water loss through vapor-transpiration. In other words, it is the amount of water needed by the various crops to grow optimally. Potential crop evapotranspiration under favorable growth conditions is estimated from ETP and crop coefficients (KC) and then compared with Penman, Blaney- Criddle, and Evanov formula. A standard evaporation pan exposure as used herein is defined as a Class A (Hargreaves, 1975). Class A pan evaporation, EV, has been widely used as an index for obtaining ETP. Figure 10 shows the comparison of the crop coefficient in studied area for both methods and the previous formula.

The Penman (1984) formula

$$E_{mass} = \frac{mR_n + \rho_a C_p \delta_e g_a}{\lambda_v (m+\gamma)} \quad (5)$$

Where m is slope of the saturation vapor pressure curve (Pa K^{-1}), Rn is net irradiance (W m^{-2}), ρa is density of air (kg m^{-3}), cp is heat capacity of air (J kg^{-1} K^{-1}), ga is momentum surface aerodynamic conductance (m s^{-1}), δe is vapor pressure deficit (Pa), λ_v is latent heat of vaporization (J kg^{-1}), γ is psychometric constant (Pa K^{-1}).

Blaney-Criddle (1960) formula

$$ET_o = \rho(0.46T_{mean} + 8) \quad (6)$$

Where: ETo is the reference evapotranspiration (mm day^{-1}) (monthly), T_{mean} is the mean daily temperature (°C) given as T_{mean} = (T_{max} + T_{min})/ 2, ρ is the mean daily percentage of annual daytime hours.

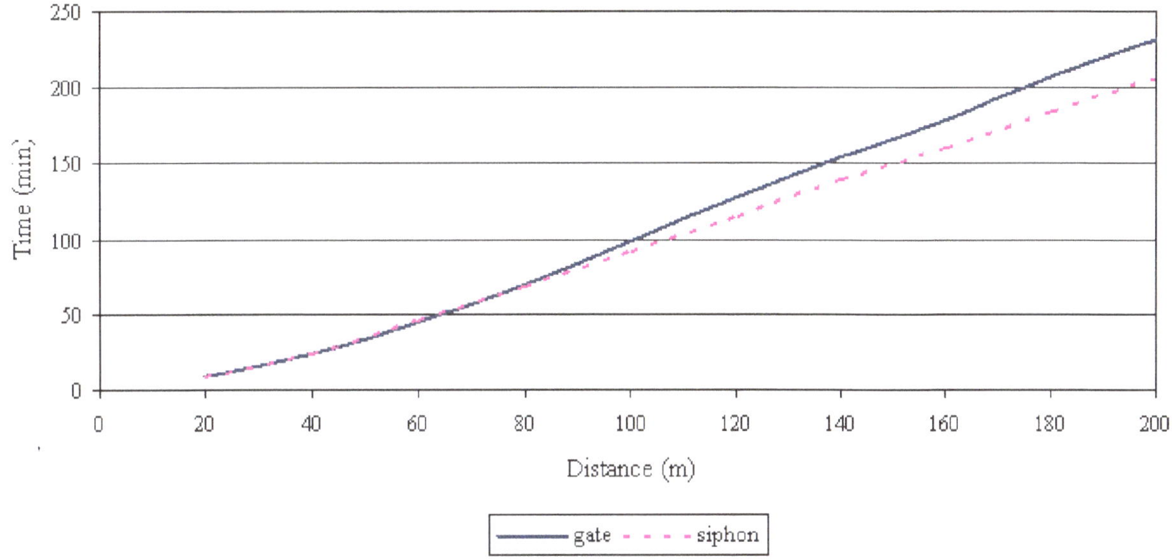

Figure 9. Time progress in both treatments.

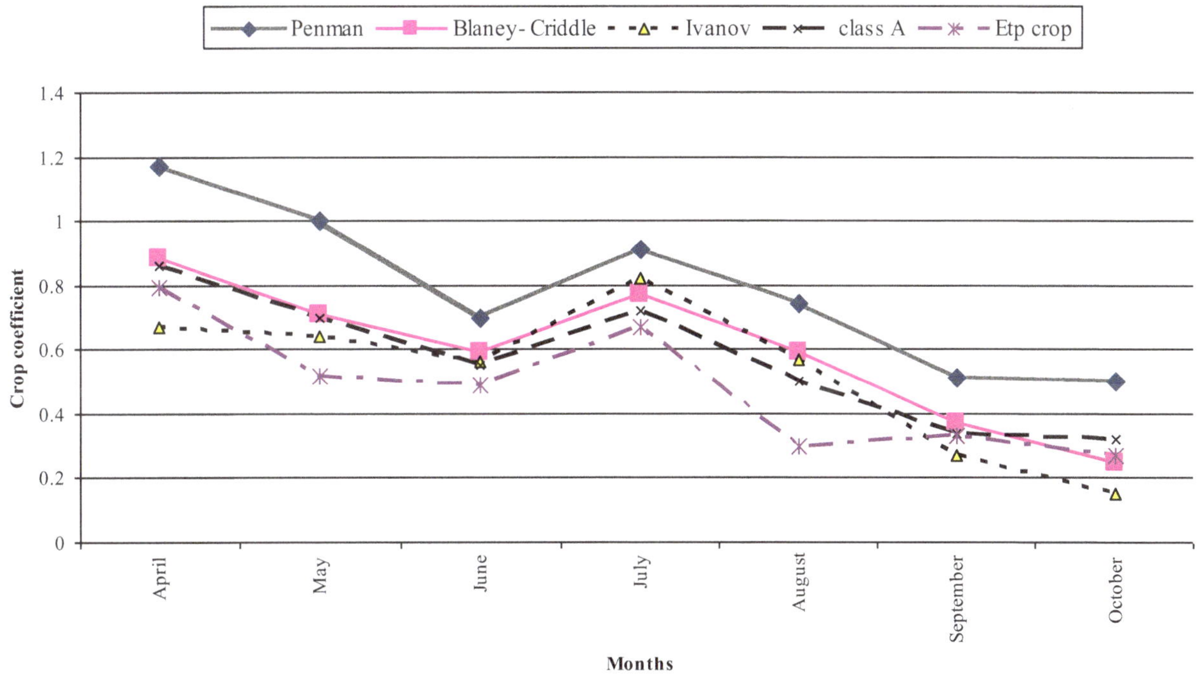

Figure 10. The comparative of crop coefficient with the previous relation.

Ivanov formula

Approach by Turc for: T > 5°C

$$ET_o = 0.0031(G + 209.4)\frac{T}{T + 15}ETPF \quad (7)$$

(Turc, 1961; Wendling and Schellin, 1986).

Where: G is global radiation in Jcm2d1, ETPF is empirical factor, T is daily mean temperature in °C.

Figure 10 shows the convergence value of the crop coefficient in all relations used in most stages of growth, except Penman relationship which gave the highest values. Figure 11 shows the average cotton yield for both siphon irrigation and gated pipe treatments. In case of siphon method, the cotton yield was 2.581 ton/ha; this

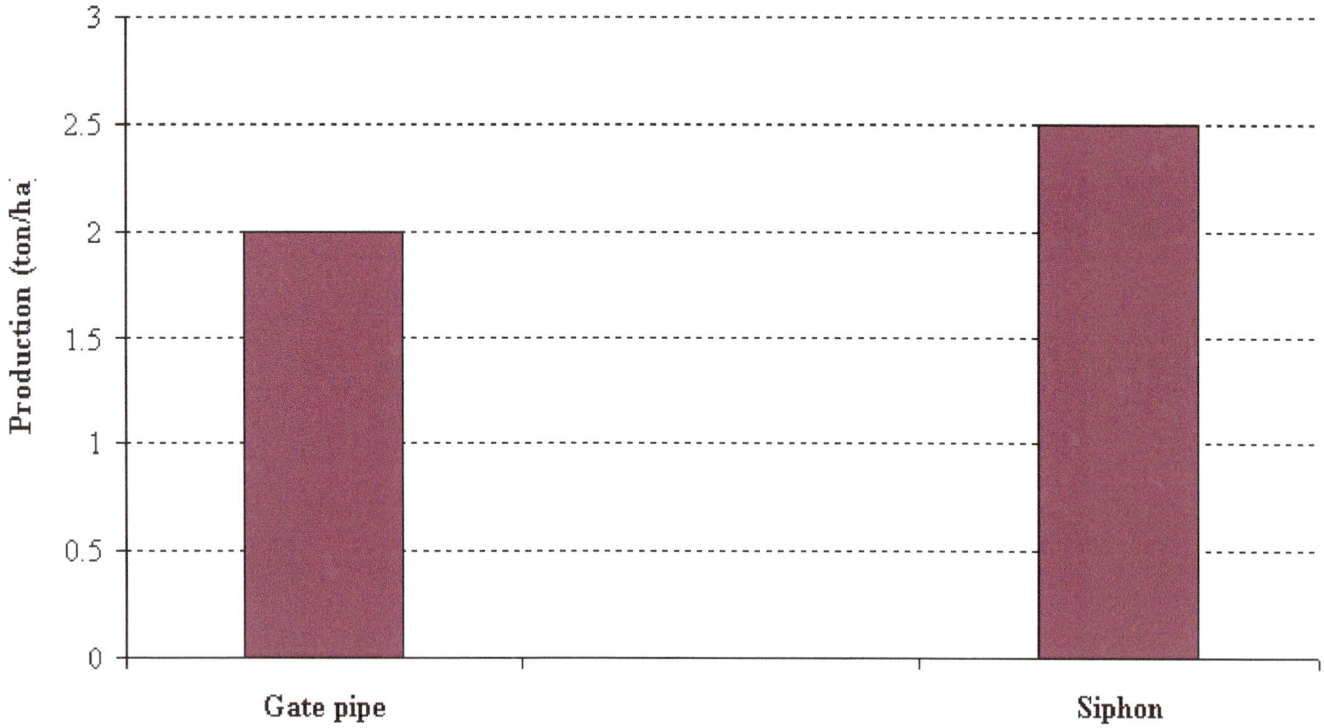

Figure 11. Crop production.

exceeds the yield of gated pipe irrigation method by 17%.

Figure 12 shows the average cotton stem length for both siphon irrigation and gated pipe treatments. In case of siphon treatment, the average cotton stem length was 74 cm; this exceeds the length of gated pipe treatment by 14%. This result due to the siphon method decease the soil salinity especially in root zone so the permeability increased and the length of the crop produced increase.

Conclusion

This paper studied the effects of gated pipe, and siphon irrigation methods on cotton yield, salinity, water table, and made sense to the water-saving irrigation technique and agricultural production. The results indicate that the soil salinity increased during the irrigation season then decreased after season. In the gated pipe method, the electrical conductivity was increased by average 13.5% at mid-season then decreased by 0.5% after the season but in siphon method, the soil salinity was increased by about 6.9% at mid-season then decreased by 13.7% after the season. The siphon method increased the salinity at 0 to 15 cm depth and decreased it at 15 to 105 cm depth. The average soil salinity was increased in siphon system by 2.7% where in gated pipe it increased by 12.9%. The sodium adsorption ratio (SAR) decreased for both methods. The siphon method decreased SAR by about 24% where the gated pipe decreased it by about 11.4%. The siphon method gives lower ECe and decrease SAR which increase the soil permeability. The gated pipe irrigation method increased the toxic salt by 9.1% where the siphon method decreased it by 12.1%. The gated pipe decreased the non-toxic salt by 14.1% where the siphon method decreased it by 24.41%. The gated pipe method decreased the water table level less than the siphon method by 7%. For the gated pipe the water consumption during the flowering phase and form roots give the largest amount which gives about 42.4% from the total amount. For the siphon the water consumption during the flowering phase and form roots give the largest amount which gives about 36.61% from the total amount. The siphon treatment need water consumption less than gate pipeline by 2.5%. The water distribution efficiency is higher in siphon 80% than in gate pipeline 72%. The storage coefficient for pipeline is 100 and 99% for the siphon. The value of the crop coefficient converges in all relations used in most stages of growth, except Penman relationship which gave the highest values. The cotton production in case of siphon method was greater than gate pipeline irrigation method by 17% and equal to 2.581 ton/ha. In case of siphon treatment, the average cotton stem length was 74 cm; this exceeds the length of pipeline treatment by 14%. Cotton production, its length, the water consumption, and water distribution efficiency correlated negatively with soil salinity, while ground water

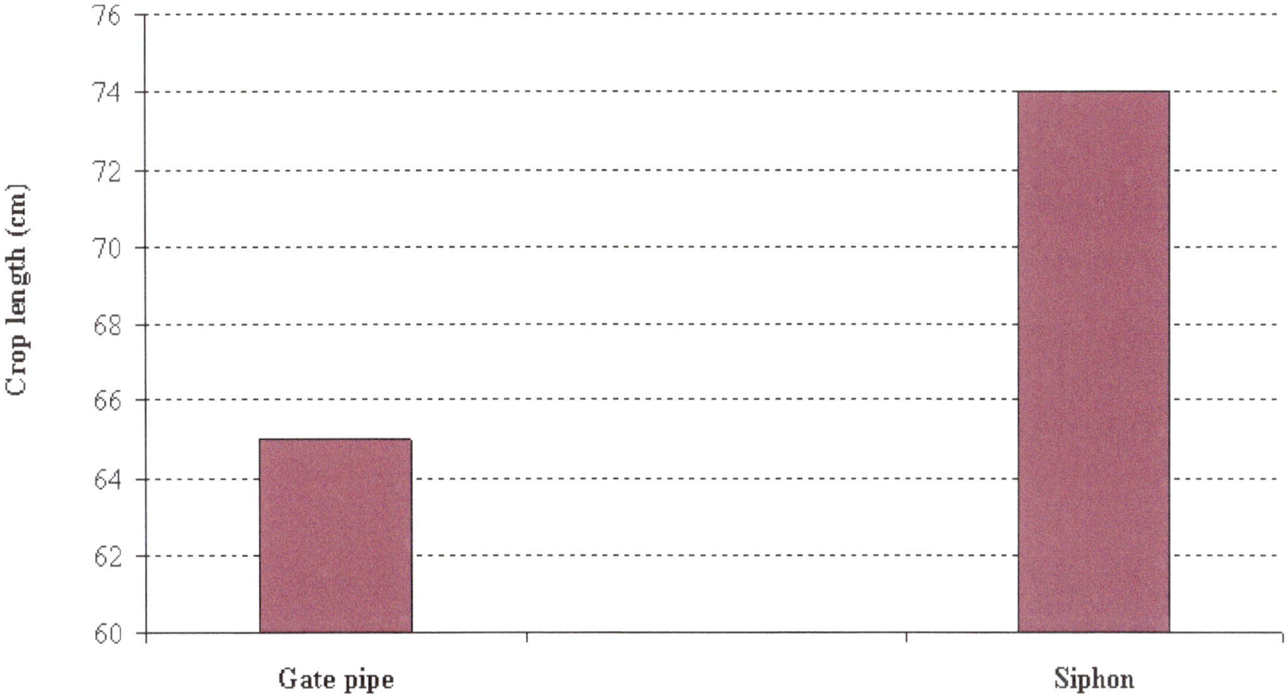

Figure 12. Stem length.

table correlated positively with the mean salinity and SAR.

REFERENCES

Abdel GE, Wahba MAS (2006). Possibility of water table management through sub-irrigation in Egypt. Tenth International Water Technology Conference, IWTC10, Alexandria, Egypt.

Afyuni M, Khoshgoftarmanesh AH, Dorostkar V, Moshiri R (2007). Zinc and cadmium content in fertilizer commonly used in Iran International. Conference of Zinc Crops, Istanbul, Turkey.

Bouksila F, Persson M, Berndtsson R, Bahri A (2010). Estimating Soil Salinity Over a Shallow Saline Water Table in Semiarid Tunisia. Open Hydrol. J. 4:91-101.

Codex Alimentarius Commission (CAC) (2008). Codex general standard for contaminants and toxins in foods. Codex STAN 193-1995. Codex Alimentarius Commission, Rome. Available at: www.codexalimentarius.org/standards.

Gafni A, Zohar Y (2001). Sodicity, conventional drainage and bio-drainage in Israel. Austr. J. Soil Sci. 39:1269-1278.

Grant CA, Clarke JM, Duguid S (2008). Chaney Selection and breeding of plant cultivars to minimize cadmium accumulation Sci. Total Environ. 390:301-310.

Hargreaves GH (1975). Moisture availability and crop production. Trans. ASCE 18(5):980-984.

Kemper WD, Heinemann WH, Kincaid DC, Worstell RV (1981). Cablegation: I. Cable controlled plugs in perforated supply pipes for automating furrow irrigation. Trans. ASAE 24(6):1526-1532.

Landwirtschaftlichen HF, Friedrich R (2011). Nitrogen management in irrigated cotton-based systems under conservation agriculture on salt-affected lands of Uzbekistan Inaugural-Dissertation. PhD of Diese Dissertation ist auf dem Hochschulschriftenserver der ULB Bonn. http://hss.ulb.uni-bonn.de/diss_online elektronisch publiziert.

Mikati G (1997). Temporal analysis of multispectral video/satellite imagery for the detection and monitoring of salinity on agricultural lands. USU Logan, Utah, pp. 95-97.

Nhundu K, Mushunje A (2010). Analysis of irrigation development post fast track land reform programme. A case study of Goromonzi district, Mashonaland East Province, Zimbabwe. Joint 3rd African Association of Agricultural Economists (AAAE) and 48th Agricultural Economists Association of South Africa (AEASA) Conference, Cape Town, South Africa, pp. 19-23.

Patel R, Prasher S, Bonnell R, Boughton R (2002). Development of comprehensive soil salinity index. J. Irrig. Drain. Eng. -ASCE 128:185-188.

Penman HL (1948). Natural evaporation from open water, bare soil and grass. Proc. Roy. Soc. London A 194(S):120-145.

Richards LA (1954). Diagnosis and improvement of saline and alkali soils. United States Department of Agriculture, Washington, DC.

Rogers ME (2002). Irrigating perennial pasture with saline water: Effects on soil chemistry, pasture production and composition. Austral. J. Exp. Agric. 42(3):265-272.

Rukuni M, Eicher CK, Blackie Eds (2006). Zimbabwe's Agricultural Revolution, Revisited. University of Zimbabwe Publications, Harare.

Wendling U, Schellin H (1986). Neue Ergebnisse zur Berechnung der potentiellen Evapotranspiration. Zeitschrift f'ur Meteorologie. 36(3):214-217.

Wigginton D (2004). Theoretical flow rates for siphons: head-discharge charts. - In Cotton CRC, WATERpak - A guide for irrigation management in cotton. Cotton Research and Development Corporation, Narrabri.

2

Improving soil surface conditions for enhanced rainwater harvesting on highly permeable soils

AMU-MENSAH Frederick K[1] , YAMAMOTO Tahei[2] and INOUE Mitsuhiro[2]

[1]CSIR Water Research Institute, Ghana.
[2]Arid Land Research Center, Tottori University, Japan.

The harnessing of runoff water from rainfall over land rather than allowing these waters to run into streams and rivers and eventually lost into the sea is attaining some popularity due to the increasing demand for scarce water resources. Sandy soils, especially in arid and semi-arid regions and along some coastal areas of Ghana (from Aflao to Tema New Town) do not generate sufficient rainwater runoff for storage because of their high permeability. This is important, as groundwater in these areas is mostly saline. To make such lands productive in terms of their ability to harvest rainwater, an experiment under laboratory conditions was conducted to apply top dressing on these land surfaces using less permeable soils on slopes of 20 and 30°. These adverse soil conditions were ameliorated with good results even though the conditions as pertains in the real world could not be completely simulated. Rather than have low or non-existent runoff from such lands from rainfall events, the application of a top dressing using a less permeable soil presented possibilities for harvesting and storing rainwater for agriculture and other uses. Harvested runoff water from rainfall of between 6 and 80 m^3/ha was obtained on the treated soils under laboratory conditions from a previous situation of zero runoff from the untreated soil. The time taken to generate runoff was drastically reduced from 4,783 min (when no runoff was generated) to less than 70 min with the treatment. This presents a hopeful mechanism for developing water resources in areas of water scarcity to improve upon poverty and livelihoods of the communities affected.

Key *words*: Ghana, rainwater harvest, Overland flow, Deep percolation losses, Soil amelioration, Tottori dune sand, Koyama mash clay, Isahaya polder clay.

INTRODUCTION

Runoff from highly permeable soils is difficult to generate because of their high infiltration rates. Due to the high permeability of sandy soils, heavy rainfall conditions in the order of 38.5 mm/h applied continuously for over four days on these soils may not generate any runoff (Amu-Mensah et al., 2013).

The rainfall regime in the Accra Plains of Ghana is the lowest in the country and rarely exceeds 850 mm/yr whereas evapotranspiration rates exceed 2,000 mm/yr. Similar conditions exist in arid and semi-arid areas of the world. Under such conditions as pertains in the Ada area of the Greater Accra Region, where local shallot cultivation is undertaken, frequent irrigation is needed to sustain crop yields because the soils lose considerable water through deep percolation and from evaporation ((Ben-Asher et al., 1978). The Ada area has a shallow water table that enables farmers to use tube wells to irrigate their farms. The efficiency of rainfall is low and without the tube wells, there would be no farming in this area. Where such water resources are not available, storage of rainwater offers a good source of water for irrigation and for other uses. When the runoff characteristics of the soils are poor, only appropriate soil amelioration measures will ensure sufficient water harvested from rainfall to meet demand.

This paper discusses an experiment conducted on a sloped field in a large glasshouse at the Arid Land Research Center of Tottori University, Japan. The purpose was to understand the process of runoff generation as influenced by the surface treatment of highly permeable sandy soils using less permeable clay and loamy sands, and the consequences to erosion as they affect the successful harvesting and collection of rainwater for storage to meet various water needs. It was also to find simple solutions to harvesting water from rainfall in the conditions mentioned, to meet especially domestic water needs. Since rainfall in the tropics is mostly relief in origin, field slopes of 20 and 30°, which represent mountain slopes where rainwater-harvesting potential exists, were used (Buah, 1998).

METHODOLOGY

In an attempt to enhance the runoff from dune sand (a highly porous soil very much like beach sand and similar to the delta soils (Wikipedia, 2013) of the coastal areas around Ada in Ghana) rainfall-runoff experiment was undertaken with treatments of Tottori Masa soil (loamy sand), Koyama Mash clay and Isahaya Polder clay. The treatments involved spreading uniformly and compacting over the dune sand soil, a 7 to 10 cm layer of the treatment soil. The properties of the soils used are given Figure 1.

Time Domain Reflectometer (TDR) soil moisture probes were inserted into the soil profile at depths of 10, 50 and 100 cm respectively in each of the soil profiles. These were connected to a Campbell Scientific CR23X data logger to log soil moisture data at the respective depths. Overhead rainfall from full-circle jet nozzles (Spraying Systems Co., Japan) provided rainfall onto the slopes for a period of 157 min. Average intensity of the spray system was 40.3 mm/h at a mean pressure of 2.75 kg/cm^3.

Data collected included rainfall intensities, rainfall amounts, runoff amounts, runoff rates, visual (photographic) changes in soil surface resulting from erosion and soil moisture variability at the various depths. The experiment was conducted in a glasshouse, on a sloped bed as shown in the schematic representation in Figure 2.

A cross-section of the soil profile showing the arrangement of the TDR probes, spraying nozzles, runoff collection bottles and soil layers is shown in Figure 3.

RESULTS AND DISCUSSION

Results show remarkable improvement in runoff generation over the layered soil surface as opposed to the bare sand runoff experiment (Table 1). Runoff from the slopes was first observed 28 min from the start of the experiment. This occurred on the 20% slope with Masa top layer dressing. It was followed by runoff from the 20% slope having as top layer, the Isahaya polder clay after 45 min of continuous rainfall. A little runoff was observed on the 30% slope with Isahaya as topsoil. This started 69 min into the experiment. The 30% Masa-topped slope did not produce any runoff however, a pool of water collected at the base of the slope 54 min into the experiment. Since this pool of water was persistent, it is probable that some runoff was generated but seeped into the soil at the base of the slope, preventing runoff to be collected into the bottles.

In all the runoff events, surface flow was enough to dislodge some soil particles. Runoff water collected was coloured in all the events however very little rill formation was observed. It is possible that the colouring of the runoff water was due mostly to silt particles suspended in the runoff flow. This is most probably due to the ramming of the slopes, which was done to compact and stabilise the soil as the top layer dressing was being applied. A more stabilised slope may not have produced any colouration in the runoff water. It is likely that heavier rainfall intensities would dislodge some soil and cause significant erosion. The Koyama clay-topped slopes did not produce any noticeable runoff. Figure 4 shows a graph of the progression rate of runoff with time for the three-runoff events that occurred.

The experimental setup showing the rainfall-runoff process in progress is shown in Figure 5. The low distribution coefficient (41 to 49%) of the overhead sprinkler system may also have contributed to the low runoff generated.

Observed runoff ratios needed for rainfall harvesting are low because of a number of factors. Firstly, the treatment soils as well as the underlying dune sand were initially dry and required a significant amount of wetting to generate some runoff. Much of the rainwater was thus used to wet the soil profile especially as the soil mass was not as stabilised as exists in real life. In spite of this, runoff was observed to occur early in the experiment as opposed to the non-treated condition when no runoff was observed from the dune sand slopes. The layer of treatment soil may be too thin and therefore does not effectively reduce the infiltration rate of the soil. If this occurred, a significant amount of the rainwater will still infiltrate into the soil profile. The fact that some runoff was generated in spite of the thin layer of treatment soil, when previously no runoff was generated, shows how effective this method can be. Additional treatment to improve the structure of the treatment material through mulching for example, could further improve runoff.

It is probable, though not confirmed that the “fingering phenomenon” (Dong et al., 2010), which is responsible for high infiltration in sandy soils overlain with a less permeable soil, is the reason for the low runoff generated. Another reason for the low runoff ratio is the short duration of the experiment. Since a significant fraction of the rainwater as well as time was required for the initial wetting of the soil, enough time was not available for the generation of actual runoff. This is especially the case for the Isahaya-treated soil on the 30° slope. In order to reduce errors, plastic rain shields were placed to prevent direct rain into the collecting gutters.

It is however significant that the experiment produced

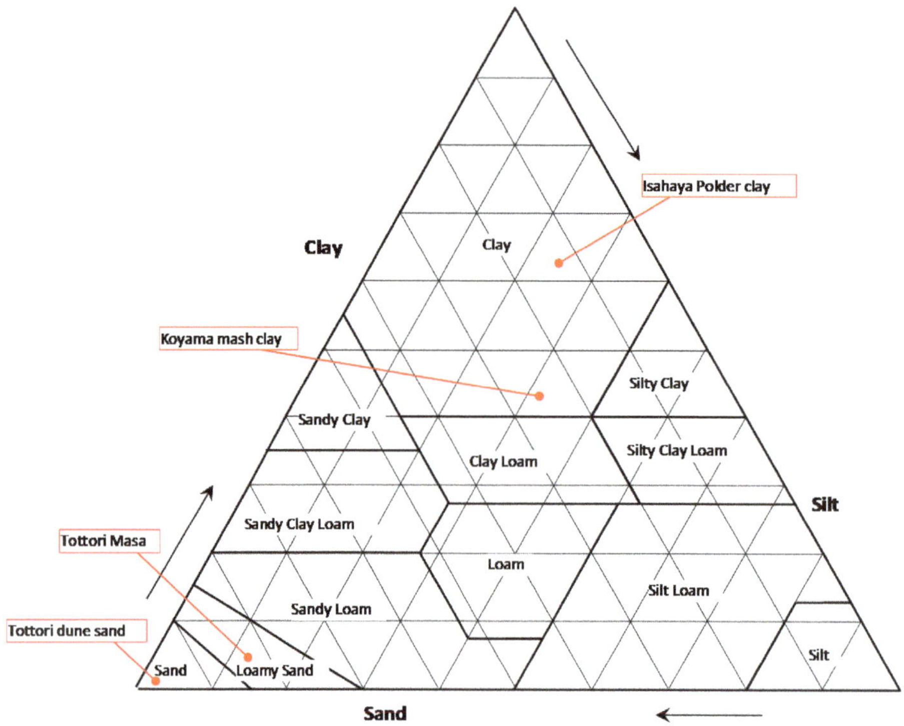

Figure 1. USDA textural classification of soils used for rainfall-runoff experiment.

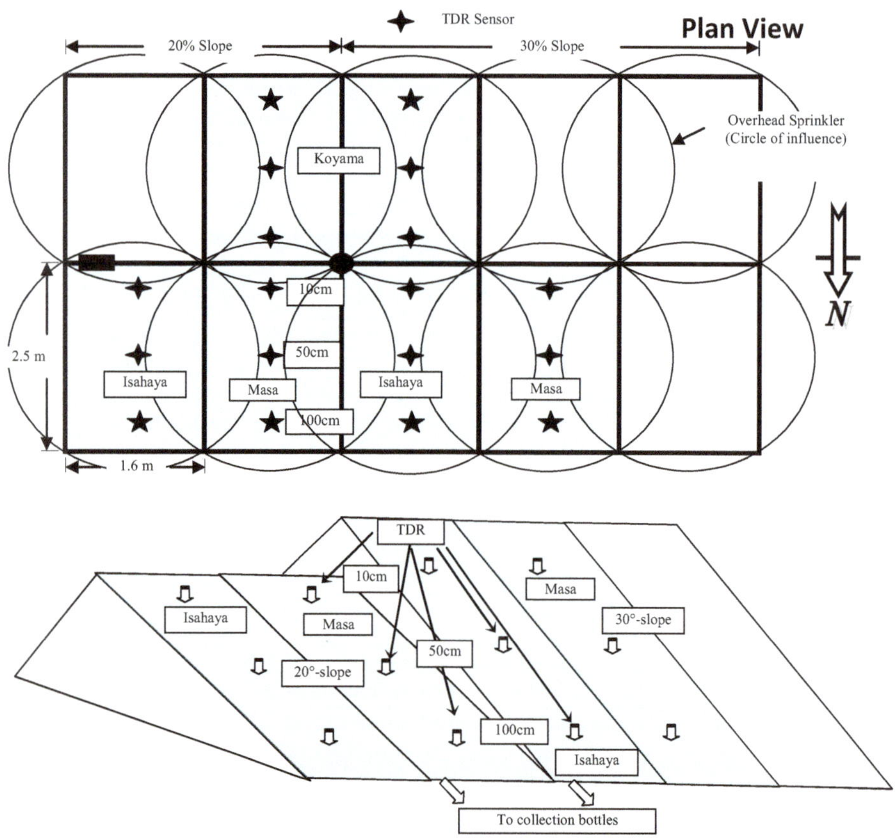

Figure 2. Arrangement of TDR probes, rainfall nozzles, runoff collection bottles and treatment soil layer.

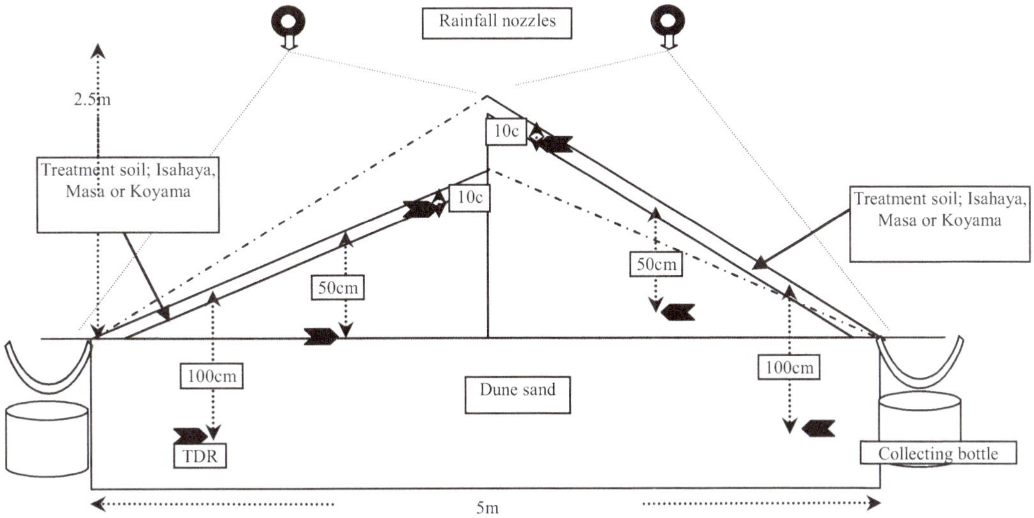

Figure 3. Cross sectional layout of layered rainfall-runoff experiment showing positions of equipment used.

Table 1. Results of runoff events on treated sand slopes.

Runoff events	Dune sand	Top layer treatment material (on dune sand) and slope		
	No treatment	Isahaya clay 20° slope	Masa 20° slope	Isahaya 30° slope
Start of runoff (from start of rainfall)	No runoff*	45 min	28 min	69 min
Total runoff (l)	0	7.95 (17,408 m^3/ha)	36.8 (80,580 m^3/ha)	2.77 (6,127 m^3/ha)
Mean runoff rate (l/h)	0	5.25	19.74	0.0006
Maximum recorded runoff rate (l/h)	0	7.16	30.6	0.0006
Runoff ratio (%)	0	1.68	7.75	0.58

*No generated runoff after 4,783 min (Amu-Mensah et al., 2013).

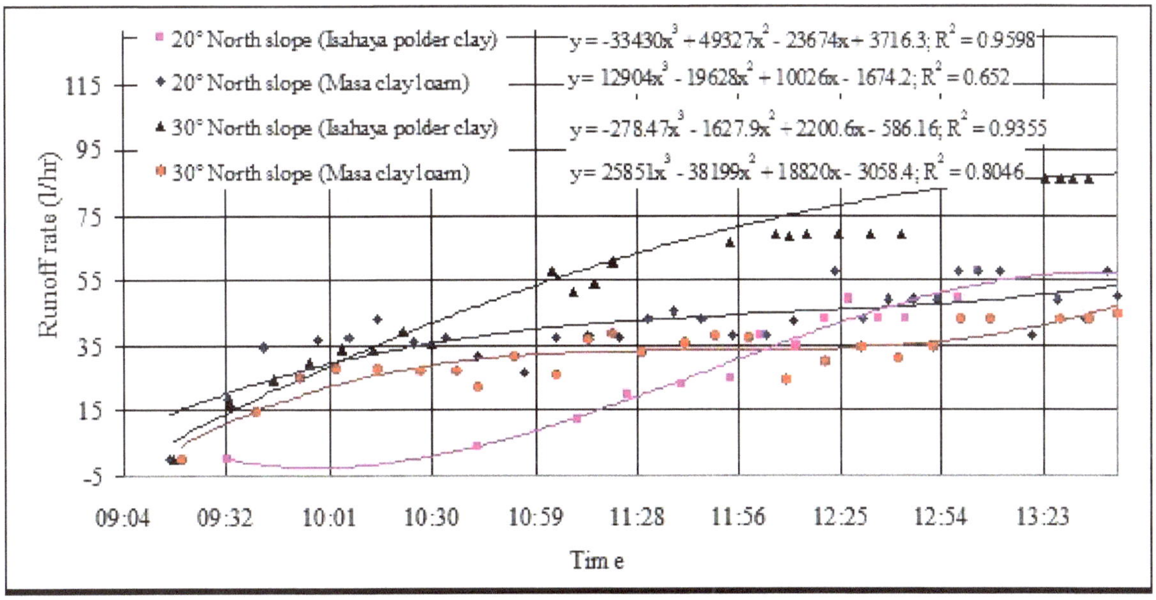

Figure 4. Runoff progression rate as measured for constant rainfall intensity of 40.3 mm/h.

Figure 5. Rainfall-runoff experimental layout showing slopes with different top layer soil treatments.

some level of runoff and therefore presents some options for improving the runoff characteristics of soils with poor rainwater harvesting potential. Based on the initial condition of the experimental plot, with its clear disparities to real land situations, which over time have established a firm and stabilised regime, it is reasonable to assume that the runoff characteristics of the experimental plot will improve with time as the plot stabilises through the wetting and drying cycles characteristics of natural lands.

Conclusions

Soil amelioration techniques for improving runoff characteristics of soils for better water harvesting are available and include using plastic sheets, bitumen spreads, concrete layers and chemical additives mixed into the top soils to improve bonding of the soils particles. These methods are quite expensive and usually interfere with the natural environment and deprive the land from being used for other activities. The use of less permeable soils as a top layer dressing to reduce infiltration and enhance runoff has been shown to be a viable option with appreciable results obtained.

This will especially be useful in areas with sand, sandy loam and loamy sand soils having high infiltration rates. Various locations in Ghana especially along the coastal stretch of the country and in some inland valley alluvial plains could benefit from this process if these areas are consciously lined with laterite and clay soils. When carefully designed and constructed, these rainwater harvesting fields could concentrate and channel rainwater into dugouts and reservoirs for storage and used for agricultural and other uses to improve the livelihoods of the people. Water, which currently infiltrate into the soil and is lost, could be harvested from the treated lands and made available for development of the riparian communities.

Sandy soils are characterised by high infiltration rates in excess of 6,000 mm/h that makes them unable to generate any sufficient runoff even when rainfall with average intensity of 38.5 mm/h are continuously administered to them. This was in spite of continuous rainfall being administered for a period of 4,783 min. Upon the treatment on these soils however, runoff generation started in as little as 28 min and in as much as 69 min after rainfall with average intensity of 40.3 mm/h was administered.

Runoff ratios observed are low and the thin layer of top dressing employed could have been the reason for the observation. This could have enabled significant amounts of the rainfall to seep into the soil profile because of the thickness of the layer. The low intensity of the rainfall event, which under tropical conditions could reach as high as 140 mm/h could also have caused these low ratios.

Considering that the rate of runoff continued to increase as the experiment progressed, it is to be expected that a well compacted top dressing which has stabilised over time, will offer better runoff generation capabilities. It is also reasonable to suppose that a thicker top dressing on the sand would yield higher runoffs as they would further reduce the permeability of the soil.

REFERENCES

Amu-Mensah FK, Yamamoto T, Inoue M (2013). Investigating rainwater harvesting on highly permeable soils - baseline conditions, Int. J. Water Res. Environ. Eng. 5(7):426-433.

Ben-Asher J, Fuchs M, Goldberg D (1978). Radiation and Energy Balance of Sprinkler and Trickle Irrigated Fields. Agro. J. 70:415-417.

Buah FK (1998). A History of Ghana. Macmillan Education Limited. London.

Dong B, Yan YY, Lia WZ, Song YC, (2010). Simulation of the Influence of Surface Wettability on Viscous Fingering Phenomenon in Porous Media. J. Biol. Eng. 7(3):267-275.

Wikipedia, (2013). Geography of Ghana. http://en.wikipedia.org/wiki/Geography_of_Ghana#Low_Plains.

Spatial and temporal variations of river water quality: A case study of River Kabini at Nanjangud in Karnataka

Sudevi Basu[1] and K S. Lokesh[2]

[1]Department of Biotechnology, Sir MVisvesvaraya Institute of Technology, Bangalore – 562157, Karnataka,
[2]Department of Environmental Engineering, Sri Jayachamarajendra College of Engineering, Mysore – 570006, Karnataka

Water pollution is alarming all over the globe and necessitates better understanding of spatial and temporal variations of pollutants in the rivers. In this study, the data sets of 8 water quality monitoring stations in River Kabini at Nanjangud which comprised of 10 water quality parameters monitored monthly over 12 years (2000-2011) are used. Temporal variation of the physico-chemical parameters in each station is represented by Box plots. High positive skewness is observed for total dissolved solids (TDS), conductivity, chemical oxygen demand (COD), and chlorides for all stations, showing rampant pollution in the river due to anthropogenic sources. The microbiological parameters in each station shows the river is polluted with total coliform and faecal coliform across baseline, trend and impact stations. The results suggest that the Kabini River is polluted from anthropogenic sources because bathing ghat exists along the river and domestic wastewater is let into the river without treatment. This study provides critical information for river conservation and its protection by suggesting possible treatment of domestic wastewater before letting in the river.

Key words: Box plots, temporal variation, anthropogenic sources, total coliform, faecal coliform.

INTRODUCTION

Rivers are vital freshwater systems of strategic importance across the world, providing main water resources for domestic, industrial, agricultural and recreational purposes. Rivers form the lifeline of human society and play a pivotal role in assimilating or carrying industrial and municipal wastewater and runoff from agricultural fields. In recent years, rivers are amongst the most vulnerable water bodies to pollution as a consequence of unprecedented development. Anthropogenic influences like urbanization, industrialization, agricultural activities as well as natural sources like erosion degrade the surface water (Najafpour et al., 2008; Bu et al., 2010). Thus the water quality of these water resources is a subject of ongoing concern and has resulted in an increasing demand for monitoring river water quality. The quality of water is described by its physical, chemical and microbiological characteristics (Venkatesharaju et al., 2010). Therefore a regular monitoring of river water quality not only prevents outbreak of diseases and checks water from further deterioration, but also provides a scope to assess the current investments for pollution prevention and control. Monitoring programs result in a huge and complex data matrix consisting of a large number of physico-chemical parameters (Alkarkhi et al., 2008). One of the methods in water resources quality assessment is the use of Multivariate statistical techniques to characterize and evaluate freshwater quality and these are useful in verifying temporal and spatial variations caused by anthropogenic factors (Shrestha et al., 2007;

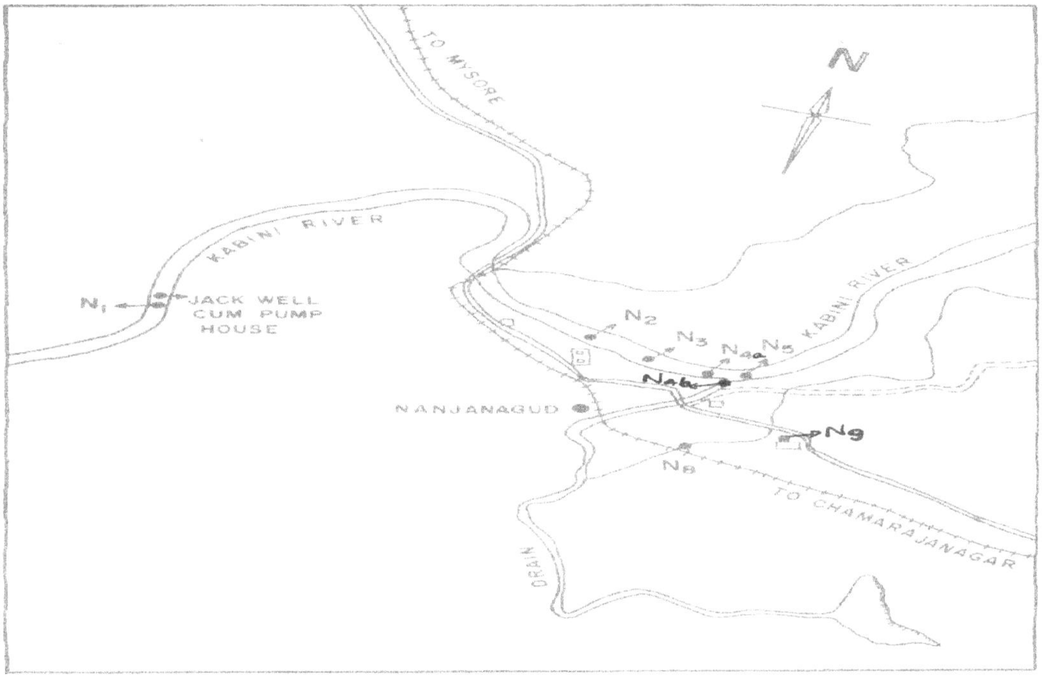

Figure 1. Map of water quality monitoring stations at Nanjangud town.

Venkatesharaju et al., 2010; Ekwere et al., 2011).

In this study, spatial and temporal variations of physico-chemical and biological parameters of water quality in Kabini River was assessed in Nanjangud town in Karnataka using multivariate statistical tools like box plots.

MATERIALS AND METHODS

Study area

Nanjangud town is situated 20 km towards the south of Mysore city. River Kabini, a tributary of the River Cauvery, flows along this town. Nanjangud is a piligrim town attracting crowds on various occasions. Several industries are located in the vicinity of the banks of this river.

There are eight stream monitoring stations in this area, six of which are on River Kabini and two on Hulla Halli Channel. A drain carrying wastewater from Nanjangud town enters the river along this stretch. The stations are shown in Figure 1.

Monitoring stations

The water quality monitoring stations along River Kabini at Nanjangud town are classified as baseline, trend and impact stations. Baseline stations are concerned with natural and unpolluted state of river basin. In these stations, there is no influence of human activities on water quality. Impact stations are used for measuring the quantity of pollutant and extent of pollution due to human interference. Trend stations show how a particular point on a water course varies over time due to influence of human activities. These stations not located on main river systems are sited on major tributaries and points just upstream of the confluence with the main river.

Baseline stations - N2 and N3

Station N2 is located on the river near the waterworks intake of Nanjangud town about 4 km downstream of station N1. Station N3 is located downstream of N2 on the river near the Travelers Bungalow.

Impact stations – N4b and N5

The station N4b is located on a drain which is carrying wastewater from Nanjangud town. The drain enters the river just downstream of the station N4a. Station N5 is a downstream station located near the water intake pump house of Gemini Distilleries Industry. Near this station the river flows at a greater velocity as the river width is small when compared with the width at the upstream station.

Trend stations - N1, N4a, N8 and N9

Station N1 is an upstream river station located near the Karnataka Industrial Area Development Board (KIADB) water intake point. Several industries are located near this station. Station N4a is located near the bathing ghat in Kabini River. Bathing and washing of clothes are very prominent along the right bank in a stretch of about 250 to 300 m length. Station N8 is an upstream station, located near the railway bridge leading towards Chamarajanagar across Hulla Halli Channel. Station N9 is located just downstream of the culvert across the Hulla Halli Channel.

Data preparation

The data sets of 4 water quality monitoring stations which comprised of 10 water quality parameters monitored monthly over 12 years (2000-2011) are used for this study. The data is obtained from the Water Quality Monitoring work of Cauvery River Basin in Mysore District, Karnataka State assigned to Sri Jayachamarajendra

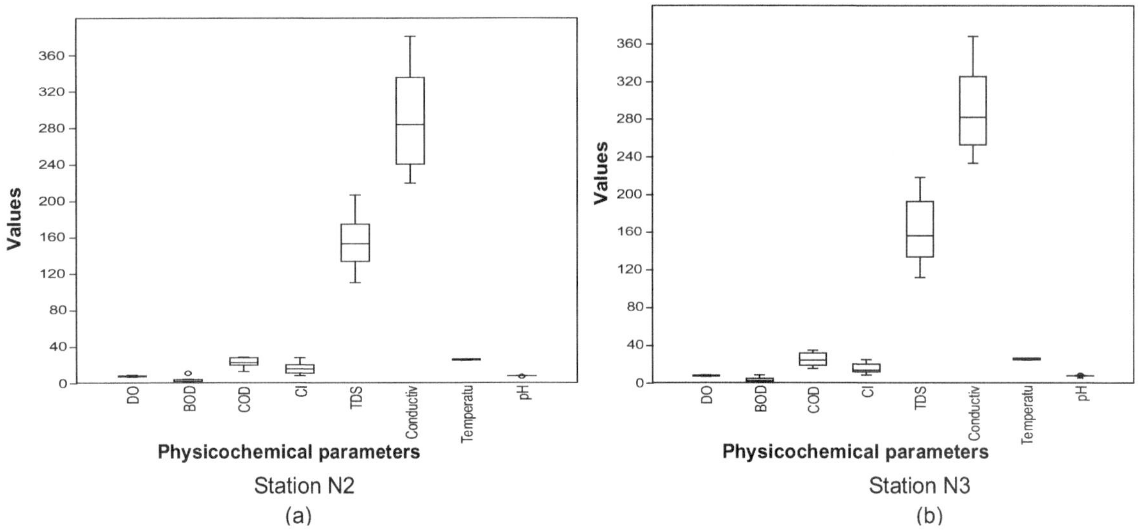

Figure 2. Physicochemical parameters of baseline stations.

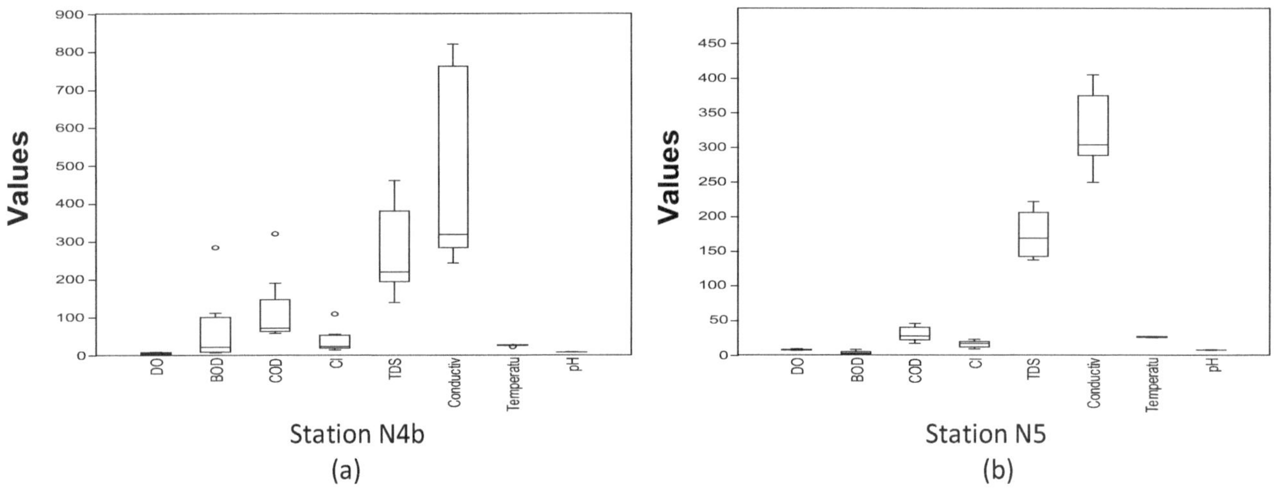

Figure 3. Physicochemical parameters of impact stations.

College of Engineering, Mysore under a nationwide River Water Quality Monitoring Project of the National River Conservation Directorate (NRCD), Ministry of Environment and Forests, Government of India, under its National River Conservation Project (NRCP). Although there are more water quality parameters in these stations, only 10 most important parameters are chosen because of their continuity in measurement through the 12 years. The 10 selected water quality parameters include dissolved oxygen (DO), biochemical oxygen demand (BOD), chemical oxygen demand (COD), chlorides (Cl), total dissolved solids (TDS), conductivity, total coliform (TC), faecal coliform (FC), temperature and pH.

Statistical analysis

The data for spatial and temporal variation of pysico-chemical parameters of baseline, impact and trend stations was analyzed using Paleontological Statistics (PAST) version 2.15 Software Package. The 12 year annual average is taken for each parameter at each station for the plot. The spatial and temporal variations are represented by box plots for various physico-chemical parameters.

Box-plot provides a visual impression of the location and shape of the underlying distributions. The line across the box represents the median, whereas the bottom and top of the box show the location of the first and third quartiles. The whiskers are the lines that extend from the bottom and top of the box to the lowest and highest observations inside the region (Razmkhah et al., 2010). Several researchers (Pejman et al., 2009; Bu et al., 2010; Razmkhah et al., 2010; Garizi et al., 2011) have used box plots to assess seasonal and other water quality variations along rivers.

RESULTS AND DISCUSSION

The spatial and temporal variations for physico-chemical parameters of baseline, impact and trend stations are shown in Figures 2 to 4, respectively.

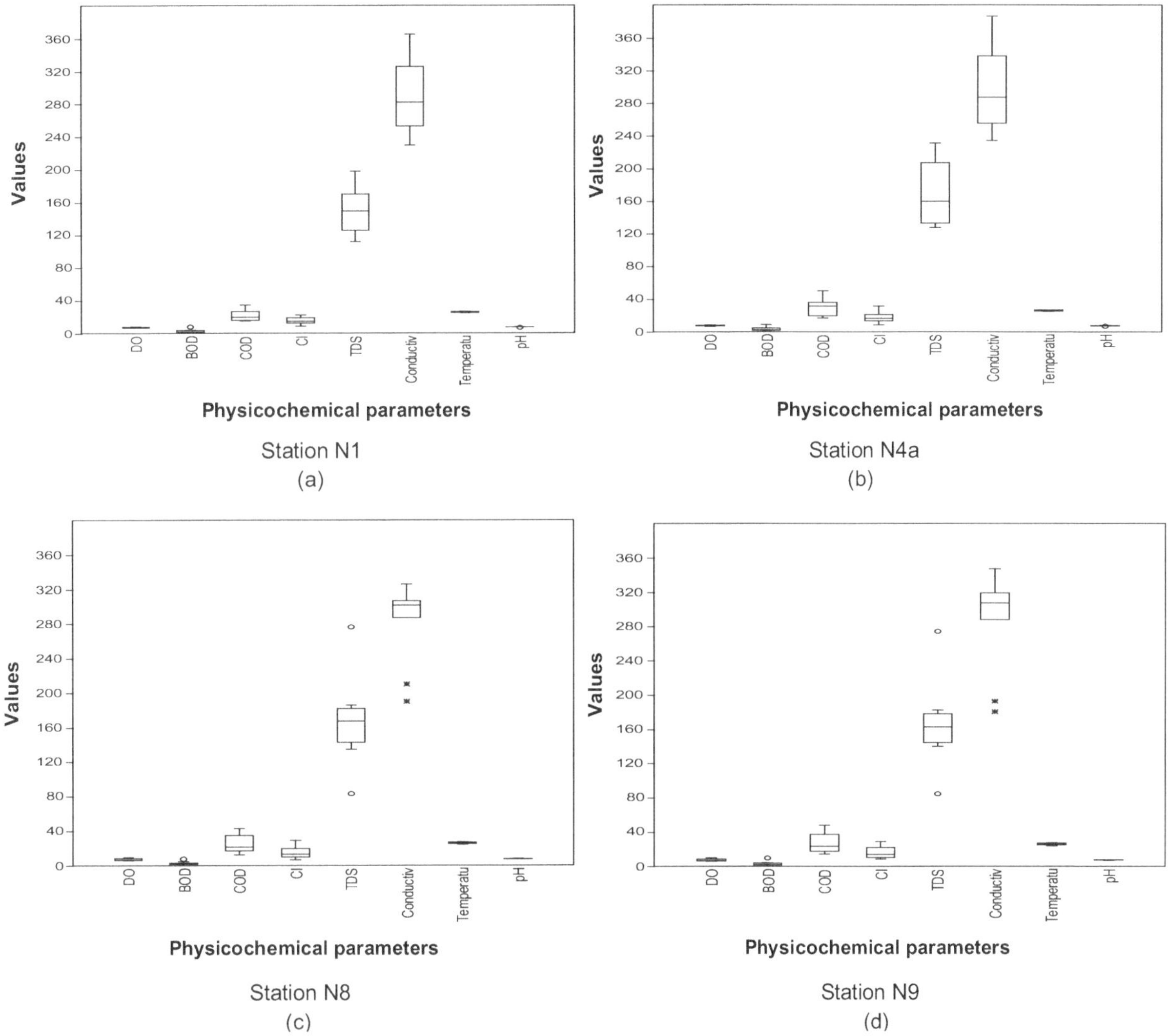

Figure 4. Physicochemical parameters of trend stations.

As shown in Figure 2a and b, the values of TDS and conductivity show asymmetric plots and are skewed with high positive values. This is because of wide variety of inorganic salts and organic matter due to natural conditions and agricultural runoff. A high value of TDS indicates effects of anthropogenic sources along the river (Charkhabi et al., 2006) and in this case also the station N3 is located near Travelers Bunglow. Conductivity variation reflects the status of inorganic pollution and is a measure of TDS in water (Garizi et al., 2011). High conductivity value observed is a result of increased agricultural land use and runoff in the river. COD and chlorides are also present in this part of the river indicating pollution. This study indicates that the baseline stations in the river are getting polluted.

Figures 3a and b show the physicochemical parameters of the impact station. It is seen that that station N4b is highly polluted with positive skewness observed for BOD, COD, Cl, TDS and conductivity. The DO value in station N4b is less. This is because TDS affect the depletion of DO in water and similar trend was observed by Charkhabi et al. (2006). The station N4b is located on the drain carrying wastewater from Nanjangud town and enters the river. In station N5 the skewness observed in all parameters is less than that at station N4b. This is because dilution is occurring near this station as the river flows at a greater velocity since the river width is small when compared with the width at the upstream station.

The physicochemical parameters of trend stations are given in Figure 4a to d. COD, chlorides, TDS and conductivity are high in these stations. Compared to station N1, N4a shows high positive skewness for COD and Cl, showing pollution from anthropogenic sources. This is because station N4a is located near the bathing ghat in Kabini River.

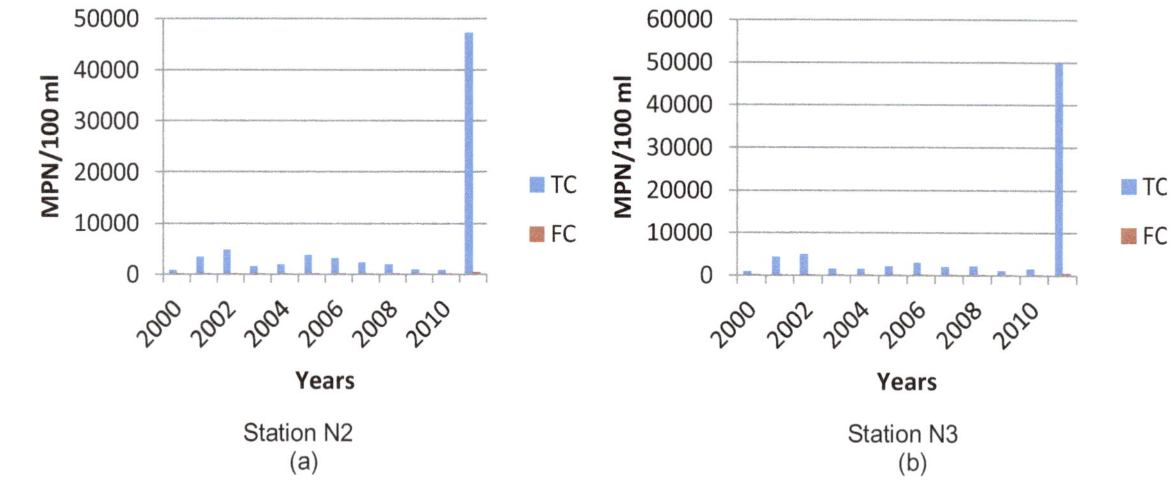

Figure 5. Microbiological parameters of baseline stations.

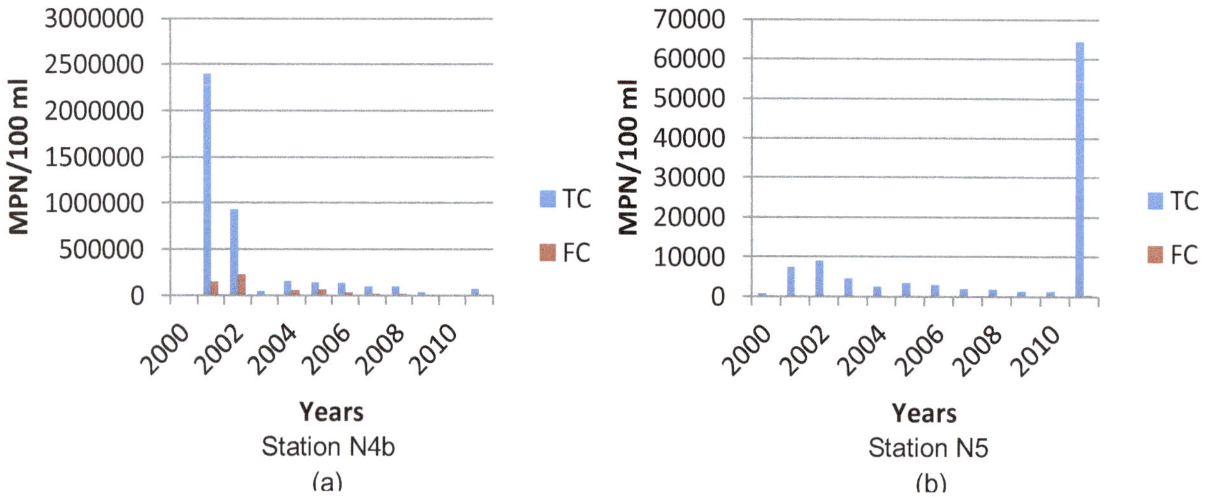

Figure 6. Microbiological parameters of impact stations.

The data for spatial and temporal variation of microbiological parameters of baseline, impact and trend stations was analyzed using Microsoft Office Excel 2007. The 12 year annual average is taken for each parameter at each station for the plot. The microbiological parameters of water quality data of baseline, impact and trend stations are shown in Figures 5 to 7, respectively.

It is seen from Figure 5a and b that TC is high in stations N2 and N3, indicating pollution of the river in baseline stations. Station N2 is located on the river near the waterworks intake of Nanjangud town and this is a matter of concern. Station N3 is located downstream of N2 on the river near the Travelers Bungalow and shows a high value for TC.

As shown in Figure 6a and b, the station N4b is located on the drain carrying wastewater from Nanjangud town and enters the river and TC value is high. In station N5 also the TC value is high the river flows at a greater velocity since the river width is small when compared with the width at the upstream station. Similar trend is observed in trend stations shown in Figure 7a to d.

Conclusion

It is alarming that the pollution in river Kabini is escalating over the years in this stretch at Nanjangud town due to indiscriminate human activities and agricultural uses along the river. The pollution level of the river is on the rise and can cause serious problem in near future. From this study of the surface water quality of the river, it can be concluded that the water of this river may represent serious threat to the ecosystem due to anthropogenic pollution. Although, some parameters may not be at critical pollution level, but high levels of COD, TDS, conductivity, total coliform and faecal coliform clearly

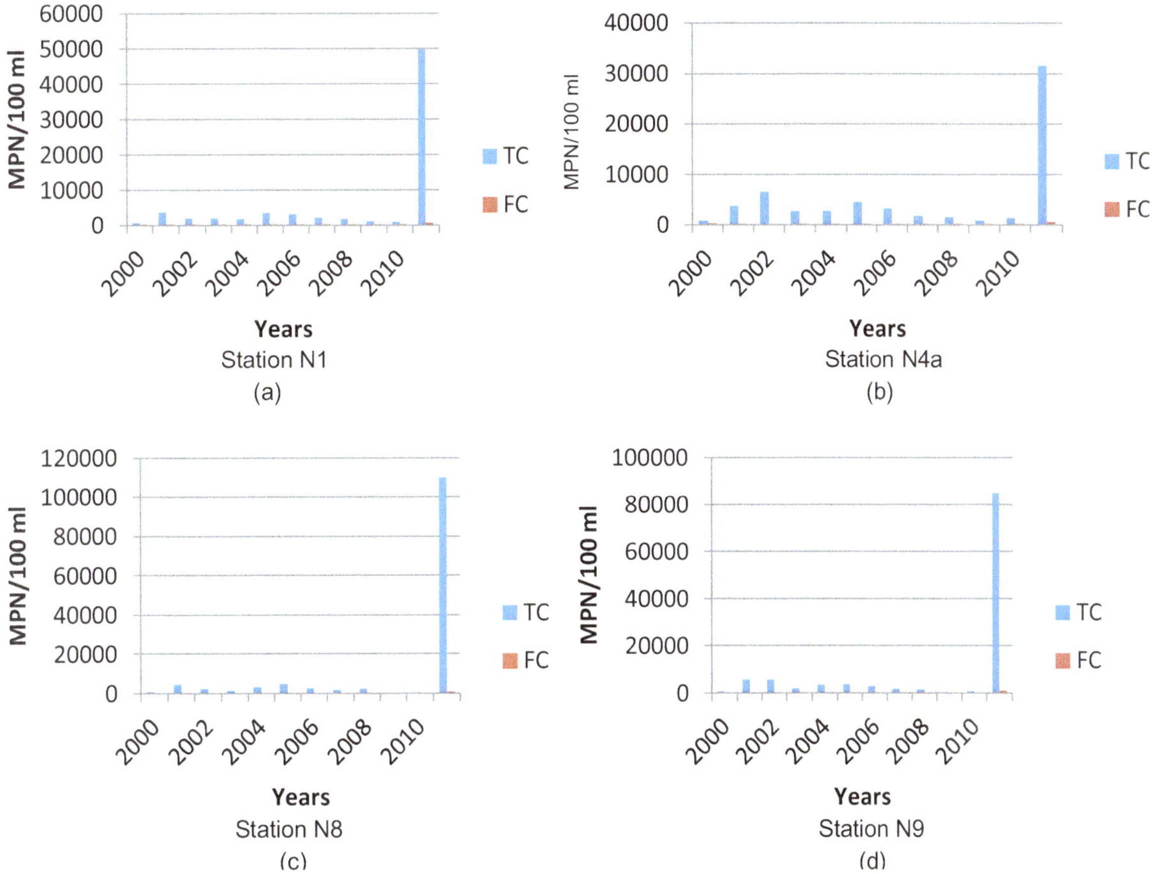

Figure 7. Microbiological parameters of trend stations.

indicates severe water pollution and there is a need for wastewater treatment plant to prevent further river water pollution.

ACKNOWLEDGEMENT

The authors thank National River Conservation Directorate (NRCD), Ministry of Environment and Forests MoEF, Government of India for the financial support under National River Conservation Directorate Project.

REFERENCES

Alkarkhi AFM, Ahmad A, Ismail N, Easa A, Omar K (2008). "Assessment of surface water through Multivariate Analysis." J. Sust. Dev. 1(3):27-33.

Bu H, Tan X, Li S, Zhang Q (2010). "Temporal and spatial variations of water quality in the Jinshui River of the South Qinling Mts., China." Ecotoxicol. Environ. Saf. 73(5):907-913.

Charkhabi AH, Sakizadeh M (2006). "Assessment of spatial variation of water quality parameters in the most polluted branch of the Anzali wetland, Northern Iran." Pol. J. Environ. Stud. 15(3):395-403.

Ekwere A, Edat A, Ukpong A, Obim V (2011). "Assessment of seasonal variations of hydrochemical signatures of surface water quality using multivariate statistical methods." J. Sust. Dev. Afr. 13(8):51-64.

Garizi AZ, Sheikh V, Sadoddin A (2011). "Assessment of seasonal variations of chemical characteristics in surface water using multivariate statistical methods." Int. J. Environ. Sci. Technol. 8(3):581-592.

Najafpour SH, Alkarkhi AFM, Kadir MOA, Najafpour GD (2008). "Evaluation of Spatial and Temporal Variation in River Water Quality." Int. J. Environ. Res. 2(4):349-358.

Pejman AH, Bidhendi GRN, Karbassi AR, Mehrdadi N, Bidhendi ME (2009). "Evaluation of spatial and seasonal variations in surface water quality using multivariate statistical techniques." Int. J. Environ. Sci. Technol. 6(3):467-476.

Razmkhah H, Abrishamchi A, Torkian A (2010). "Evaluation of spatial and temporal variation in water quality by pattern recognition techniques: A case study on Jajrood River, Tehran, Iran." J. Environ. Manage. 91:852-860.

Shrestha S, Kazama F (2007). "Assessment of surface water quality using multivariate statistical techniques: A case study of the Fuji river basin, Japan." Environ. Model. Softw. 22(4):464-475.

Venkatesharaju K, Somashekar RK, Prakash KL (2010). "Study of seasonal and spatial variation in surface water quality of Cauvery river stretch in Karnataka." J. Ecol. Nat. Environ. 2(1):001-009.

4

Investigating rainwater harvesting on highly permeable soils - baseline conditions

AMU-MENSAH Frederick K.[1], YAMAMOTO Tahei[2] and INOUE Mitsuhiro[1]

[1]CSIR Water Research Institute, Ghana.
[2]Arid Land Research Center, Tottori University, JAPAN.

Dune sand was subjected to varying rainfall intensities in order to generate runoff under laboratory conditions on slopes of 20 and 30°. Soil moisture probes were inserted into the soil at varying depths to investigate the movement of water through the profile. Results indicate that in spite of continuous simulated rainfall administered for over 4,000 min, no runoff could be generated over the sand surfaces and across the different slopes. It was deduced that for this to occur, rainfall intensities in excess of 6,000 mmh^{-1} would be required. The results also indicated that to generate runoff from such sandy soils, treatment of the upper 0 to 10 cm of the soil profile would be required to reduce significantly the infiltration rate of the soil profile, which then will translate into runoff generation.

Key words: Ghana, rainwater harvest, overland flow, deep percolation losses, soil amelioration, sand

INTRODUCTION

Rainwater harvesting is becoming increasingly important as a means of securing water for various uses. Though Ghana has sufficient water resources, it is becoming increasingly evident that the quantity and quality of the surface water resources of Ghana is decreasing (WARM Study, 1998).

Rainwater harvesting is very crucial to the sustainable development of water resources for domestic, agricultural, industrial, aquaculture and ecosystem protection water needs. Since rainfall incidence and amount is limited and unpredictable in many areas of its occurrence, any process that maximises the retention and collection of rainwater through minimising of the losses incurred is beneficial to the success of such resource development process (Amu-Mensah et al., 2006). Not all soil types aid water harvesting from rainfall due to a number of factors including slope, permeability and vegetative cover. This paper discusses an experiment conducted on a sloped field in a large glasshouse, at the Arid Land Research Center of Tottori University, Japan for the purposes of understanding the process of runoff as they affect the successful harvesting and collection of rainwater for storage to meet various water needs. Since rainfall in the tropics is mostly relief in origin, field slopes of 20° and 30° that represent mountain slopes typical of Ghana where rainwater-harvesting potential exist, were used.

RAINFALL RUNOFF PROCESS

Rainwater runoff is that part of precipitation, as well as any other flow contributions, which appears as surface streams of either perennial or intermittent form. This is the flow collected from a drainage catchment or watershed, and it emerges at an outlet of the catchment.

It occurs when rain falls with intensity greater than the rate at which it is able to infiltrate the soil and penetrate into the water table. It may also occur when low intensity rainfall occurs over extended time as to saturate the soil profile and therefore generate excess water that runs off the land. The rate at which water seeps into the soil is known as the infiltration rate ($f(t)$; function of time, t) of the soil. It depends on the permeability of the surface of the soil and on the antecedent rainfall, which is important because of its effect on the moisture level of the soil profile. Initially, when the soil is dry, infiltration rate is high (fo) and the soil absorbs virtually all the rainfall. As the upper layers of the soil fill up with water, the infiltration rate reduces exponentially until it reaches a constant rate (fc).This relation, developed by Horton is expressed in Equation.

$$f(t) = fc + (fo - fc)e^{-kt}$$

k is a decay constant $[t^{-1}]$ Equation 1 Horton's infiltration relation

Horton's infiltration model is one of several models including Philip, Swartzendruber, Kostiakov (with its variants) and the SCS models, developed to describe the infiltration of water through various soils. Each has its strengths and weaknesses, however the Horton model represents well, the soil infiltration process as compared to other models (Zolfaghari et al., 2012). The Horton's infiltration model is assumed in this adopted for this work.

Total runoff from a catchment comprises direct and indirect runoff, which together end up at the outlet of the catchment. Direct runoff consists of surface flow and infiltrated rainfall moving laterally through the upper horizons of the soil and latter reappearing at the surface of the soil. Surface or overland flow arises from excess rainfall that is not able to infiltrate into the soil but flows over the land surface in channels. It travels over the ground surface and through channels to reach the outlet of the catchment. Indirect runoff on the other hand is formed by delayed subsurface and groundwater flow formed from seepage of rainwater into deeper layers of the soil profile. This accumulates and flows laterally in deeper horizons and latter emerges as springs in riverbeds and other low elevations to contribute to stream flow. During dry seasons, most of the flow in rivers and streams are a result of this indirect flow and is referred to as the base flow of the river or stream.

Rainwater harvesting assumes that there is no flowing water resource and therefore all water harvested is mainly from surface flows, principally direct runoff. This means that for the target soil fc must be low and k must be high. The result of this is that the terminal or steady infiltration rate of the soil is low enough to generate sufficient direct runoff and the decay is rapid enough to ensure the attainment of the steady state infiltration in the shortest time. This is illustrated in Figure. Soil compaction, chemical additives, bitumen coating and covering with plastic sheeting are a few ways to attain this state. This paper examines the baseline conditions of rainfall runoff existing in dune sand subjected to varying intensities of rainfall under different slope conditions.

The dynamics of water movement and storage in the soil, evaporation of moisture from the soil surface, infiltration rate of moisture in the soil as well as rainfall intensity and duration, affect the amount of water to be harvested from a land surface (Bouwer, 1990). The soil amelioration proposed in this study is therefore not a panacea for improving the rainwater harvesting properties of any land surface.

RUNOFF FROM A SLOPED FIELD UNDER DUNE SAND CONDITIONS

In order to understand the rainfall-runoff process for improving rainwater harvesting, experiments were conducted on a sloped field inside a glasshouse at the Arid Land Research Center of Tottori University, Japan. Two pairs 20° and 30° slopes, each pair facing north and south, were used in the experiment. Each 20° field measured 2.7 m × 1.8 m on the surface, while each 30° slope measured 2.9 m × 1.8 m and had Tottori dune sand packed to a bulk density of 1.6 gcm^{-3}. Amplitude Domain Reflectometer (ADR) Theta soil moisture sensors (Cambridge Delta-T™ ML-1® probes) were inserted into the soil at depths of 10, 50 and 100 cm for each field. These sensors measure the dielectric constant of the soil matrix through the amplitude difference between a 100 MHz incident sinusoidal wave transmitted into the soil, and a reflected wave received by the sensor. This is output through internal circuitry as voltage within a range of 0-1000 mV. The moisture content of the soil is directly related to the output voltage recorded by the ADR. An initial plot of sets moisture content against measured dielectric constant provides calibration for converting the measured dielectric constants to equivalent moisture content in the soil. A schematic of the experimental layout is shown in Figure 2. Overhead sprinkler nozzles placed 2.5 m above the slope provided a spray of "rain" water at an average intensity of 38.5 $mmhr^{-1}$. The measured distribution properties of the rainfall system and the layout of the process for measuring uniformity index are shown in Figure 3. The measured characteristics of the rain system are presented in Table 1.

The ADR theta probes, temperature and humidity sensors, wind speed sensor and solar radiation sensor (Epply Pyranometer, EKO™ MS-62) were then connected to a Campbell Scientific™ CR23X® datalogger. This were used to log soil moisture information and all other datasets at 1 s intervals and the average in a 1 min interval was output to final storage.

Since the ADR sensor readings are dependent on moisture variation as well as on soil type, they were calibrated for Tottori dune sand. The process involved adding known quantities of water to a mass of initially dry soil sample. Several samples, from dry to saturated conditions were prepared and thoroughly mixed. The probes

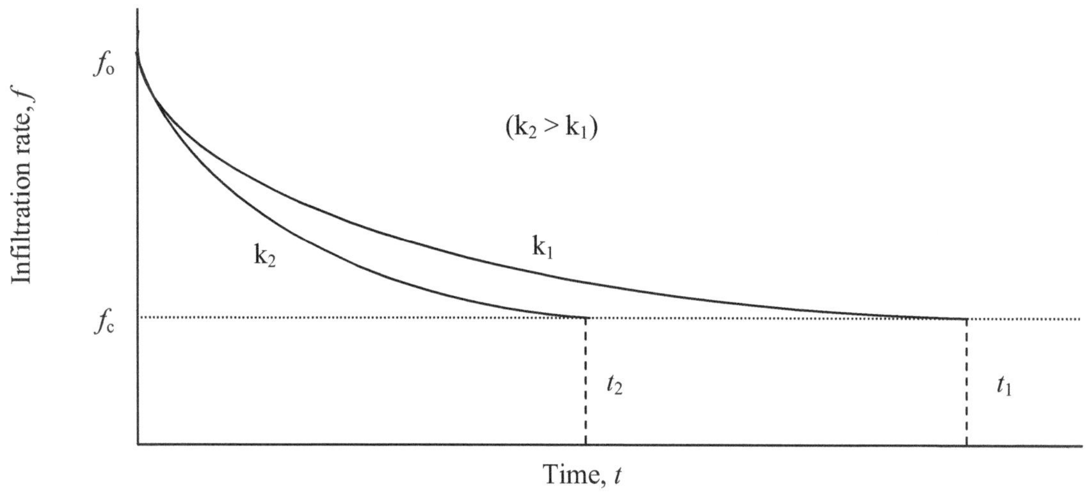

Figure 1. Effect of decay constant k, on time of runoff incidence (as k increases, t decreases).

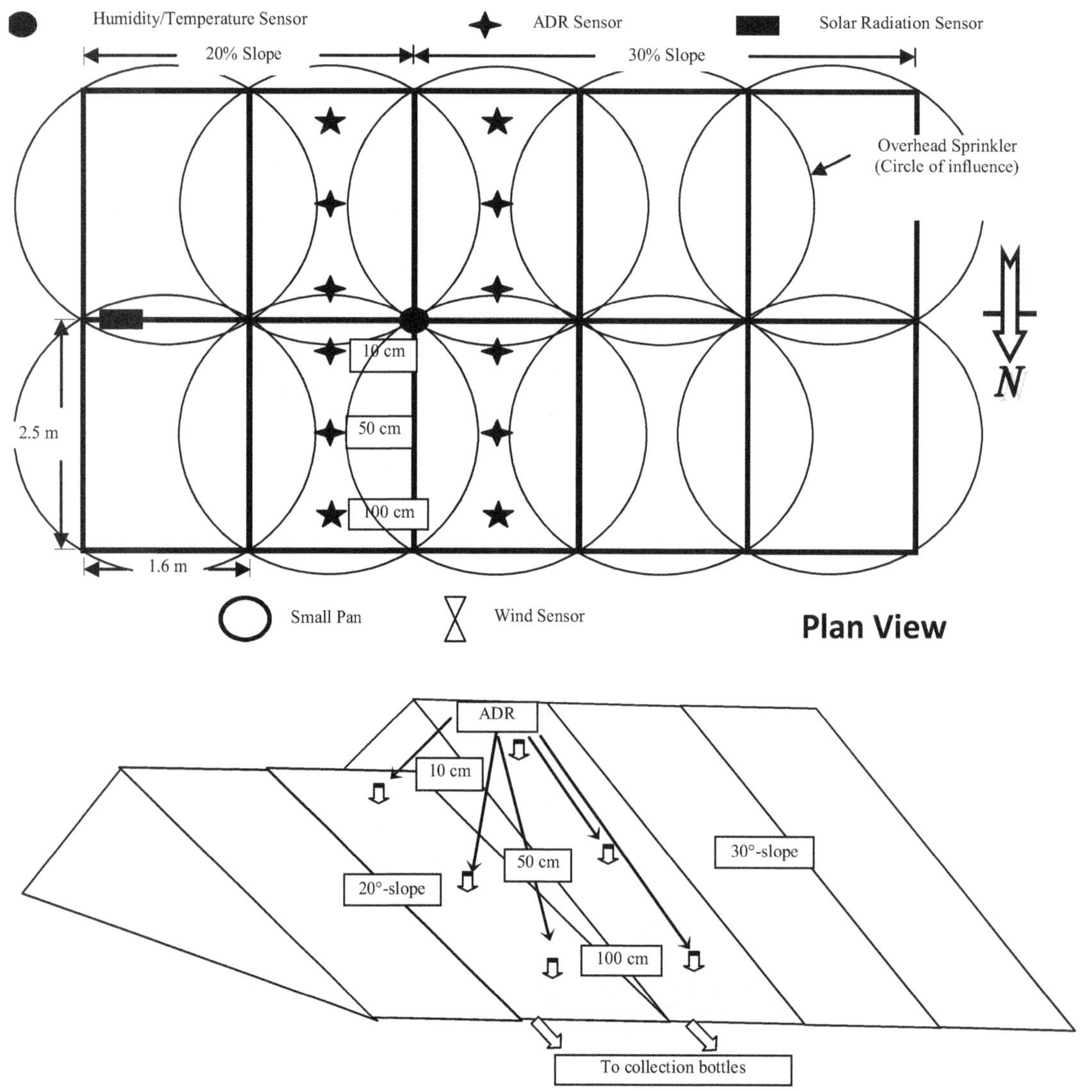

Figure 2. Schematic of sloped rainfall-runoff experimental layout showing location of ADR probes.

Figure 3. Setup for measuring uniformity and spray characteristics of overhead rainfall system (cups collect rain water.

Table 1. Characteristics of rainfall system for sloped rainfall-runoff experiment.

Mean sprinkler intensity = 38.5 mmh^{-1}	20° N slope	30° N slope	20°S slope	30° S slope
Christiansen's uniformity index (%)	50.2	56.8	53.3	54.7
Distribution coefficient (%)	41.9	40.3	47.4	49.8

were then successively inserted into the moist soil and readings were logged onto the CR23X datalogger for 5 s. A sample each of the moist soils was weighed and placed in a constant temperature oven (105°C) for 24 h until the sample was oven dry. Analysis of the gravimetric moisture content was done and converted to volumetric moisture content using the dry bulk density of the soil. The 5n s average ADR readings stored on the datalogger were then plotted against the volumetric moisture content. Regression analyses of the lines of best fit were obtained and the results are represented in Figure 4. Correlation coefficients of more than 0.99 were obtained using a third order polynomial equation as shown in the figure. With this data, it was possible to monitor the moisture variation in the slope experiment by applying the regression equation of each sensor. The experiment was carried out for 79 h and 43 min and the data collected were analysed.

Though the overhead sprinkler system was continuously operated for almost five days, there was no runoff from either the 20° or 30° sand plots in both north and south inclinations. This may be attributed to the high infiltration rate of sandy soils though some runoff might have occurred if higher rain intensities had been applied. Figure 5. illustrates these observations.

Immediately after the start of rain, soil moisture is seen to rapidly rise at all the depths starting at the surface and moving downwards. The rise in moisture content is steep indicating a rapid movement of water through the dry soil profile. Moisture content distribution through the profile is not uniform but the movement of the waterfront is identified from the sudden change in the moisture content at subsequent depths. In all the fields, the sequence of moisture content change is from the 10 cm-depth to 50 cm and finally to 100 cm. The probes were not placed in the same vertical plane but were horizontally displaced to cover the field as shown in Figure 2. It is therefore difficult to attribute the increase in moisture content at the lower depths only to vertical movement of water alone. It is probable that lateral movement of water contributes to this

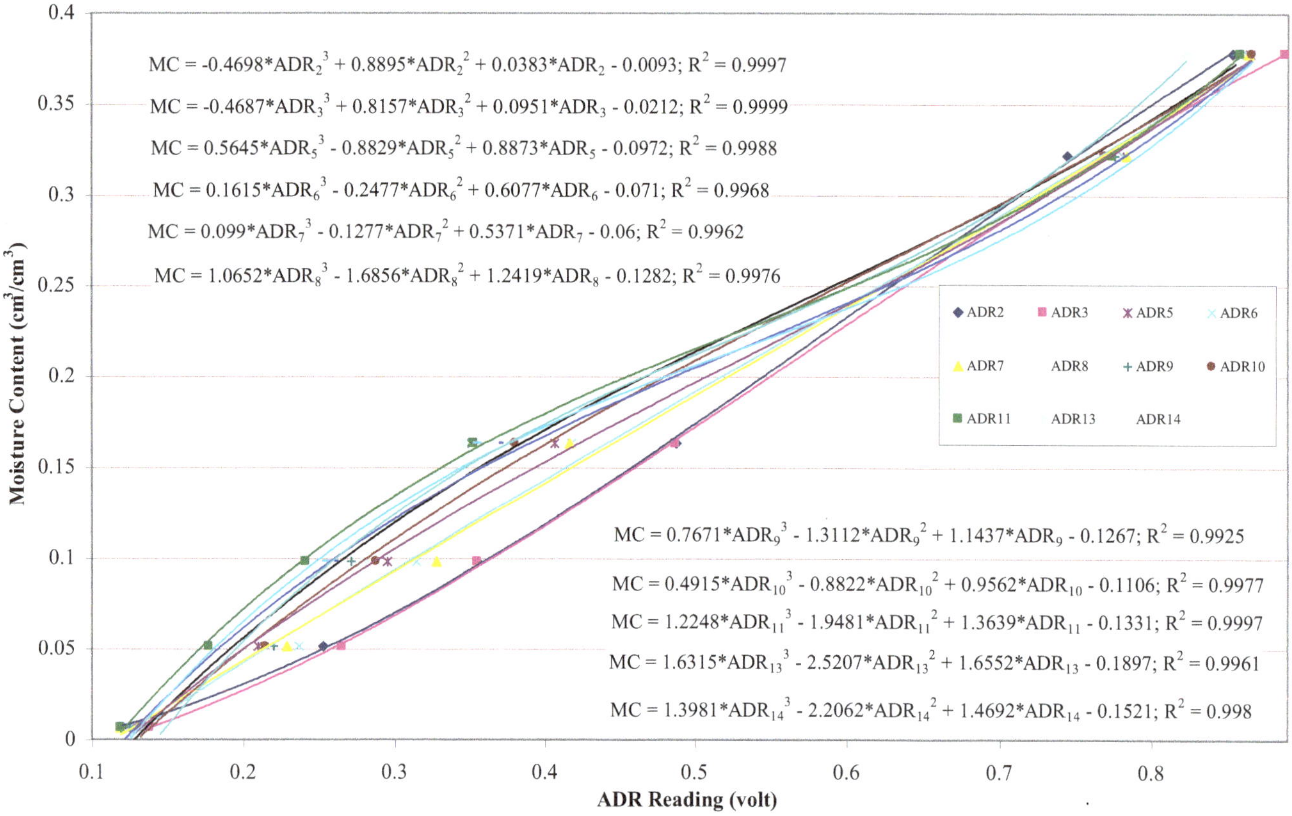

Figure 4. Calibration curves for ADR sensors used in rainfall-runoff experiment.

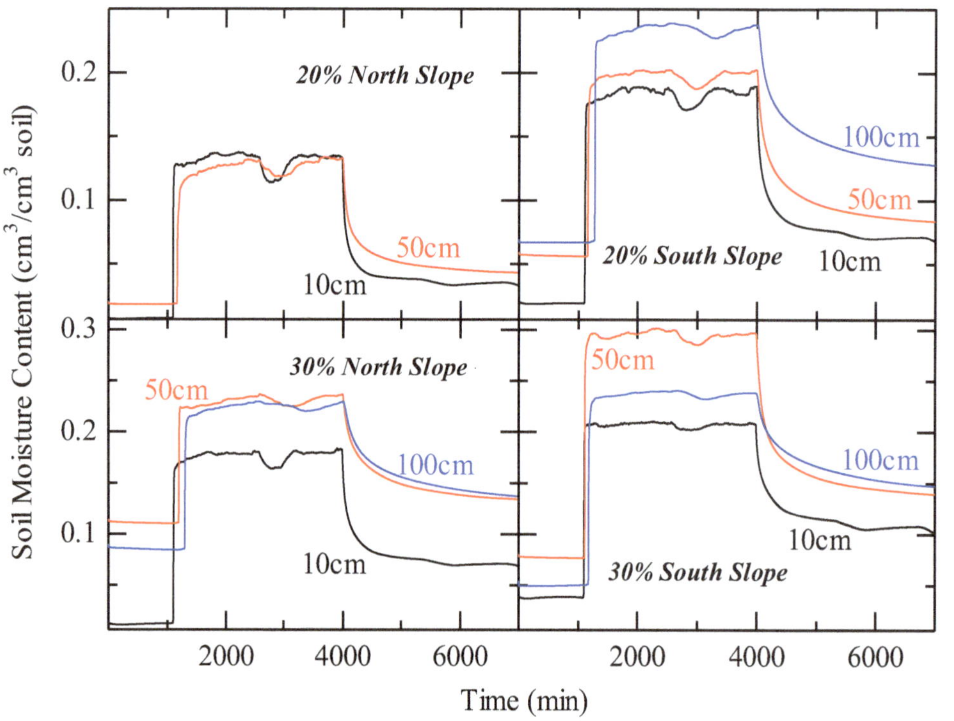

Figure 5. Soil moisture variations in sloped sand fields at different depths, before, during and after rainfall.

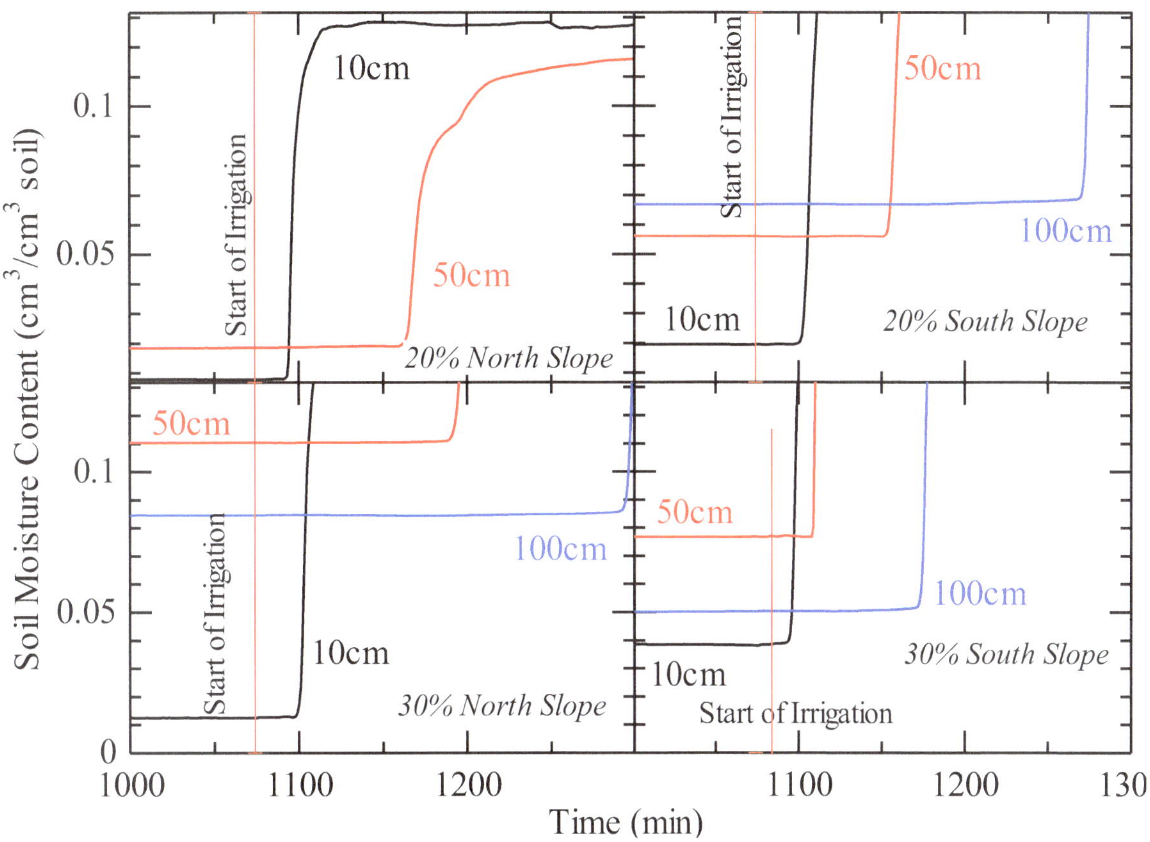

Figure 6. Rapid increases in moisture content of soil profile due to high infiltration rate.

phenomenon. Water entering the soil accumulates until its weight is more than the attractive forces between the soil and water. The water then moves to lower layers as a result of the pull of gravity, wetting the soil profile in the process. The time taken for the waterfront to move from one depth to another is used to determine the rate of movement of the water when the soil is being wetted. Infiltration rate is approximated to the rate of movement of the wetting waterfront. The infiltration rate for each layer of soil is obtained from Equation 2, which approximates the slope of the wetting front curve, n

$$If = \frac{\Delta d}{\Delta t}$$

where If is Infiltration rate (mm/h)

Δd and Δ are depth of water movement (mm) and time taken (h) respectively.

When the rainfall system is switched off, moisture content in the soil profile is seen to decrease rapidly. This shows that the soil is very permeable and infiltration rate is high. The rate of water movement is approximate to the rate at which moisture content at the different depths, change. In order to estimate the rate at which water infiltrates in the soil, the time taken for moisture content change from one depth to the next is calculated. Since the depth between the reference points are known, the rate in terms of distance per unit time can be deduced. Figures 6 and 7 respectively show enlarged portions of the wetting and drying phases of the moisture content curves at the three depths.

The infiltration rates calculated during the initial wetting are presented for the three depths in Table 2. Infiltration rates calculated soon after rainfall is stopped are also presented in Table 3.The high infiltration capacity of Tottori dune sand means that runoff is only possible with rainfall intensities in excess of 6,000 mmh^{-1}.

Runoff are also achieved when an impermeable surface exist at a shallow depth beneath the soil profile. This is the case in the rice plains of the Oda River basin of Ghana where sandy soils are underlain at average depths of 1 to 1.5 m by heavier clay and loam soils (Opoku-Duah et al., 1999). Though most of the soil profile is sandy in nature, runoff occurs at the slightest rain event enabling the fields to be irrigated.

Water harvesting from sand slopes may be achieved when the entire soil depth is saturated because of the presence of a high water table. This is the case at the Aframso rice plains of Ghana where the main problem is

flooding during the rainy season (Opoku-Duah et al., 2000).

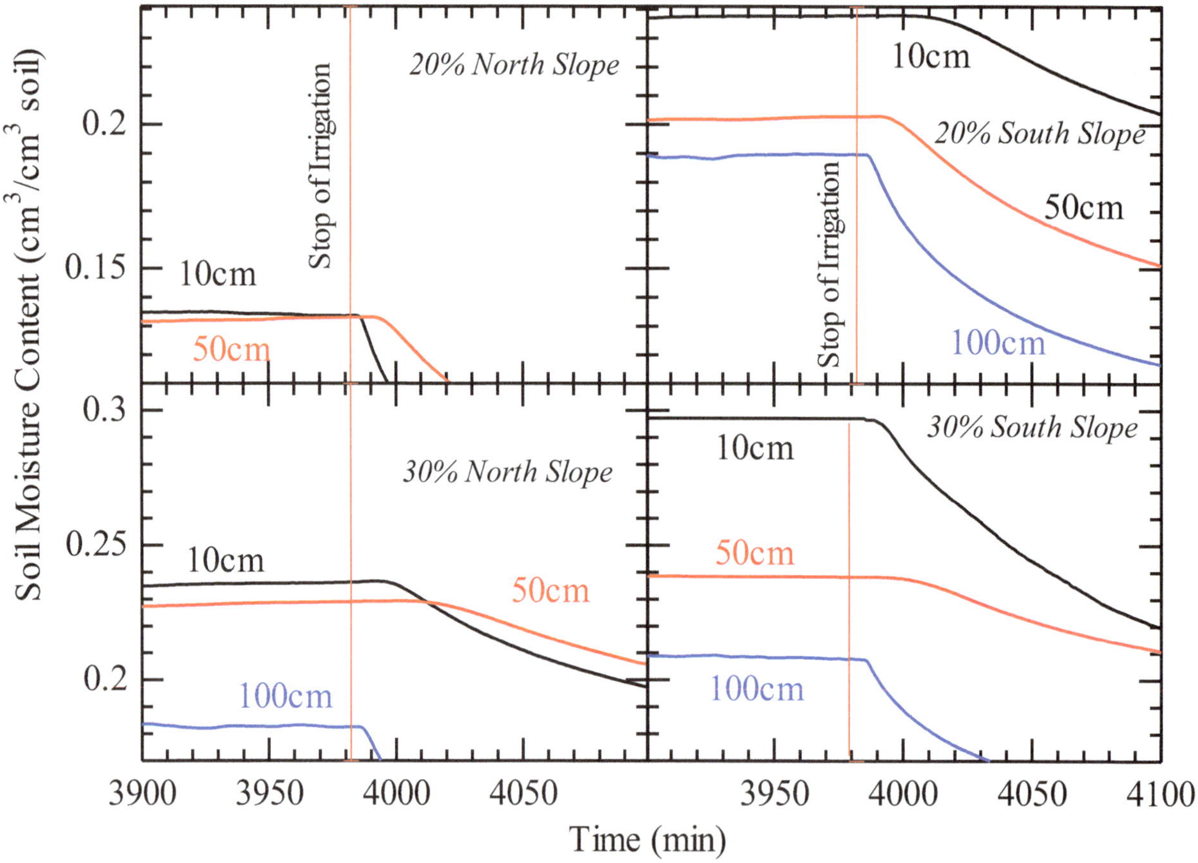

Figure 7. Rapid depletion in moisture content of soil profile due to high percolation rate.

Table 2. Initial infiltration rates calculated at start of rainfall.

Profile range (cm)	Δ*d* (cm)	20° North slope		20° South slope		30° North slope		30° South slope	
		Δ*t* (min)	*If* (mmh^{-1})	Δ*t* (min)	*If* (mmh^{-1})	Δ*t* (min)	*If* (mmh^{-1})	Δ*t* (min)	*If* (mmh^{-1})
0-10	10	16	375	22	273	23	261	17	353
10-50	40	69	348	55	436	88	273	16	1500
50-100	50	N/A	N/A	116	259	104	288	60	500
0-50	50	86	353	77	390	111	270	33	909
0-100	100	N/A	N/A	193	311	192	313	93	645

N/A – data not available.

Table 3. Terminal infiltration rates calculated after rainfall is stopped.

Profile range (cm)	Δ*d* (cm)	20° North slope		20° South slope		30° North slope		30° South slope	
		Δ*t* (min)	*If* (mmh^{-1})	Δ*t* (min)	*If* (mmh^{-1})	Δ*t* (min)	*If* (mmh^{-1})	Δ*t* (min)	*If* (mmh^{-1})
0-10	10	3	2000	3	3333	3	2000	3	2000
10-50	40	6	4000	6	4000	8	3000	4	6000
50-100	50	N/A	N/A	13	2308	20	1500	11	2727
0-50	50	9	3333	9	3333	11	2727	7	4286
0-100	100	N/A	N/A	22	2727	31	1935	18	3333

N/A – data not available.

During the rainy season, runoff from higher slopes flow into the plains, which are sandy and form a reservoir storing and releasing water by lateral flow into the river. At the Gbi-Godenu rice plains of Ghana, runoff is obtained through seepage flow at the base of the slope caused by lateral flow as a result of reduced infiltration caused by iron-pans. Alternatively, the sand slopes can be lined with cheep rubber sheeting, clay or loamy soils or planted with Vertiver grass having dense root network. This would also likely generate some runoff for agricultural use.

CONCLUSIONS

Soil amelioration techniques for improving runoff characteristics of soils for better water harvesting are available and include using plastic sheets, bitumen spreads, concrete layers and chemical additives mixed into the top soils to improve bonding to the soils (Prinz et al., 1997). These methods are quite expensive and usually interfere with the natural environment and deprive the land of other possible uses. The use of less permeable soils as a top layer dressing to reduce infiltration and enhance runoff presents an interesting and viable option that could give appreciable results.

This will especially be useful in areas with sand, sandy loam and loamy sand soils having high infiltration rates. Various locations in Ghana especially along the coastal stretch of the country and in some inland valley alluvial plains could benefit from this process if these areas are consciously lined with laterite and clay soils. When carefully designed and constructed, these rainwater harvesting fields could concentrate and channel rainwater into dugouts and reservoirs for storage and used for agricultural and other uses to improve the livelihoods of the people. Water that currently infiltrate into the soil and is not available for use could be harvested from the treated lands and made available for development of the riparian communities.

The water harvesting options presented are for agricultural water uses only as quality issues need to be addressed to enable such water to be potable. Currently serious perceptions and norms exist in the minds of most users that make the deployment of roof rainwater harvesting a problem (Hans et al., 1999, 1999a).

Soils with infiltration rates in excess of 6,000 mmh^{-1} did not generate any runoff when simulated rainfall with average intensity of 38.5 mmh^{-1} was continuously administered on them. This was in spite of continuous rainfall being administered for a period of 4,783 min. It is recommended to investigate the effect of soil surface treatment on runoff occurrence using less permeable soils for top dressing of dune sand under similar conditions.

ACKNOWLEDGEMENT

The authors acknowledge the kind sponsorship of the Japanese Ministry of Education and Science (Mumbusho) that enabled the research to be undertaken.

REFERENCES

Amu-Mensah FK, Sabi EB, (2006). Rainwater Harvesting; Methods of Capture, Storage, Treatment and Uses: Challenge to Ghanaian Scientists, Technologists and Architects, in Rainwater Harvesting; a sustainable solution to water shortage problems in Ghana. Proceedings of the 10th Biennial Workshop. Ghana Science Association. GNAT Hall, Accra.

Amu-Mensah FK, Yamamoto T, Inoue M (2000). Planning for Sustainable Agriculture in Ghana using Tank Irrigation. Trans. JSIDRE 205:75-84, Japan.

Bouwer H (1990). Water Conservation in Arid Zones. In Water Saving Techniques for Plant Growth, (Eds. H.J.W. Verplancke, E.B.A. De Strooper and M.F.L. De Boodt), NATO ASI Series E. Appl. Sci. 217:21-31.

Doorenbos J, Pruitt WO (1977). Guidelines for predicting Crop Water Requirements. FAO Irrigation and Drainage, Paper 24, Rome, 179.

Feddes RA, Bastiaanssen WGM (1990). Forecasting Soil-Water-Plant-Atmosphere Interactions in Arid Regions, In: Water Saving Techniques for Plant Growth, (eds. H.J.W. Verplancke, E.B.A. De Strooper and M.F.L. De Boodt), NATO ASI Series E. Appl. Sci. 217:57-78.

Hans H (1999). Attitudes towards Rainwater Harvesting. Rainwater Harvesting Research Group: FAKT, Warwick University, Lanka Rainwater Harvesting Forum, IIT Delhi

Hans H (1999a). Domestic Rainwater Harvesting: Perceptions of Water Professionals and the way forward. Rainwater Harvesting Research Group: FAKT, Warwick University, Lanka Rainwater Harvesting Forum, IIT Delhi

MWRWH (1998). Water Resources Management (WARM) Study Synthesis Report. Ministry of Water Resources, Works and Housing, Accra.

Opoku-Duah S, Kankam-Yeboah K, Amu-Mensah FK (2000). Determination of Water Requirements for Producing Irrigated Rice and other crops in the Afram River valley bottom at Aframso, Ghana. Ghana J. Sci. 40:15-24.

Opoku-Duah S, Kankam-Yeboah K, Amu-Mensah FK (1999). Water Quality for Irrigated Rice-based cropping system in the Oda river valley bottom at Besease, Ghana. Ghana J. Sci. 39:63-72.

Prinz D, Wolfer S, Siegert K, Dessalegne Agaze T (1997). Water Harvesting for Crop Production, FAO Training Course on CD-ROM (September 1997) Rome.

Zolfaghari AA, Mirzaee S, Gorji M (2012). Comparison of Different Models for Estimating Cumulative Infiltration. Int. J. Soil Sci. 7(3):108-115.

5

Management of urban water for domestic and industrial uses and sustainability in Anambra State, Nigeria

A. C. C. Ezeabasili[1] , O. L. Anike[2] and A.U. Okonkwo[3]

[1]Heriot Watt University, England, UK.
[2]Nnamdi Azikiwe University, Awka, Anambra State, Nigeria
[3]Anambra State University, Uli, Anambra State, Nigeria.

Nigeria has abundant water resources. This natural endowment is evident in the yearly rainfall and large surface bodies of water-rivers, streams as well as abundant reservoirs of underground water. However, because of lack of good developmental policy, the rational use of the water resources still poses problems and challenges in most states of the country. For some time now these water problems have been left with the River Basin development authorities with little success, primarily because of lack of good management and best practices, despite huge sum of money spent. Although there are surplus surface and underground water resources within the Anambra basin, basic water supply for domestic and industrial use have been in very short supply for the teeming 5 million population despite the fact that government through international programmes like the Millennium Development Goals (MDGs), UNICEF, EU has done a lot to aid water supply. This paper is timely because of a recent introduction of a major regional water supply project earmarked to take off in Onitsha, the industrial and commercial hub of Anambra. It is expected that the quantity and quality of water would improve with this project to be financed by the government in urban areas of Onitsha, Awka and Nnewi. Published as well as unpublished secondary sources were used to present the access to drinking water in Anambra State. Reform efforts are currently going on in the state and a review of the reform reveals the effects of political and economic challenges on the existing strategies.

Key words: Urban water, water reforms, sustainability, sectors.

INTRODUCTION

Nigeria has not been able to guarantee the most important need of safe drinking water for its citizens and small population of industries. Every part of Anambra State is battling with insufficient of this basic necessity. The local government headquarters of Nnewi, Awka, Ekwuluobia, Ihiala, Ogidi, Abagana and Onitsha have been without sufficient domestic and industrial water for the past forty years since after the Nigeria Civil war. Major sources of water were rivers and stream for the first twenty years while the development of private water for individual homes and industries by the private sector, local governments and international agencies have recently impacted on the provision of water. The problem of raising the level of drinking water from low to medium term and in cost effective ways is imperative in the state. The objective of the paper is to assess the ongoing water sector reform in Anambra State of Nigeria, to determine the key barriers to more effective reform for urban water provision with respect to water quality and water sector efficiency. Ongoing attempts by government including those of private sectors, to reform the water sector and other necessary recommendations are included in this review.

In the Urban areas of Anambra State piped water supplies does not exist in the seven (7) major cities of Onitsha, Awka and Nnewi, Abagana, Ihiala, Ogidi Ekwulubia. Virtually 95% of the population and industries receive their required water from water boreholes. It is only the rich that can afford the luxury of providing personal water boreholes for personal uses and industries. The present Administration in the state is planning major water schemes for Nnewi, Awka and Onitsha which is presently being executed. Other small town water scheme is either completed or ongoing.

THE CURRENT STATE OF URBAN WATER SUPPLY

Water availability and quality

The water supply in the urban cities was only available sparingly and the water is of questionable quality. Private and Public water borehole are common sites and supply most of the rected water. Intermittent water supply, and unpredictable service from boreholes and vendors impose both financial and health costs on the three major cities of Nnewi, Onitsha and Awka.

Many households within the urban cities were found to have undertaken long-term investments in the form of water tanks, handpumps or boreholes. Households with water tanks install booster pumps and pump water directly to water tanks. This increases the risks of contamination of the general water supply. Water tanks though popular are sources of various biological contaminants. Water pump (submersible) fixed to clevated water tanks as well. The WHO Guidelines for Drinking-water Quality (WHO 1998, 1997, 1993) are assessing the health risks posed by contaminants in drinking water. This provides a primary health requirement for a sufficient water supply, which the Government of Nigeria takes to mean about 40 L per person per day. The second requirement is that the water be microbiologically safe (FMoWR, 2006).

In most developing countries, the primary contaminant of surface and ground waters is human and animal waste (McKenzie, 2007). The WHO guidelines suggest that *Escherichia coli* should not be detectable in a 100-ml sample of water for the water to be considered to be of moderate good quality. The Government of Nigeria accepts these guidelines but has been unable to ensure that they are met. Water-borne diseases from faecal contamination are some of the biggest public health risks in the country. The incidence of typhoid fever diagnosed rightly or wrongly remains contentious till now. Monitoring of water quality in Nigeria cities is haphazard or nonexistent. Standards for drinking water that are actually enforced could have enormous positive impacts on public health, but for this to occur, the procedures for water testing and data sharing have to be made regular, standardized and public.

Water resources

Anambra State is drained principally by a number of rivers which include River Niger, Anambra River, Idemili and Mamu River Nkisi Obizi etc (Figure 1). In some areas the formation have confined aquifers and their depths depend on the nature of the formation relative high rainfall which occurs within 7 months of the year ensures the sustenance of perennial streams and rivers traversing the state including effective re-charge of groundwater acquifers (Muoghalu and Okonkwo, 1998). The majority of small water supply systems in the state are ground water based while the big water supply schemes are surface water based. The Nkissi stream is the main source of raw water for the existing Greater Onisha Water Supply Scheme, but for Awka and Nnewi, there is no scheme yet in existence. The greater Onitsha water scheme presently not functional is being considered as a major concession for the private sector. Before this when the public sector was responsible for the water works, epileptic water supply is available as provided by the state water cooperation, the water corporation was highly subsided by the Anambra State government but was finally starved of the subsidy, between 1999 and 2003 when its activities collapsed (AWSSP, 2009).

AREAS FOR URBAN REFORM

Pricing reform

Recovering at least part of the cost of a new water system or of upgrading and maintaining an existing water system is the primary rationale for pricing reform everywhere. Several studies have argued that poor people will pay for water if it is conveniently and reliably supplied and that appropriate pricing reform can promote both efficiency and equity (Boland and Dale, 2000; Mcintosh, 2003). Existing evidence suggests that many low-income households in Nigeria can afford to pay more for water, particularly if the increase in prices is accompanied by better service.

One might question whether metering in very low-income areas is cost-effective for the water utilities. In general, piped connections to homes in slums were not metered; therefore we do not have the data. Metering is

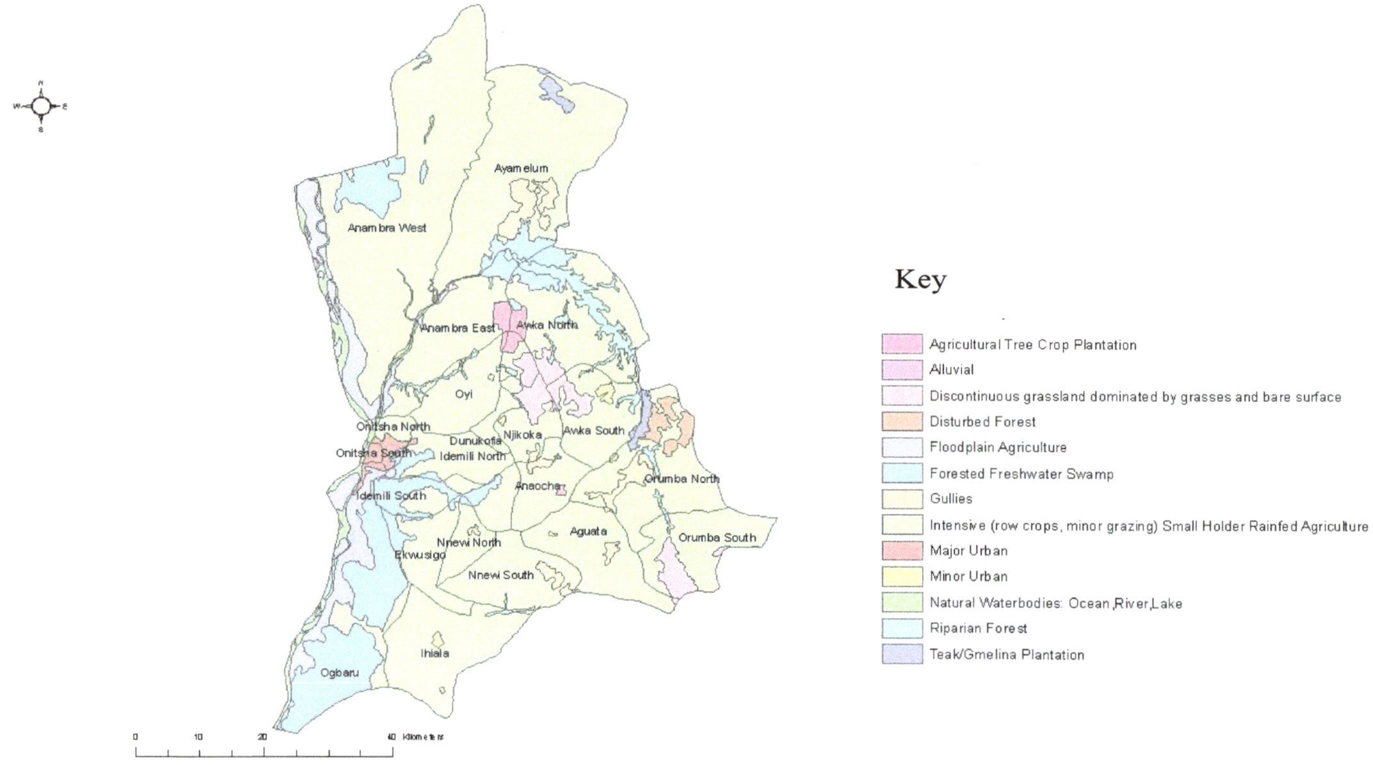

Figure 1. Anambra State Local Governments.

arbitrary done and the public does not have the ability to check the excesses of the water corporation.

However, even among low-income households there may be scope for metering with some adaptation in the future. Provision of metering devices and its functionality programme implementation in the state would likely improve with the modern water works that are envisaged for the future.

Financial reforms

A traditional mechanism for raising the capital needed for water and sewerage system expansions and upgrades is the bond, usually issued without guarantees from the federal government. Through such bonds, private credit markets lend money to state governments for a fixed period of time and at predetermined interest rates. These best practices in the developing countries have not worked well in Nigeria where state and Federal governments had been into direct participation. This practice has been variously abused in the rare instance where it was permitted by legislation in the states and the local governments especially between 1999 and 2007.

Private sector participation in urban water delivery

Over the last decade, privatization to a greater or less degree has been seen as one of the primary ways to infuse capital into the urban water sector and to overcome some of the inefficiencies of management. It has been urged upon developing countries by international lending agencies as an essential component of water sector reform.

However, the 2002 National Water Policy of the Government of Nigeria for the first time called for the encouragement of private sectors participation (PSP) in water resources. Private sector participation should be encouraged in planning, development and management of water resource projects for diverse uses, wherever feasible. Private sector participation incorporates a wide range of private sector involvement. The method of participation is coming on new in the state and would be in adapted forms. At one end lies contracting out of services to the private sector, such as mains repair, billing and collecting. Such arrangements are relatively straightforward and involve short-term (5 years) renewable contracts. More private involvement occurs under longer-term (20 to 35 years) concession and Build-own-transfer (BOT) contracts. This entails that under a concession, a private firm manages and operates the whole utility at its own commercial risk. BOT contracts are used for major investment in new facilities. At the other end of the public-private spectrum lies full divestiture, whereby the Government sells the assets of the water supply company to a private firm, who runs it on a permanent basis subject to government regulation.

Evaluation of the effects of PSP on urban water supplies has been constrained by the overall poor quality of data available and the small number of cases from which to draw conclusions. As in India, in many parts of the world the public utility does not release regular information on costs, water quality, the size of the network, etc., making it difficult to measure the pre-privatization trends in access, costs, and quality, and thus to determine what would have happened in the absence of private involvement. However, much of the criticism regarding privatization is that it will result in large increases in water tariffs, making water unaffordable to the poor (Sridhar, 2003b). But it is argued that it is better to have costly potable water than none. Overall, the experience of other countries suggests that PSP in the urban water sector may or may not, improve efficiency and provide better service to the poor. While it is too soon to evaluate the effects of private involvement in urban water in Anambra State, several projects are now underway. We comment here on three ongoing efforts. Nigeria's furthest step towards full privatization of water supply is the Build-Own-Operate-Transfer (BOOT) contract which is planned to be carried out by the Government at Onitsha.

Further private sector engagement such as concessions and leasing agreements will be difficult without prior pricing reform. Successful PSP therefore requires public awareness campaigns of the true costs of the current policies. It is equally important not to oversell the reforms possibly under private sector involvement through unrealistic goals and exaggerated public statements.

With the unattractiveness of peri-urban areas to the formal private sector (Cairncross, 2003), more city governments should consider recognizing, contracting with, and regulating local water entrepreneurs as mainstream rather than interim delivery mechanisms. In the absence of official recognition, water vendors will continue operating anyway, but without quality controls, price monitoring or accountability.

Political economy of reform

In light of the widespread inefficiencies in the water sector in developing countries, there is clearly ample scope for reform. However, as in most democratic nations, any major reform needs to survive the political process, while even small changes in prices require public approval. As Noll et al. (2000) argued, several features of urban water systems make reform difficult.

The political benefits of water reform are often low, as reform may affect employment and investment in the public enterprise. Changes in prices and staff layoffs are more visible to the public than improved operating efficiency, reductions in state subsidies, and small improvements in quality.

The longer reform is delayed, the more difficult it becomes. In particular, critics viewed the price rises as laying the ground for privatization of water, with little apparent recognition that even an efficient public utility would require prices substantially higher than those prevailing pre-reform.

Promoting changes in the water sector therefore requires finding a way to raise the political benefits of reform efforts. Public awareness could be further increased by reporting the results of water quality tests, along with information on hours of service. This information could be coupled with accessible data on how much water subsidies take from the state budget, and price increases could be explicitly linked to targeted improvements in key sectors. This would work for the informed members of the society and those with different positive political inclination.

EXISTING SITUATION OF THE WATER SUPPLY AND ENVIRONMENT IN ANAMBRA STATE

Access to water supply services

There is very little data available to inform the drafting of the policy and its implementation strategy. However, the Federal Ministry of Agriculture and Water Resources (FMAWR) appointed a consultant to carry out a baseline survey of Water Supply and Sanitation (WSS) in the state. The findings of these study shows that access to safe water from public water supply in all the Local government of the state ranges from as low 0.2% to above 40%. It is estimated that on the average, access to potable water is not more than 20% across the state.

Condition of the existing government water supply infrastructures in Anambra State

Recent study by Water Supply and Sanitation Program (WSSP) shows that the State water supply infrastructures are in a very poor state. Out of 62 systems owned by the State Water Corporation only three of them are functional. Water supply infrastructure identified under the Federal Government baseline survey for Anambra State comprised 497 motorized boreholes, 72 hand pump boreholes and 7 surface water schemes. The total installed capacity excluding the Nkissi surface Water Scheme Which Serve Onitsha North, Onitsha South, Idemili North, Idemili South and Ogbaru LGAs is 30,795 m^3/day.

Greater Onitsha Water Supply Scheme (Nkissi Water Works) which was destroyed by massive siltation of the intake facilities is being considered for concession and rehabilitation. The State Government had entered into concession with a Jordanian Firm for the provision of water and an independent power distribution at Onitsha.

The take off is still awaited. This was abandoned for a South African firm with the mode of procurement changing to the traditional procurement route. The Greater Onitsha Water Scheme was the live wire of Anambra State Water Corporation generating over 90% of its revenue. The corporation has been out of operation since the past 15 years due to inefficient operation, corruption and other ills associated with public installations in Nigeria.

Capital funding and development of the sector

The sector has been poorly funded. According to the Anambra Sate Economic Empowerment Strategy (SEEDS) document, it was estimated that over N8.0 billion was required to improve the state water supply between the year 2004 and 2007. In reality less that 20% of that was actually spent in the sector during the period. Meanwhile water demand continued to increase annually due to increase in population and economic activities in the state, thus the water coverage has actually continued to decline during the period.

Supply-demand gap

It has been estimated that the demand for water in Anambra State in the year 2005 was 213,952 m^3 per day. This will rise to 27,313 m^3/day in 2015. The percentage of water available for the state or the theoretical percentage of water supply against demand was also found to be 7.2% (Anambra State WSSSP, 2009). This means that on a state wide basis only 7.2% of safe water requirement is met in 2005 or is available in Anambra State and this is very poor. Clearly then, extensive work would have to be embarked upon to find new water supply sources and establish new water schemes to close up the demand gap. This has been done and is found in the creater Onitsha water scheme, proposed greater Nnewi and Awka water scheme, the small town water schemes and water boreholes projects scattered over the state.

Water supply promoters in the Anambra State

According to the recent baseline survey on WSS conducted by the Federal Government under the Urban Water Sector Reform, Anambra State has 102 small water scheme or 17.7% of the federal government promoted water. And 28 of these schemes or 27.5% are functional.
The State Government promotes 52 of the schemes which is 9% of the total and only 13 of them or 25% are functional. Local Governments in Anambra State promoted 8 or 1.4% of all the schemes in the state and 3 of these or 37.5% are functional. Donors promoted 148 schemes or 25.6% of the total number of schemes of which 122 or 82.4% are functional. Members of the various communities promote 257 schemes or 44.5% of all schemes out of which 131 or 89.9% are functional. Together Donors and members of the various communities provided 405 schemes or 70.2% of all schemes out of which 353 or 87.2% are functional. Donors and Communities constitute the backbone of potable water supply in Anambra State and the functionality of these schemes is superior to other providers (AWSSR, 2009; WSP, 2001).

The Local Government Areas (LGAs) appear now to be stepping up involvement in water supply in this state. Clearly then, the most sustainable way forward is for the government to take the path of developing water supply in the state through community managed systems, this will best be done through government strong financial support to community water supply projects.

Private sector participation on water supply

Over 80% of water supply services provision in the State is in private hands and the service providers charge more than ten times the expected public sector charging rates.

Water pricing challenges

Bottled water: *Prices for bottled water are set in the market. Retail prices vary widely between countries, brands, bottle sizes and place of sale (supermarket, restaurant). They can be as high as ₦600 per cubic meter.*

Tanker trucks: Prices for water sold by tanker trucks in bulk, which is common in cities, for households without access to piped water supply. Prices for trucked water vary between about ₦1, 500,000 per cubic meter.

Utility tariffs: *Prices for piped water supply provided by utilities, be they publicly or privately managed, are determined administratively. They vary from ₦100 when it was available per cubic meter.*

Irrigation: *Prices for irrigation water that is being provided by a public agency are also typically determined administratively.*

The water rate charges by the water corporation were for the following:

1. Metered Supply: M_1 -domestic; M_2 - industrial/commercial.
2. Big Hotels; Bakeries, breweries; block moulding, car washing; laundry services, banks, filling stations.
3. Small hospitals; maternities.

One does not foresee what the new rate would come to and the success of selling these rates to the public since some individual and community schemes have made some organization forget the public water works.

Cost recovery and sustainability

From the tariff approved for the state water corporation, it is evident that the domestic connection charges are lower than the production cost by almost 50% while the Anambra State Government is required to produce and sell water in a way as to cover its operational cost. When a utility earns less revenue than its expenses in a particular year, it run short of funds to properly run its services in the following year and this leads to lower water production with the resultant low sales due to the low quality of service. The resultant effect is the dearth of the consumer's willingness to pay. The utility will then have even less revenue to operate in the following year. The vicious cycle will continue leading to the continued deterioration or service delay. This policy is addressing this by suggesting sweeping reforms.

Subsidy and access to WSS services

Common argument usually attributed to the application of flat subsidy to domestic consumers is to say that water is being made affordable as part of the government commitment to its citizens. Unfortunately the opposite is being achieved in this case. The reality is that majority of people in urban areas of Anambra State do not have water connections in their houses. Those connected are mostly those that can afford to pay higher price. Instead, they are benefiting from government subvention originally intended to assist the poor gain access to service. Furthermore the poor are the ones that purchase water from vendors and tanker operators at the exorbitant rates. The alternative is for them to get water from contaminated sources that are unhealthy. Sickness leads to low human productivity which in turn increases poverty.

Public/private sector participation

The private water supply operators are in business because the consumers are ready to pay for their services. This leads to the conclusion that in spite of the reported high level of poverty, large section of the population is willing and ready to pay higher prices for water supply if the service is reliable. This finding opens a whole new opportunity to carry out two sweeping reforms; the first is to produce a level playing field for both government and private companies. This would release latent potentials to compete favourably so that both can generate sufficient revenue to expand the services.

The second is the opportunity to redefine roles of all the actors in the provision of services especially in separating policy, service delivery and regulations. Experience has shown that reforms like this brought tremendous improvement in water supply in both developed and developing countries.

Design and application of appropriate tariff in Anambra State

Sufficient data on the production cost of water supply for urban areas is not available to enable proper costing of water production. This is perhaps because the major water treatment plant has been out of operation for a number of years. It is not clear what the guiding principles for setting tariff lower than production cost without giving details on how the shortfall in the revenue will be financed. The tariff structure should take into account the following:

1. To allow the water supply system to have resource and liquidity maintenance that consumers are able and willing to pay as failure to do so will mean to them spending more money to get potable water from other sources.
2. Put into consideration the poor who may be denied access to clean water if the rates are too high.
3. Be assured that it will not cause social unrest and is politically acceptable to the Government.

Unaccounted-for water

Unaccounted for water refers to water losses in the form of system leakages in the distribution network (which includes physical losses due to pipe bursts and leakage, and water theft through illegal connection) and losses in revenue due to poor billing and collection system and weak tariff structure. Unaccounted for water has not been assessed due non-availability of water in the network. Unaccounted for water can be taken as the difference between the account of water produced or purchased and the amount of water sold to all customers. Unaccounted for water includes underground leakage, unauthorized use, unavailable leakages, inaccurate master, industrial, commercial and domestic water and unusual causes. Underground leakage is caused by age of the pipe, soil conditions, traffic loading, pipe movement, poor installation practices and electrolysis.

Listening surveys and water audit was used in checking this in the past. Improved various types of the surveys and audit processes are imperative for the new water works scheme expected.

Other problems facing the state WSS sector in urban areas

The following problems are found to be the major causes of inadequate water supply and sanitation in the state:

1. There is no coordination and focus and because of this, planning, activities were "reactive" rather than "proactive" until the setting up of the new Ministry of Public Utilities, Water Resources and Community Development in August 2008, there was no central agency

to develop policy direction of Government."
2. There is lack of a philosophy, objectives and strategy guiding government activities in this sector.
3. Approach to plans for service delivery were "top bottom supply driven" rather than "bottom up demand driven."
4. The very strong community structure and vibrant private sectors, for which the state is well known, was not exploited in developing an efficient water delivery system for the people.

ANAMBRA STATE WATER CORPORATION (ANSWC)

Anambra State Water Corporation (ANSWC) established by Edict No 3 of 7th May 1999 has as its main function; to supply and mange water supply in the whole of the State (including urban, semi-urban and rural water supply). It was established to develop, manage, control, provide, conserve and distribute water in the state for public, domestic and industrial purposes. It had power to enter into contract or other agreements for the purpose of expedient performance of its function.

As at the year 2006, ANSWC had 12 Zonal Offices covering the twenty one LGAs of Anambra State for managing their water scheme of the 62 water schemes operated by ANSWC, 58 (93.5%) were non functional while only 4 (6.5%) are functional. The record is indeed an issue of great concern. Even more disastrous for ANSWC is the collapse of the Greater Onitsha Water Supply Scheme since 2001. When operational, the scheme provided 90% of ANSWC's internally generated funds. With so many public water boreholes and private ones in private building the people of the state would not be said to be short of water supply but the portability remains the question.

Water policy

The existing level of service of water supply and sanitation in the state is far from satisfactorily improving priorities of Anambra State Government. There was need for a comprehensive WSS services in a sustainable way. The institutional arrangement for Water Supply and Sanitation were not properly streamlined to have proper focus in a coherent and sustainable way. In a situation like this, one finds gaps and overlaps and lack of clarity on the responsibilities of the state actors. An institutional framework defines strategies and programmes that the sector must implement the organizations to implement these policies, and regulatory framework needed to support the policies as well as the institutional actions. The institutional framework looks beyond organizations and their functioning and takes into account the policy, legal and regulatory environment that supports the delivery of effective and efficient services which is the ultimate objective of any reform process. One of the primary impacts of the policy is the creation of the enabling environment for the implementation of the principles of the policy. This conducive working environment is created by the policy, legislature and financing structure Policy development gives an opportunity for setting objectives for managing water resources and water service delivery within a framework of overall development objectives.

Legislative framework sets the rules to follow in order to active policy objectives and goals. One of the main duty here is the development of the Water Law. The law covers the ownership of water schemes, the use of permit operation and maintenance and the regulatory norms. Also necessary is the need to repeal and reform the existing legislatures, edicts with the development of the State Water and Sanitation Policy. A component of the enabling environment is the financing and incentive structures that deals with the allocation of financial resources to meet the water needs. The financing needs of the water sectors are huge, and water projects tend to be capital intensive and require investment policy with financing options.

The fundamental objective of the policy is to bring reform to institutional roles and responsibilities for better water governance. The policy document defines the roles of resource managers, service providers, regulators, the roles of private sector, local authorities, civil society organization (CSOs) community based organizations (CBOs) and other water sector stakeholders. This forms part of the overall institutional framework.

Capacity building or human resources development involves the upgrading of the skills and understanding of all levels of capacities: Public decision makers, water professionals, regulatory bodies and other bodies like civil society groups.

An important aspect in the policy development is the overall application of management instruments. A primary aspect of this is a good understanding of resources and the needs in water resources assessment. This assessment starts with the collection of hydrological, physiological, demographic, and social-economic data. This will enable development in the sector to be modeled in line with the integrated water resources management (IWRM) principles. A very important concept here is the demand management. It is a process of using water more efficiently by balancing supply and demand and requires the following:

1. Improved efficiency of use.
2. Recycling and use.
3. Improved efficiency of supply.

Demand management application will work with good social change supply and sanitation systems as a matter of priority, the government is already planning for subsequent challenges with regards to expansion of the water supply throughout the state, there is a need to accompany any rehabilitation works with a good structure of operation and maintenance system. This is strongly emphasized in the policy. Lack of operation and maintenance was the bane of the dysfunctional water schemes. This aspect of the reform and indeed all aspect of the policy are necessary to ensure sustainable growth and development of the service delivery in the State.

CONCLUSIONS AND RECOMMENDATION

The literature on drinking water in Nigeria is characterized by an overall sense of policy failure and barriers to access, private sector participation, of fiscal reform, willingness to pay, of civil society participation. It remains a challenge to answer the key question for designing an affordable and sustainable urban drinking water program. With respect to private sector participation, relative to the efficiency or prices, analyses of the kinds of contracts, regulatory regimes and citizen oversight that can ensure accountability and the inclusion of low-income communities, are less common. It is agreed that governments should not be in the water provision business, but should ensure that private providers are regulated with respect to price structures and water quality, and should provide incentives for these providers to serve the poor. This new role for government translates to developing partnerships with the private sector and with civil society for water delivery.

A major barrier to the design of appropriate policies is the lack of reliable, up-to-date and publicly accessible information on many aspects of the Anambra water system. Baseline information is necessary in order to evaluate various reforms in progress, and in order to allow for benchmarking against government targets, that is, other states and nearby countries. A lack of transparency over the true costs of under-priced and inefficient municipal systems dampens public support for major reforms that may be needed.

Scanty information on groundwater withdrawals make urban and Semi-Urban drinking water interventions unsustainable. While efforts are underway to carry out some benchmarking of financial performance of several large utilities, regular and comparable data need to be made available on, water quality, subsidization, metering, groundwater levels, and infrastructure maintenance. The abundance of groundwater resources in Anambra State remains a big boast to solving the problems of domestic and industrial water with ease since exploration and exploitation is easily feasible through the state.

REFERENCES

Anambra State Water Supply and Sanitation Sector Policy (2009). First Edition.

Boland J, Dale W (2000). The political economy of water tariff design in developing countries, in Ariel Dinar (ed.) The Political Economy of Water Pricing Reforms. Oxford University Press: New York pp. 215-236.

Cairncross S (2003). Water supply and sanitation: Some misconceptions. Trop. Med. Int. Health 8:193-195.

David JM, Barja G, Miguel U (2005). Capitalization and Privatization in Bolivia: An Approximation to an Evaluation. J. Nellis and N. Birdsall (eds.) Reality Check: The Distributional Impact of Privatization in Developing Countries, Center for Global Development: Washington.

McIntosh AC (2003). Asian Water Supplies: Reaching the Urban Poor, Asian Development Bank and IWA Publishing: London.

Mckenzie R (2002). Urban Water, India: Urban Water Supply in India, Status reform options and possible lessons

Sridhar L (2003a). Piped dreams, www.indiatogether.org, [accessed 2/2004].

Sridhar L (2003b). Water: The privatization debate, http://www.indiatogether.org/2003/nov/env-wtrdebate.htm [accessed 03/2004].

Water and Sanitation Program (WSP) (2001). Nagari "Thirteenth Meeting of the Urban Think Tank: Private Sector Participation in Urban Water and Sanitation: Managing the Process and Regulating the Sector, December 5-6, 2001, UNDP-World Bank Water and Sanitation Program-South Asia.

WHO (1993). Guidelines for drinking-water quality. Recommendations. 2(1), Geneva.

WHO (1997). Guidelines for drinking-water quality: Surveillance and control of community supplies. 2(3) Geneva.

WHO (1998). Guidelines for drinking-water quality: Health criteria and other supporting information. 2(2), Geneva.

Cost differentials and scale for newer water treatment technologies

Mohammed H. I. Dore[1]*, Rajiv G. Singh[1], Arian Khaleghi-Moghadam[1,2] and Gopal Achari[3]

[1]Brock University, St. Catharines, Ontario, Canada.
[2]Bloomsburg University of Pennsylvania, United States.
[3]University of Calgary, Alberta, Canada.

Although a number of jurisdictions in Europe have abandoned chlorine as a disinfectant, most water utilities with conventional water treatment plants in North America seem to rely on chlorine for disinfection. However there are a number of newer treatment technologies that offer advantages and appear to be cost effective at a number of scales. It is the objective of this paper to consider 5 newer treatment technologies and work out the cost differentials as a function of scale. We use nonlinear least squares regression to estimate the average unit cost in US dollars as a function of scale and demonstrate the extent to which these technologies differ in cost. We show that at a certain scale the differentials narrow and project the switch point at which some of the technology costs cross over, indicating the flow rates at which some technologies become cost effective.

Key words: Drinking water treatment technologies, costs of treatment, small water systems, disinfection, economies of scale, ultra violet (UV), ozonation.

INTRODUCTION

In Europe, some countries have moved away from using chlorine as a primary disinfectant in treating drinking water. European Union (EU) member countries in general are guided by the EU Drinking Water Directive (European Commission, 1998) but many adopt their own stricter measures with respect to disinfection of drinking water. Some countries like the Netherlands have discontinued the use of chlorine as a primary as well as a secondary disinfectant without compromising the microbiological safety of the drinking water (Smeets et al., 2009). The successful use of non-chlorine based treatments arises mainly from: (1) using safe microbiological groundwater as main source water (2) using sedimentation, filtration, ozone and/or UV disinfection as treatment processes (3) avoiding the use of chlorine altogether, (4) producing biologically stable drinking water to prevent microbial growth later in the distribution systems and (5) implementing proper and effective monitoring systems (Smeets et al., 2009). In Denmark, the city of Copenhagen has abandoned the use of chlorinated water from lakes after 37 years of use and instead has turned to groundwater use. The Lovo Waterworks of Stockholm Water Company in Sweden which serves drinking water to 1 million people recently switched to a combination of UV and preformed chloramines in order to achieve a more efficient disinfection and to prevent the formation of disinfection byproducts (DBPs) and biodegradable organics or assimilable organic carbons (AOCs) (Eriksson and Nordwall, 2002). The preformed chloramines also did not form measurable amounts of trihalomethanes (THMs) when used in combination with the ultra violet (UV) treatment (Eriksson and Nordwall, 2002). UV based treatments are likely to be used as chlorine or chlorine dioxide treatments in France while in Belgium the focus has shifted away from chlorine to UV as the primary treatment (Connell, 1998).

In North America, however, the emphasis has been on the use of traditional treatment technology which is of course chlorine. However, one of the major disadvantages of

*Corresponding author. E-mail: dore@brocku.ca.

Table 1. Disinfection by-products found in chlorinated and UV treated tertiary effluent (EPA, 1996).

Compounds detected	Before disinfection (ug/L)	Chlorine (ug/L)	UV (ug/L)
4-methyl-2-pentanone	5.4	5.9	4.5
Chloroform	2.7	21.0	2.9
Dichlorobromomethane	0.9	22.0	1.0
Dibromochloromethane	1.1	<1.0	0.8
Methyl bromide	2.8	<1.0	<1.0
Bromoform	<0.5	3.1	<0.5

using chlorine as the primary disinfection treatment for drinking water is the formation of DBP precursors, which poses long term health problems. Table 1 provides a summary of the effectiveness of UV in reducing or preventing the formation of DBP precursors compared to chlorine according to an EPA (1996) study. Although the information is based on wastewater treatment, it provides some insight into the comparative efficacy between UV and Chlorine.

UV disinfection either reduced or only slightly increased the above mentioned disinfection by-products while chlorine increase some DBPs significantly compared to UV. Similar results were shown for THMs (EPA, 1996). However, unlike chlorine and chloramines, UV will not be able to prevent the growth of biological contaminants in the distribution system.

The long term health risks of consuming water with DBPs are well known and over 260 million people were exposed to DBPs as of 2005 in the US. As a result, the US EPA implemented Stage 2 of the Disinfection and Disinfection Byproducts Rule (DBPR) in 2005 which would regulate total trihalomethanes (TTHM - chloroform, bromoform, bromodichloromethane, and dibromochloro-methane) and haloacetic acids (HAA5 - monochloro-, dichloro-, trichloro-, monobromo-, dibromo-). Most of these TTHMs and HAA5 are formed as a result of chlorination disinfection used in drinking water treatment. The rule was expected to reduce bladder cancer cases by 280 per year, when 26% of bladder cancers resulted in fatalities (EPA, 2005a). Based on bladder cases alone, Stage 2 of DBPR was expected to provide reduction in health costs in the region of $0.8 to 1.5 billion annually while the average cost of implementing the rule was only $79 million per annum (EPA, 2005a). In the US, 76% of community water systems and non transient community water systems which use disinfection receive their drinking water from ground water sources; ground water source serve a total of 94 million or 37 % of the total population (EPA, 2005b). Surface water sources serve the remaining 161 million people and 85% of ground water plants disinfect with chlorine only; a further 13% disinfect by chloramine only. The remaining 2% use other types of disinfection methods. Some 54% of surface water plants disinfecting uses chlorine while 41% disinfecting uses a combination of chlorine, chloramines and chlorine dioxide (EPA, 2005b).

In Canada, chlorine is also the major disinfectant used by treatment plants. According to the Environment Canada survey of Municipal Water and Wastewater (Environment Canada, 2005), there were 2042 reporting drinking water systems in 2004, of which 481 gave information on at least one of the type of technology they use for drinking water treatment. The treatment technologies considered in the Municipal Water and Wastewater Survey (MWWS) are: micro straining, coagulation, sedimentation, slow sand filtration, pH control, corrosion control, fluoridation, membrane filtration, granular filtration, chlorination, chloramination, ozone, ultra violet (UV) and chlorine dioxide treatment. Of the 481 systems which reported at least one type of the aforementioned treatment technologies, 167 or 35% used chlorine alone that is, not used in conjunction with any other treatment technology such as filtration device, sediment removal system etc. Therefore, chlorination is the dominant treatment technology for systems which depend on a single treatment type. Only 1 used ozone alone[1] while 9 used chloramines alone. Chloramination would provide free chlorine in the distribution and there would be no need for secondary disinfection by chlorine. If we extrapolate, this would mean that 715 out of 2402 systems in Canada use chlorine as the primary and only treatment technology. Figure 1 shows the number of systems which use at least one treatment type.

The five most commonly used treatments are chlorination, coagulation, granular filtration, pH Control, and sedimentation. Other filtration and/or sediment removal treatment were micro-straining, coagulation, sediment removal, slow sand filtration, membrane filtration or granular filtration. 129 out of 481 systems use at least one type of filtration or sediment removal treatment. Of these 129 systems, coagulation, sedimentation and granular filtration were the most common types used while micro straining and membrane filtration was the least used.

[1] We assume that residual chlorine is used as a secondary treatment regardless of the treatment technology used. However, those systems which reported chlorine alone were assumed to use chlorine as the primary disinfectant.

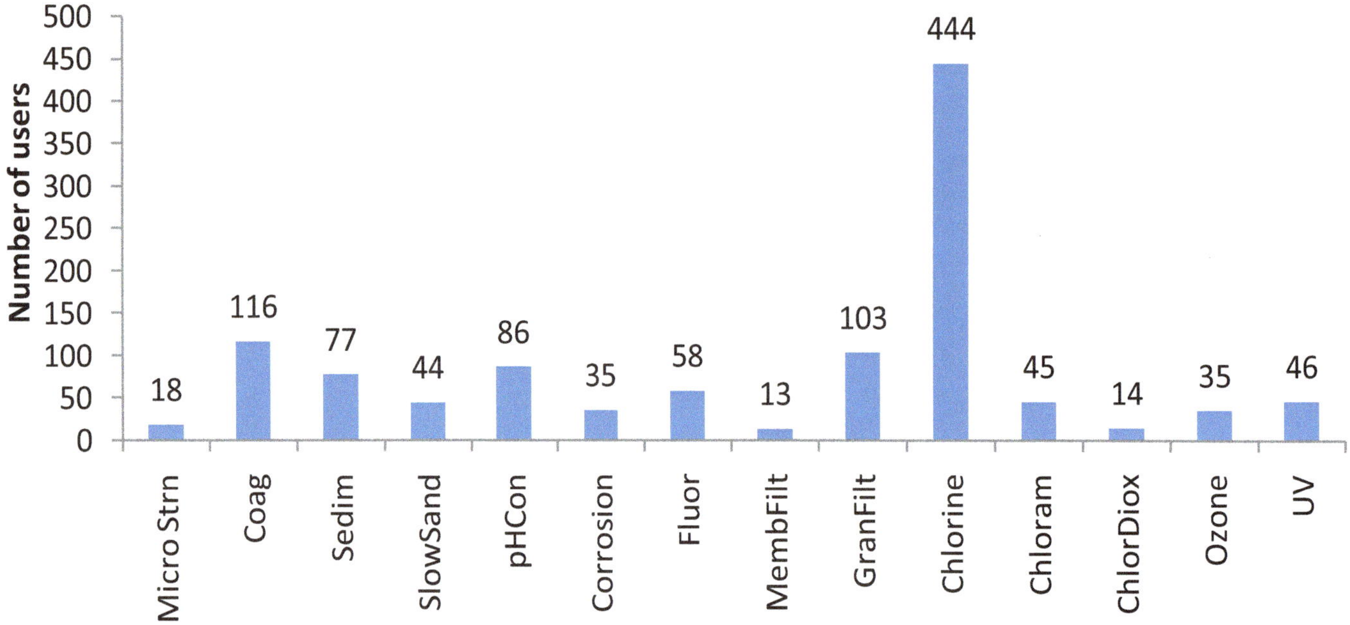

Figure 1. Type of treatment technology used and the number of users in Canada. Micro Strn=Microstraining, Coag=coagulation, Sedim=sedimentation, SlowSand=slow-sand filtration, pHCon=pH control, Fluor= fluoridation, MembFilt=membrane filtration, GranFilt=granular filtration, Chloram=chloramination, ChlorDiox=chlorine dioxide, UV=Ultraviolet treatment.

Although chlorination can provide effective disinfection of many contaminants for some source waters, the discovery of new and chlorine resistant pathogens reduces its effectiveness considerably. Cryptosporidiosis and hepatitis are two examples of water borne illnesses which chlorine is unable to treat. Newer technologies such as UV and ozonation would provide effective treatment of these pathogens at dosages which are commercially available. For instance, UV dose of 40 mJ/cm^2 can achieve greater than 4 log reduction in Cryptosporidium and Giardia (EPA, 2010). UV at 120 mJ/cm^2 can achieve up to 3 log reduction in human adenovirus type 2 which is the most resistant enteric virus to UV light, while a dose of 118 to 124 mJ/cm^2 can achieve 4 log inactivation of adenovirus Type 40 and 41 which are the UV-resistant viruses that are of most concern in drinking water disinfection according to the EPA (2005c). Ozone at 1.0 mg-min/ml can achieve up to 3 log reduction in Giardia Lamblia depending on pH and temperature conditions (EPA, 1991). In addition several studies highlighted in an EPA (2005c) study showed that UV disinfection produces little or no DBPs. Similar results were found for Ozone in the same EPA study; chlorinated DBPs are not formed under ozonation but may form bromate if bromide is present in the water prior to disinfection (EPA, 2005c). Ozone can affect the formation of DBPs if natural organic matter (NOM) is present but a granular media biofilter would be suitable for the removal of NOM as well as the prevention of re-growth of organisms later in the distribution system (EPA, 2005c). Newer technologies are also more effective than chlorine in removing physical and chemical contaminants. The EPA has identified 16 polycyclic aromatic hydrocarbons or PAHs as priority contaminants; the mutagenic and carcinogenic properties of PAHs are discussed in Luch (2005). Ali and Tarek (2009) have shown that ozone reduces considerably more PAHs than chlorine and chlorine dioxide. Kornmüller and Wiesmann (2003) have also shown that ozone also reduces PAHs although this experiment was confined to water and oil emulsions.

The literature has highlighted the disadvantages of using chlorine as a primary and often at times the only disinfection type used for treating drinking water. However, other alternatives are available which can remove more contaminants than chlorine alone while at the same time reduce or have no significant impact on DBP formation. In this paper, we examine 5 treatment technologies which can be of use for drinking water systems of most sizes (except the smallest) and show the cost differentials between them. Many of these technologies have cheaper average costs over a wide range of daily capacities than systems which use chlorine alone. Not only can these technologies provide lower cost drinking water treatment but they reduce health costs associated with DBPs in drinking water, treat newly discovered pathogens, physical and chemical contaminants and are therefore adaptable to future regulations.

Table 2. Treatment technologies.

Technology	Description
Actifloc	(i) Consists of a clarification system (Actiflo) and filtration system (Dusenflo Mixed Bed Filters) (ii) Reduces turbidity, color, suspended solids, algae, taste and odour (T&O), metals and total organic carbon. (iii) The resulting filtered water from the Dusenflo gravity filter can contain little or no Giardia and Cryptosporidium cysts.
UV Swift SC	(i) Utilizes the ability of ultra violet rays to deactivate microorganisms. (ii) This system on its own is chemical free and produces no disinfection by-products. (iii) However, it can also be used in conjunction with other treatment processes forming a "multi-barrier" approach for treating water for drinking purposes (iv) UV Swift SC will inactivate bacteria, viruses and protozoa, including Giardia and Cryptosporidium with a dose of 40 mJ/cm^2. (v) We assume some filtration system to remove sediments (for example, sand filtration) would be required and is included in the cost.
MF-UF	(i) Micro filtration and ultra filtration involves separating water from organic and inorganic matter contained in the water by forcing it through a micro porous membrane (ii) Pore sizes in microfiltration membranes are 0.1 to 10 microns thick while ultra filtration membranes are between 0.001 and 0.1 microns (iii) Microfiltration will remove Giardia and Cryptosporidium cysts, bacteria, and some viruses; however not all viruses can be removed via this process. (iv) Microfiltration is also used in sterilization of beverages and pharmaceuticals, clearing of fruit juices, wine and beer, separation of oil-water emulsions and pre-treatment of water for nano filtration and reverse osmosis (v) Ultra filtration removes all viruses, bacteria and suspended solids between 0.001 and 0.1 µm. Ultra filtration is used in paint treatment, oil-water emulsion separations, the food industry and textile industry.
Ozonation (SCOR)	(i) SCOR utilizes the ability of ozone to inactivate microorganisms through oxidation. (ii) The system consists of an ozone pretreatment unit, a BioSand filter and a BioCarbon filter. (iii) The roughing filtration system removes suspended solids and coliforms as well as some Cryptosporidium. (iv) The BioSand Filter is used to treat parasites, color, cysts, manganese, mercury, iron and turbidity while the BioCarbon Filter treats dissolved organic carbon, tannins, pesticides, iron, bacteria, color and odors.
UV Phox	(i) A UV-oxidation process designed to provide disinfection and T&O treatment; it destroys Geosmin and 2-methylisoborneol. (ii) Also removes pharmaceutical, personal care products, pesticides and trace contaminants (iii) System consists of a UV reactor, H_2O_2 dosage and storage system. We assume some filtration system to remove sediments (for example, sand filtration) would be required and is included in the cost.

Contaminant removal for UVSWIFT SC, MF-UF, ACTIFLOC, SCOR, and UV PHOX

In this part of the work, we present five technologies that are currently used for drinking water treatment: Actifloc, UV Swift, MF-UF, UV Phox, and SCOR. UV based technologies UV Swift SC and UV Phox would most likely require that source water be pretreated using a filtration or sediment removal process before being disinfected by UV. Also, which technology is appropriate is determined by site-specific factors and some remote communities may need to rely on Point-of-Use or Point-of-Entry treatment devices. But for centrally treated water plants, the five technologies discussed here should be considered as possible candidates. For example, UV Phox, which is an advanced oxidation process (AOP), may be appropriate for specific source-water characteristics. Nevertheless new and more stringent regulations on drinking water quality in the future may make it worthwhile for many systems to consider using UV-oxidation based treatment technologies. Some small communities are already using AOPs for ground water remediation, even at a small scale[1].

Each technology is described in Table[2] 2. All the technologies considered here produced municipal standard drinking water.

Table 2 shows that some technologies offer greater contaminant removal than other. For instance UV Phox would remove pharmaceuticals, personal care products and pesticides which Actifloc would not remove while at the same time removing the target contaminants that Actifloc would remove. Actifloc is perhaps the least effective of the technologies described in Table 2 but would remove turbidity, color, suspended solids, algae, taste and odour (T&O), metals, total organic carbon and some Giardia and Cryptosporidium. Actifloc will therefore provide an even higher level of disinfection than chlorine alone which is the most commonly used disinfection method among small systems.

METHODOLOGY

Non-linear least squares estimation and differential costs

Nonlinear least squares estimation

We use the non-linear least squares (NLLS) estimation process since it can capture a wider range of functional forms than the ordinary least squares method. Simple linear models may not describe certain data generating processes very well especially if the functional form changes over its domain. The NLLS technique also has the added advantage of yielding better estimates when the amount of data is limited.

Table 3 shows the estimated cost functions for the various technologies[3] described in Table 2 with the functional form $y_i = \beta_1 X_i^{\beta_2} + \varepsilon_i$ where y_i is the average cost per cubic meter, X_i is the capacity in cubic meters per day and ε_i is the error term which satisfies the standard Gaussian assumptions.

RESULTS AND DISCUSSION

Using the above estimation procedure, we obtain the cost estimates for a few discrete plant sizes; this is shown in Table 4.

Cost estimates in Tables 3 and 4 are treatment costs only and do not include cost of residual chlorine require by law in the US and Canada. However, we assume that this cost of residual chlorine cost would be the same for all technologies holding all other factors constant and was therefore left out.

Our results show that for smaller capacities (100 and 500 m^3/day), UV Swift SC, MF-UF and Actifloc are the cheapest treatment technologies. Both the MF-UF and Actifloc drinking water treatment can cost on average 10 cents per cubic meter for a 100 m^3 sizes plant while UV Swift SC can cost on average 7 cents per cubic meter for a similar sized plant. For a 100 m^3 plant SCOR (ozonation) and UV Phox (advanced oxidation processes) would cost on average 38 and 21 cents per cubic meter respectively. Although SCOR and UV Phox are more expensive than Actifloc, MF-UF and UV Swift SC at this capacity, they will remove more contaminants. UV Swift SC seems to be cheaper and provides a greater range of basic contaminant removal than Actifloc at smaller sized plants. It also has a similar range of contaminant removal as MF-UF but will cost a few cents per cubic meter cheaper at very small scales. However, for UV, some additional costs must be added for suspended solid removal, such as sand filtration, which could add up to 5 cents per cubic meter and has already been included in Table 3. At larger scales (5000 m^3 and over) however, MF-UF and Actifloc are cheaper than UV Swift SC by a few cents per cubic metre. In fact, SCOR is also cheaper than UV Swift SC at capacities greater than 5,000 m^3 per day although it was at least 30 cent per cubic meter more expensive for a plant with capacity 100 m^3 per day. If we consider medium sized plants as plants with daily capacity between 1000 and 5000 m^3, then Actifloc, Mf-UF and UV Swift SC can cost on average between 3 and 6 cents per cubic meter. SCOR and Phox therefore would cost between 9 and 16 cents per cubic meter under the unofficial 'medium' sized plants category. Our statistical modeling estimations which are based on data supplied by manufacturers also suggest that the Advanced Oxidation Process (UV Phox) is cheaper than ozonation (SCOR) at smaller capacities, roughly the same at 'medium' sized capacities and more expensive at larger capacities (greater than 5000 m^3/day) even though AOPs will remove a larger number of micro-pollutants (see description in Table 2). However, the degree to which AOPs are cheaper than ozonation at smaller capacities is much larger than the degree to which AOPs are more expensive than ozonation at larger capacities. For UV Phox, our information indicates that the cost per cubic meter is around $0.45 (in 2008 Canadian dollars) at a scale of 3800 cubic meters per day[4] when actual plant-specific costs are taken into account.

SCOR ozonation seem to be the most expensive at smaller capacities but has a higher contaminant removal level than the other disinfection technologies in Table 2. Ozone treatment plants[5] are expanding rapidly in small systems across Saskatchewan and Manitoba in Canada. Some examples of such plants are located in: Peebles, Macoun, Echo Valley, Glen Ewen, Manor, the Resort Village of Aquadeo, Jenner, Town of Craik, Town of Fleming, Village of Dunrea, Benson, Antler, Guernsey, Macoun, Goodeve, and Osage (site of first pilot plant).

There are currently about 30 ozone plants in operation (at the end of 2010). Compared to a UV-based treatment plant, this is more expensive but nevertheless it is proving to be attractive to a number of smaller communities, mainly in Saskatchewan and Manitoba.

Cost differentials

We can reproduce graphically the set of estimated cost functions from Table 3 and show the associated cost differentials between technologies at different scales.

Table 3. Estimated Cost functions using Nonlinear Estimation $y_i = \beta_1 X_i^{\beta_2} + \varepsilon_i$ where y_i is the average cost per cubic meter, X_i is the capacity in cubic meters per day and ε_i is the error term which satisfies the standard Gaussian assumptions.

Coefficients	Actifloc	UV Swift	MF-UF	Ozonation (SCOR)	UV Phox
β_1	0.3226 (0.077)	0.2653 (0.025)	0.4171 (0.060)	2.2107 (0.000)	0.7576 (0.4357)
β_2	-0.2503 (0.027)	-0.6003 (0.000)	-0.3048 (0.003)	-0.381 (0.000)	-0.3394 (0.1397)
R^2	0.946	0.975	0.853	0.999	0.759
$\bar{R}^2$	0.919	0.968	0.829	0.999	0.638
S.E. β_1	0.115	0.076	0.180	0.065	0.784
S.E. β_2	0.042	0.048	0.061	0.009	0.142

p-values in parenthesis.

Table 4. Estimated average cost functions for Actifloc, UV Swift, MF-UF and Ozonation in 2008 CDN dollars.

Disinfection technology	Average cost function	Predicted cost per cubic meter based on plant with capacity in m³ per day					
		100	500	1,000	5,000	10,000	50,000
Actifloc	$y = 0.3226x^{-0.2503}$	0.10	0.07	0.06	0.04	0.03	0.02
UV Swift SC*	$y = 0.2653x^{-0.6003}$	0.07	0.06	0.05	0.05	0.05	0.05
MF-UF	$y = 0.4171x^{-0.3048}$	0.10	0.06	0.05	0.03	0.03	0.02
SCOR	$y = 2.2107x^{-0.381}$	0.38	0.21	0.16	0.09	0.07	0.04
UV Phox*	$y = 0.7576x^{-0.3394}$	0.21	0.14	0.12	0.09	0.08	0.07

*Filtration/sediment removal cost of 5 cents per cubic meter included in predicted cost.

Figure 2 shows the estimated cost curves based on manufacturers rated costs for Actifloc, UV Swift SC, MF-UF, SCOR and UV Phox.

From Figure 2, we observe that some technologies are cheaper than others but for only a particular range of daily output. SCOR and UV Phox provide a significantly wider range of contaminant removal than Actifloc, UV Swift SC and MF-UF; hence their costs are always between 10 to 40 cents per cubic meter greater than the unit costs of Actifloc, UV Swift SC and MF-UF depending on scale. Figure 3 shows the cost differentials between SCOR and UV Phox respectively.

Although SCOR is more expensive than UV Phox for capacities below 3705 m^3 per day, it still seems a viable option for small systems in Saskatchewan and Manitoba as mentioned earlier. However, UV Phox is used in the City of Cornwall which has a daily capacity of 3800 m^3/day.

Actifloc, UV Swift SC and MF-UF will remove fewer contaminants than both UV Phox and SCOR and their unit cost are similar over a very wide range of daily capacity (Figure 2). From Figure 4, we observe that UV Swift SC is cheaper than Actifloc and MF-UF for capacities[6] less than 780 m^3/day. It is also cheaper than Actifloc but more expensive than MF-UF between capacities 780 m^3/day and 1300 m^3/day and more expensive than Actifloc and MF-UF at capacities greater than 1300 m^3/day. Actifloc seem to be cheaper than MF-UF at capacities less than 110 m^3/day but more expensive at larger capacities. Actifloc therefore seem to be a cost effective option for very small plants, and UV Swift SC seems to be cost effective for small to medium sized plants. Finally, MF-UF seems to be cost effective for medium to large sized plants. Note, however, that UV Phox and SCOR can be very competitive at very large capacities and have the added advantage of removing more contaminants.

We found that the cost per unit for some *actual* small systems in British Columbia which use older treatment technologies such as chlorination range from $1.00 to $3.75 per cubic meter for system with a daily capacity of 100 m^3 and between $0.35 and $0.75 cents per cubic meter for a system with a daily capacity of 500 m^3. Most of the above mentioned technologies are cheaper per cubic meter than those costs even when we add an additional $1.00 per cubic meter for distribution and mark up costs. Therefore the choice of treatment technology may not be due to cost considerations alone and other factors must explain their choice. Perhaps the use of older treatment technologies such as chlorination simply

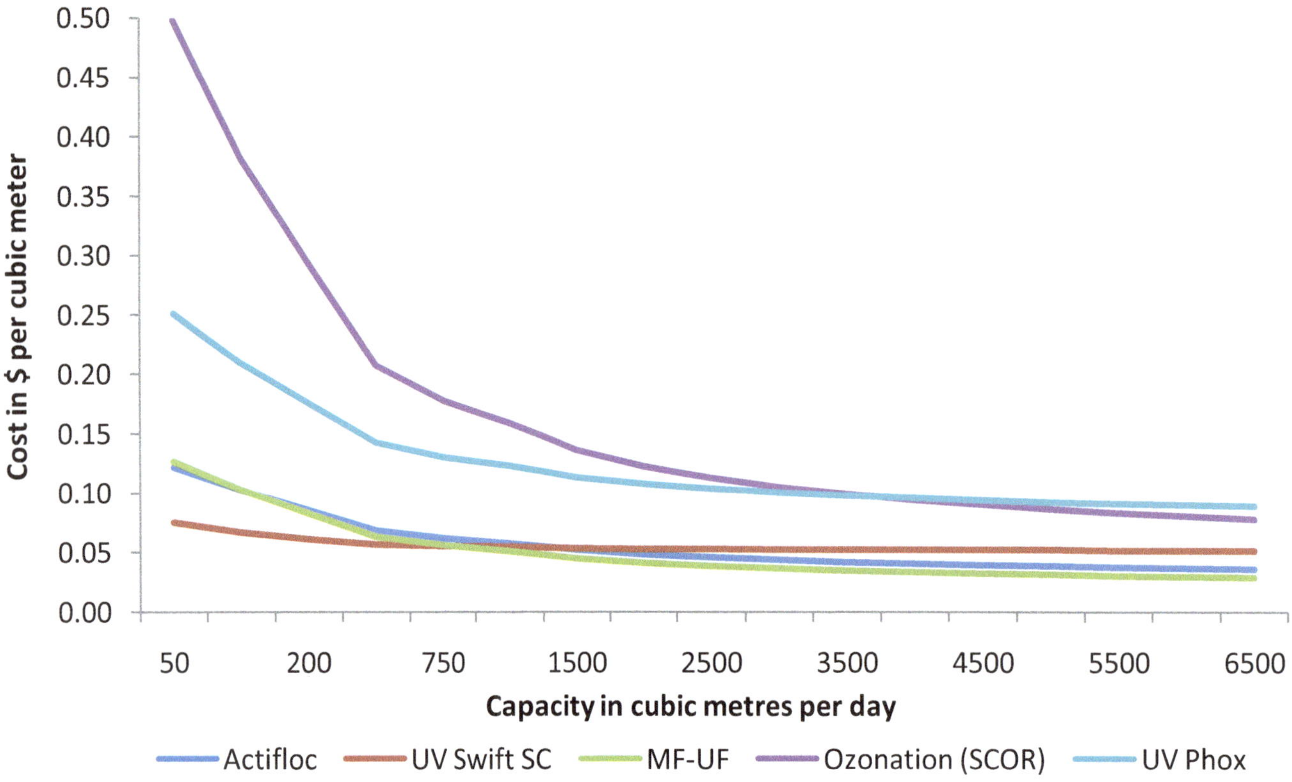

Figure 2. Estimated cost curves: Actifloc, UV Swift SC, MF-UF, SCOR and UV Phox.

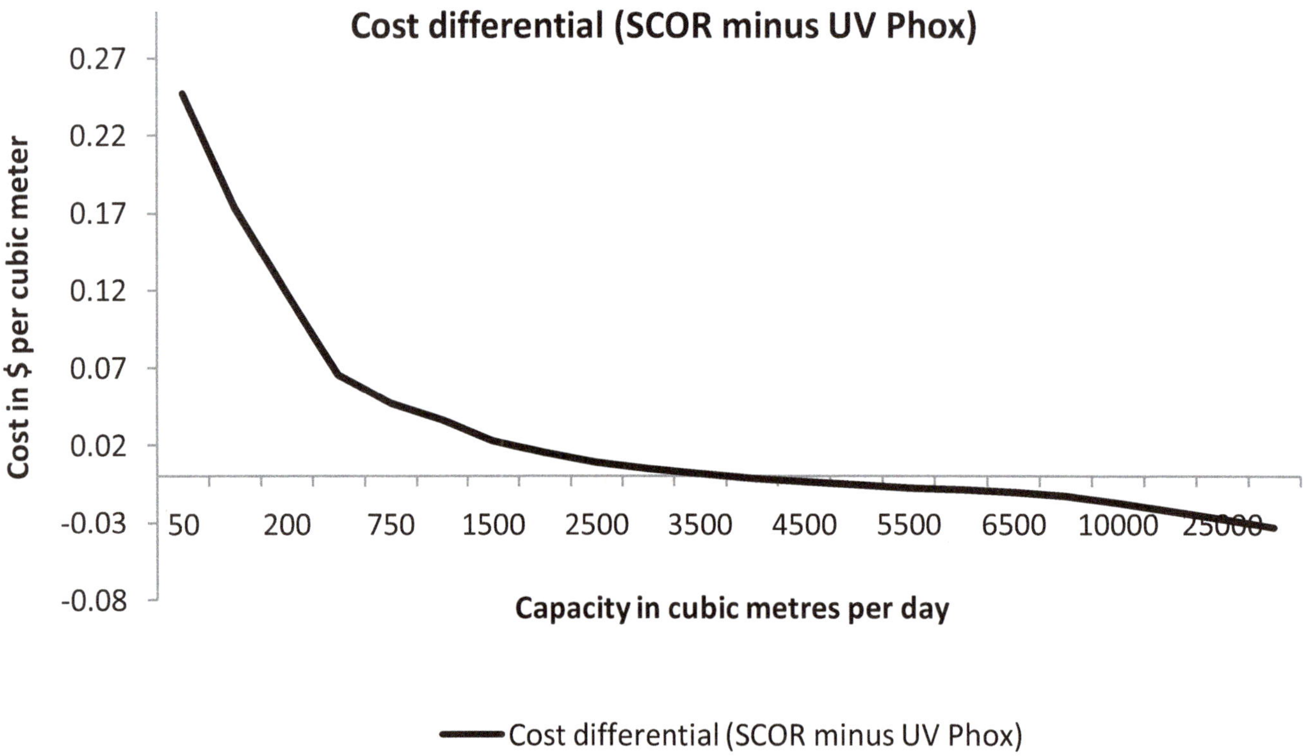

Figure 3. Cost differential between SCOR and UV Phox.

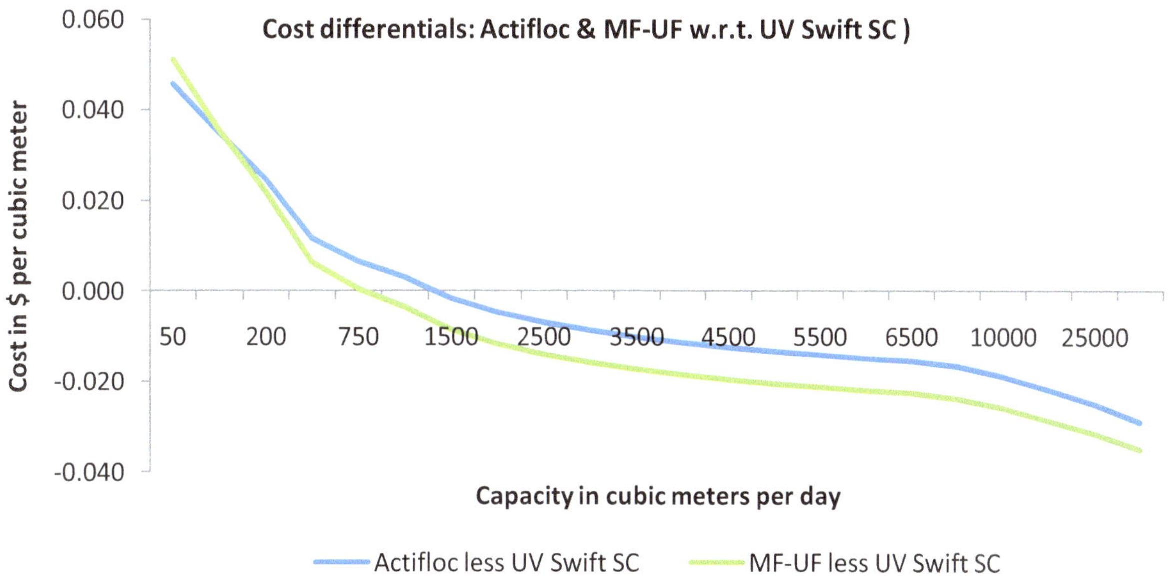

Figure 4. Cost differentials: Actifloc and MF-UF.

indicates that these plants have not been modernized over the last 150 years as there appears to be no regulatory incentive to do so in the public sector as meeting the bare minimum regulatory requirements for treating water for drinking purposes is enough.

Conclusions

Based on the above analysis and discussion, we provide the following conclusions: The estimated cost functions show that in general the greater the contaminant removal of a treatment technology the higher the cost per cubic meter:

1. For treatments which have roughly the same contaminant removal such as MF-UF and UV Swift SC, the latter is cheaper than the former at capacities smaller than 780 m^3/day. Therefore UV Swift SC is probably more suitable for small systems while MF-UF is suited for medium sized to larger systems;
2. Actifloc is cheaper than UV Swift SC at capacities greater than 1300 m^3/day but would remove less contaminants;
3. MF-UF would remove more contaminants than Actifloc but is cheaper at capacities greater than 110 m^3/day. Perhaps Actifloc is suited for very small systems;
4. The estimated cost curves show that systems, especially smaller ones, could achieve a higher removal of contaminants at a lower cost than their currently used technology;
5. If a small publicly owned system adopts Actifloc or MF-UF technology then they can incur daily treatment costs of about 9 to 11 cents per cubic meter (based on a production capacity of 100 m^3 per day);
6. UV Swift SC appears to be the least expensive for systems which produce less than 780 m^3 at only 7 cents[7] per m^3 for a plant with capacity of 100 m^3/day which shows that the competitive advantage remains even when costs of sediment removal is included. If a community is concerned with the removal of micro-pollutants, then a UV based Advanced Oxidation Process would still be cheaper that ozonation.
7. Our estimates indicate that on average SCOR is more expensive than UV Phox for capacities below 3705 m^3 per day. However, SCOR is used in small systems in Saskatchewan and Manitoba indicating that it is very competitive in that region.
8. Trojan UV Phox on average is more expensive than SCOR at capacities greater than 3705 m^3 per day but it is used in the City of Cornwall which has a daily capacity of 3800 m^3 per day.

In general, manufacturers' rated costs tend to be lower than actual plant-level average costs as they do not include some plant-specific costs, such as higher labor, energy and transportation costs due to remoteness from large urban areas. However, small water systems are dominated by the use of chlorine as the disinfectant although there exist new technologies that are cheaper and are more cost effective without some of the disadvantages associated with the use of chlorine. For UV treatment, the cost can be as low as 7 cents per cubic meter for a plant which produces 100 m^3 per day while for higher levels of treatment such as ozone and AOPs the cost can be between 20 to 40 cents for a similar sized plant. In our research we found that some small systems in British Columbia use older treatment technologies such as chlorination and which show their costs ranging from \$1.00 to \$3.75 per cubic meter for system with a daily capacity of 100 m^3. While any specific water treatment facility will need to take account of raw source water quality, the *actual* target quality for small systems seems to be to meet only the minimum regulatory requirements in most small systems. Our results show that unless the raw water is high in color and in turbidity, a UV-based plant would be the most economical and cost-effective even when the additional cost of sediment removal is added. Our finding of the cost effectiveness of UV is in agreement with EPA (1996), Gadgil (1998) and Perotta and Bekdash (1998).

ACKNOWLEDGEMENTS

In gathering the data and in writing this paper we are grateful to Brent Alspach, Irfan Gehlen, Adam Holbrook, Madjid Mohseni, Adam Festger, Bill Cairns, Craig Hoover and Douglas Price. This research was financed in part by the National Science and Engineering Council of Canada through a grant to Res'eau-Waternet and the Social Sciences and Humanities Research Council of Canada. This revised version has also benefited from the Editors and anonymous referees of this journal. However, the authors alone are responsible for any remaining deficiencies.

REFERENCES

Ali OA, Tarek SJ (2009). Removal of polycyclic aromatic hydrocarbons from Ismailia Canal water by chlorine, chlorine dioxide and ozone. Desalination Water Treat. 1:289-298.

Connell GF (1998). European water disinfection practices parallel U.S. treatment methods. Chlorine Chemistry Council. Drinking Water Health Quart. 4(3).

Eriksson U, Nordwall I (2002). Drinking Water Disinfection by Ultraviolet and Preformed Monochloramine: The Ultraviolet-installation at Lovo waterworks and effects in the distribution net. Obtained from: http://www.svarog-uv.org/drinking.htm. Retrieved January 2, 2013.

Environment Canada (2005). Municipal Water Use, 2004 statistics. Available at: http://www.ec.gc.ca/eau-water/default.asp?lang=En&n=ED0E12D7-1.

European Commission (1998). European Union Drinking Water Directive. Council Directive 98/83/EC. See: http://ec.europa.eu/environment/water/water-drink/index_en.html Retrieved Jan 2, 2013.

Gadgil A (1998). Drinking water in developing countries. Ann. Rev. Energy Environ. 23:253-286.

John Meunier Inc (2010). Water and Wastewater treatment: Actiflo® ACP package plants. Retrieved on October 01, 2008 from: www.johnmeunier.com.

Kruger USA (2010). ACTIFLO® for Drinking Water Treatment. Retrieved on May 16, 2011 from: http://www.krugerusa.com/en/files/16362.htm.

Lenntech (2010). Micro filtration System (MFS): Membrane Technology. Retrieved on May 16, 2011 from: http://www.lenntech.com/microfiltration.htm

Luch A (2005). The Carcinogenic Effects of Polycyclic Aromatic Hydrocarbons. Imperial College Press, London

Mainstream Water Solutions Inc (2010). Municipal Systems. Retrieved on May 16, 2011 from: http://www.mainstreamwater.com/topmenu_municipal.html#1.

Parrotta MJ, Bekdash F (1998). UV disinfection of small groundwater supplies. Am. Water Works Assoc. 90(2):71-81.

Siemens (2010). MEMCOR® CP - Ultrafiltration System. Retrieved on May 13, 2011 from: http://www.water.siemens.com/en/products/membrane_filtration_separation/ultrafiltration_membranesystems/Pages/memcor_product_cp_membrane_system.aspx

TrojanUV Technologies (2010). TrojanUVSwiftSC. Retrieved on May 17, 2011 from: http://www.trojanuv.com/solutions/municipal/drinkingwater/products/trojanuvswiftsc.

Koch Membrane (2010). Municipal Water Ultrafiltration (UF) Systems for Small Communities. Retrieved on May 16, 2011 from: http://www.kochmembrane.com/mww_prepacked.html.

Kornmüller A, Wiesmann U (2003). Ozonation of polycyclic aromatic hydrocarbons in oil/water-emulsions: mass transfer and reaction kinetics. Water Res. 37:1023-1032.

Smeets PW, Medema GJ, van Dijk JC (2009). The Dutch secret: How to provide safe drinking water without chlorine in the Netherlands. Drink. Water Eng. Sci. 2:1-14.

U.S. Environmental Protection Agency (2010). Comprehensive Surface Water Treatment Rules Quick Reference Guide: Unfiltered Systems. Office of Water (4606M) EPA 816-B-10-001. See: http://www.epa.gov/ogwdw/mdbp/pdfs/qrg_mdbp_surfacewatertreatment_unfiltered.pdf Retrieved January 2, 2013.

U.S. Environmental Protection Agency (1996). Ultraviolet Light Disinfection Technology in Drinking Water Application—An Overview. *Office of Ground Water and Drinking Water*: Washington, D.C. EPA/811-R-96-002.

U.S. Environmental Protection Agency (2005a). "Fact Sheet: Stage 2 Disinfectants and Disinfection Byproducts Rule", Office of Water, EPA 815-F-05-003. Retrieved on June 1, 2011 from http://water.epa.gov/lawsregs/rulesregs/sdwa/stage2/regulations.cfm.

U.S. Environmental Protection Agency (2005b). Occurrence Assessment for the Final Stage 2 Disinfectants and Disinfection Byproducts Rule. Office of Water, EPA 815-R-05-011. Retrieved on June 1, 2011 from http://water.epa.gov/lawsregs/rulesregs/sdwa/stage2/regulations.cfm.

U.S. Environmental Protection Agency (2005c). Technologies and Costs Document for the Final Long Term 2 Enhanced Surface Water Treatment Rule and Final Stage 2 Disinfectants and Disinfection Byproducts Rule. Office of Water, EPA 815-R-05-013. Retrieved on June 1, 2011 from http://water.epa.gov/lawsregs/rulesregs/sdwa/stage2/regulations.cfm

U.S. Environmental Protection Agency (1991). Guidance Manual For Compliance With The Filtration And Disinfection Requirements For Public Water Systems Using Surface Water Sources March 1991 Edition. Office of Drinking Water, EPA/CN 68-01-6989.

ENDNOTES

[1] Stockton California remediates its groundwater using Trojan UV ECT, an AOP; their flow rate is 1,100 cubic meters per day.
[2] A hyperlink to companies which produce each technology is provided for further review , Lenntech (2010), Siemens (2010).
[3] Data were obtained from John Meunier Inc (Actifloc), Kruger USA (Actifloc), Trojan Technologies (UV Swift SC), KOCH Membrane (MF-UF), US Filter Memcor (MF-UF) and Mainstream Water Solutions Inc (SCOR technology). Data for Actifloc and MF-UF were in US dollars and were converted to Canadian dollars. However, all data were converted to a base year (2008) in Canadian dollars for proper comparison.
[4] Personal communication from Mr Morris McCormick, Drinking water treatment plant, City of Cornwall, Ontario.
[5] These Ozone treatment plants are supplied by Mainstream water solutions Inc. Their brand name is SCOR.
[6] Rounding to the nearest $10m^3$ and solving simultaneously the equations for (1) UV Swift SC and Actifloc, (2) UV Swift SC and MF-UF and (3) Actifloc and MF-UF.
[7] Includes 5 cents for filtration or sediment removal.

Critical appraisal of an irrigation command and water productivity based on satellite remote sensing

S. Shah* and H. J. Dalwadi

[1]Department of Civil Engineering, Birla Vishvakarma Mahavidyalaya Engineering College (BVM, Vallabhviyanagar), Gujarat India.

Irrigation policy makers and managers need information on the irrigation performance and productivity of water at various scales to devise appropriate water management strategies, considering dwindling water availability, further threats due to continually rising population and food demand. It is often difficult to access sufficient water supply and use data to determine crop water consumption and irrigation performance. Energy balance techniques using remote sensing data have been developed by various researchers, and can be used as a tool to directly estimate actual evapotranspiration that is, water consumption. Study demonstrates how remote sensing-based estimates of water consumption and water stress combined with secondary agricultural production data, which provide better estimates of irrigation performance, water productivity. A principle benefit of the approach is that it allows identification of areas where agricultural performance is less than potential, there by providing insights into how irrigation systems can be managed to improve overall performance and increase water productivity in a sustainable manner. To demonstrate the advantages, the approach is planned to be applied in North Gujarat, India based Dharoi Irrigation Scheme. Remote sensing-based indicators reflecting equity, adequacy, reliability and water productivity can be estimated. This paper is purely the extensive research review and conceptualized concept to be used for future research in above said region.

Key words: Actual evapotranspiration, equity, adequacy, reliability, water productivity, surface energy balance, SEBAL.

INTRODUCTION

While irrigation has greatly increased global and regional food security, further rapid increases in agricultural production will be required to meet future food and fiber demands. This goal can be achieved either by bringing more area under irrigation or by increasing the yields of existing cropped area whilst using similar or even reduced water resources. Prospects for finding new water sources in these areas are relatively slim, because most of the surface and groundwater resources have already been exploited. Therefore, further expansion in irrigated area is often limited by water availability. Thus it is important to find ways to increase agricultural production by careful evaluation of existing irrigated lands. This strategy will not only help to meet future food demands but may also ease competition with other sectors and help to ensure water availability for nature. The science of evaluating irrigation systems has undergone major development during the last 30 years, moving from a focus on classical irrigation efficiencies to performance indicators (Bos et al., 1994; Clemmens and Bos, 1990). Public domain internet satellite data and scientific development makes remote sensing an attractive option to assess irrigation performance from individual fields to scheme or river basin scale (Bastiaanssen and Bos, 1999; Bos et al., 2005; Akbari et al., 2007). Such spatial information is increasingly important for large irrigation

*Corresponding author. E-mail: samta_parag@yahoo.co.uk.

systems and river basins such as the Sabarmati basin irrigation system of North Gujarat, India to help identify suitable water management strategies across different scales. This is particularly important because the means to improve water productivity lie substantially at farm scale in terms of crop management and at system scale in terms of irrigation water distribution and possible allocation.

There are a number of well known and emerging irrigation management objectives and needs in North Gujarat, India. The region being scanty rain fed. Remote sensing, geographic information systems (GIS) and geo-statistics approaches, along with limited field data, can be used to solve distributed water balance and estimate net groundwater use. Government can launch a new program to maintain a computerized database for irrigation releases to improve irrigation management, reduce rent seeking, increase transparency and demon-strate which users are getting what quantity of water. It is expected that these initiatives will improve data management and availability of surface supplies.

But to work, information on overall water consumption (surface and groundwater) at various scales will be essential for judicious and efficient water resources management in the region. There is a need to study water distribution and consump-tion patterns and the impacts of this on productivity. Remotely sensed estimates of actual evapotranspiration take account of factors that reduce water consumption below potential, such as salinity, crop condition and poor irrigation scheduling all of which are important, specifically in region like North Gujarat,India. The objectives of this paper are: application of the surface energy balance algorithm for land (Dharoi in North Gujarat, India.) to estimate actual evapotranspiration (ET_a), quantification of agricultural water consumption in different irrigation subdivisions of Dharoi scheme; understanding the seasonal and spatial patterns of evapotranspiration in North Gujarat, India and their linkage to groundwater consumption, completion of irrigation and water use performance diagnoses in terms of crop water productivity and water consumption.

MATERIALS AND METHODS

This study is proposed on for the upper Sabarmati river basin irrigation system in North Gujarat, India with particular focus on and around Mehasana District, a land unit lying between the Unjha, Visnagar, Vijapur, Sidhpur and Kheralu villages. The gross area of the right bank irrigation is about 142321 ha. Summers are long and hot, lasting from April through September with maximum daytime temperatures ranging from 40 to 45°C. The winter season lasts from December through February with maximum temperatures ranging between 25 and 27°C. Mean annual precipitation is about 850 mm (Dharoi irrigation department). SEBAL can be used for actual evapotranspiration calculations, which is a well-tested and widely used method to compute ET_a (Bastiaanssen et al., 1999). SEBAL results have been validated in a number of countries using field instruments including weighing lysimeter, scintillometer, Bowen ratio and Eddy-correlation towers, indicating daily ET estimates have errors on the order of 16% or lower at 90% probability (www.waterwatch.nl). At longer time scales, such as a crop season, E_T errors cancel out to 5% (Bastiaanssen et al., 2005). SEBAL is an image-processing model, which computes a complete radiation and energy balance along with resistances for momentum, heat and water vapor transport for each pixel (Bastiaanssen et al., 1999; Bastiaanssen, 2000). From the reflectance and radiance measurements of the bands, first land surface parameters such as surface albedo vegetation index, emissivity and surface temperature. The key input data for SEBAL consists of spectral radiance in the visible, near-infrared and thermal infrared part of the spectrum. In addition to satellite images, the SEBAL model requires routine weather data parameters (wind speed, humidity, solar radiation and air temperature). With this data, evapotranspiration is then calculated from the latent heat flux, and the daily averaged net radiation, Rn_{24}.The latent heat flux is computed from the instantaneous surface energy balance at satellite overpass on a pixel-by-pixel basis.

$$\lambda E = R_n - (G_0 + H) \quad (1)$$

Where λE = (W/m^2) is the latent heat flux (λ is the latent heat of vaporization and E is the actual evaporation), Rn (W/m^2) is the net radiation, G_0 (W/m^2) is the soil heat flux and H (W/m^2) is the sensible heat flux.

The latent heat flux describes the amount of energy consumed to maintain a certain evapotranspiration rate. The energy balance Equation (1) can be decomposed further into its constituent parameters. Rn is computed as the sum of incoming and outgoing short wave and long-wave radiant fluxes. G_0 is empirically calculated as a G_0/R_n fraction using vegetation indices, surface temperature and surface albedo. Sensible heat flux is computed using wind speed observations, estimated surface roughness, and surface to air temperature differences that are obtained through a self-calibration between dry (see=0) and wet (H =0) pixels. The dry and wet pixels are manually selected, based on vegetation index, surface temperature, albedo and some basic knowledge of the study area. The need for this technique makes SEBAL somewhat subjective and difficult to automate. SEBAL uses an iterative process to correct for atmospheric instability caused by buoyancy effects of surface heating. More details and recent references/literature on SEBAL is available at www.waterwatch.nl. Then instantaneous latent heat flux, λE, is the instantaneous evaporative fraction (Bastianssan et al., 2000).

$$\lambda = (\lambda E / \lambda E + H) = \lambda E / (R_n - G_0) \quad (2)$$

λ expresses the ratio of the actual to the crop evaporative demand when the atmospheric moisture conditions are in equilibrium with the soil moisture conditions. The instantaneous value can be used to calculate the daily value, because the evaporative fraction tends to be constant during daytime hours, although the H and λE fluxes vary considerably. The difference between the instantaneous evaporative fraction at satellite overpass and the evaporative fraction derived from the 24 h integrated energy balance is often marginal and in many cases be neglected for time scales of 1 day, G_0 is relatively small and can be ignored and net available energy / ($Rn-G_0$) reduces to net radiation (Rn). At daily timescales actual evapotranspiration, ET_{24} (mm/day) can be computed as:

$$ET_{24} = (86{,}400 \times 10^3 / \lambda \rho w) \times \lambda R_{n24} \quad (3)$$

Where Rn_{24} (W/m^2) is the 24 h averaged net radiation, λ (J/kg) is the latent heat of vaporization, and ρw(kg/m^3) is the density of water. For timescales longer than 1 day, actual evapotranspiration can be estimated using the relation proposed by Bastiaanssen et al. (2002). The main assumption is that specified in Equation (3) remains constant over the entire time interval between capture of each remote sensing image so that:

$$ET_{int} = dt\ (86,\ 400 \times 10^3/\ \lambda\rho w) \times \lambda\ Rn_{24t} \quad (4)$$

Where ET_{int} (mm/interval) is the time integrated actual evapotranspiration, and Rn_{24t} (W/m^2) is the average Rn_{24} for the time interval dt measured in days. Rn_{24t} is usually lower than Rn_{24}, because Rn_{24t} also includes cloudy days.

Remote sensing-based irrigation performance indicators quantify the efficiency of the project. A wide range of irrigation and water use performance indicators are available (Rao, 1993; Bastiaanssen and Bos, 1999; Bos et al., 2005) to assist in achieving efficient and effective use of water by providing quantification aid in such study, indicators representing equity, adequacy, reliability, and water productivity can be used to evaluate the performance of irrigated agriculture in the North Gujarat, India based Sabarmati basin. Irrigation policy makers and managers need information on the irrigation performance and productivity of water at various scales to devise appropriate water management strategies, considering dwindling water availability, further threats due to continually rising population and food demand. It is often difficult to access sufficient water supply and use data to determine crop water consumption and irrigation performance.

Energy balance techniques using remote sensing data have been developed by various researchers, and can be used as a tool to directly estimate actual evapotranspiration that is, water consumption. Study demonstrates how remote sensing-based estimates of water consumption and water stress combined with secondary agricultural production data, which provide better estimates of irrigation performance, water productivity. A principle benefit of the approach is that it allows identification of areas where agricultural performance is less than potential, there by providing insights into how irrigation systems can be managed to improve overall performance and increase water productivity in a sustainable manner. To demonstrate the advantages, the approach is planned to be applied in North Gujarat, India based Dharoi Irrigation Scheme. Remote sensing-based indicators reflecting equity, adequacy, reliability and water productivity can be estimated. This paper is purely the extensive research review and conceptualized concept to be used for future research in the above said region Reservoir project. Traditionally, equity is calculated from the supply side.

But in water scarce system, like in the North Gujarat, equity in water consumption is more relevant from the farmer's perspective and can be computed from remote sensing-based ET_a maps of an irrigation system. Adequacy is the quantitative component, and is defined as the sufficiency of water use in an irrigation system. In contrast, reliability is the time component and defined as the correspondence of water supply upon request. Both, adequacy and reliability of water supplies to cropped area can be assessed using the evaporative fraction maps as they directly reveal the crop supply. In this study, adequacy is defined as the average seasonal evaporative fraction and reliability as the temporal variability, that is temporal coefficient of variation of evaporative fraction in a season. Evaporative fraction values of 0.8 or higher indicate no stress (Bastiaanssen and Bos, 1999), and below 0.8 reflect increases in moisture shortage to meet crop water requirements as a result of inadequate water supplies. Similarly, the lower values of coefficient of variation represent the more reliable water supplies throughout the cropping season. The term water productivity is defined as the physical mass of production or the economic value of production measured against gross inflows, net inflow, depleted water, process depleted water, or available water (Molden, 1997; Molden and Sakthivadivel, 1999). For systems comprised of multiple agricultural enterprises, the water productivity is often computed in monetary terms.

Remote sensing data to evaluate the irrigation performances (Findings from case studies)

Remote sensing data to determine actual evapotranspiration and crop water stress for managing irrigation systems was suggested by several authors. However, it was also showed, how high resolution landsat images could be used to map crops and crop phenology stages in two Italian irrigation districts, and to detect crop-specific effects such as water stress. Menenti et al. (1989) made a list of indicators that could be quantified with remotely sensed data. They looked at equity by evaluating actual flow per irrigated area at different spatial scales. The irrigated area was taken from satellite images, supplemented with records of flow from the field. They also used the interesting marginal benefit indicator as defined by being the difference between actual evapotranspiration in situations with and without irrigation divided by the volume of water that is supplied to that unit. This shows the portion of irrigation water used to enhance actual evapotranspiration.In a way, data on flow and actual evapotranspiration reveals also valuable information on system losses between the off-take and the root zone.

If evapotranspiration under rain-fed conditions is negligible, or the crop selection in the irrigated areas is entirely different from the neighboring rain-fed agricultural practices, the benefit indicator can be re-defined as actual evapotranspiration over actual flow delivered. Menenti et al. (1989) have used remote sensing data to generate maps of irrigated area and potential crop water requirements based on remote sensing derived k_c factors. Operational algorithms to compute actual evapotranspiration were not yet available at that time. Instead, a hydrological simulation model was suggested to compute actual evapotranspiration. Moran (1994) described the results of a farm irrigation management study in Arizona where the maximum possible cloud-free SPOT and Landsat images were used in conjunction with airborne remote sensing data from an altitude of 100 m above ground level. Alfalfa and cotton were continuously monitored in terms of vegetation index (SAVI) and surface temperature to deduce the water deficit index (WDI) for adequacy evaluations. It became obvious that satellite data could not always be obtained due to technical shortcomings in the satellite receiving systems in that case aircraft data was found more reliable and faster available.

Data collection and preprocessing

This study can be conducted for the period representing fall and spring cropping seasons. Daily meteorological data on temperature, humidity, wind speed and sunshine hours for and regional can be collected from the North Gujarat Meteorological department for prescribed period. Rainfall data can be collected from number of gauging stations in and around North Gujarat and be interpolated to obtain gridded values of seasonal and annual rainfall. Considering the type of vegetation and bunds around field, interception and runoff losses from cropped areas are minimal and therefore ignored in this analysis (Ahmad et al., 2006). Data describing surface flow diversions at the heads of the main irrigation canals can be collected for the same time period from the irrigation and power department for the main canal systems of North Gujarat. For the gross value of production estimation, data on crop area, production and output prices can be collected from secondary sources including the Economic Wing of the Ministry of Food, Agriculture and Livestock. Data on cropped area and production are available at district level with for all fall and spring crops. District level GVP can be transformed to the scale of irrigation subdivisions based on the fraction of district area falling in a specific irrigation subdivision using overlay analysis in GIS. This was mainly done to overcome the limitation of district boundaries, which do not match those of the irrigation subdivisions. GVP can be computed on real prices and wholesale price index (WPI) can be used to convert the current/nominal prices into constant/real prices with a base year considered.

Framework for remote sensing and irrigation performance indicators

Summary of definitions

This section is not intended to standardize performance indicators that can be quantified by use of satellite data, but merely a summary of technically possible and workable indicators. There is not enough experience worldwide to arrive at a consensus. More awareness and case studies are needed to demonstrate the advantages of remote sensing-based indicators as compared to the data collection based on classical field surveys. From the literature review, it is concluded that a larger set of indicators exists for adequacy and equity than for reliability and productivity. This is a mere consequence of the surface energy balance and the partitioning of energy, being direct indicators for adequacy and equity. Note that all surface energy balance studies rely on thermal infrared measurements. Productivity has basically two components: one for the physical biomass or yield per unit of water and a second which describes the use of the water resources through the comparison of water consumed by crop evapotranspiration versus the volume of irrigation water supplied. The latter 'overall consumed ratio' is bio-physically related to the former due to the strong relationships between actual evapotranspiration and crop yield. More research is recommended to comply with a remote sensing determinant of reliability in the irrigation service. The indicators against which the performance of irrigation can be assessed vary widely with legal aspects, related regulations and with traditions of local irrigation communities. All groups of performance indicators has its own requirements for frequency of images and time constraints between image acquisition and availability at the door of the irrigation manager. For example, in rigid water distribution systems, preseason planning can be based on interpretations of remotely sensed water distribution data from the proceeding season combined with the recorded deviations between actual and intended water supply. The protective rotational water supply systems aim at equity and some of these systems have an increasingly severe problem of sustainability (salinity). The water distribution with pressurized irrigation systems with sprinklers and drip systems demand high investments, productivity evaluations are more important than issues such as equity or reliability.

Overall consumed ratio (e_p)

Overall consumed ratio quantifies the degree to which crop irrigation requirements are met by irrigation water in the irrigated area. The ratio is defined as follows:

$$e_p \, ¼ = (ET_p - P_e) / V_c$$

Where "ET_p is potential evapotranspiration, in mm, "P_e is effective precipitation, in mm" and "V_c is volume of irrigation water diverted" from resource and/or groundwater in mm. Overall consumed ratio should be set within an existing irrigated area and compared to the actual ratio on a monthly and seasonal basis .

Relative water supply (RWS)

The relative water supply used as an indicator of adequacy of irrigation water delivery compares supplied water with that demanded. It is calculated as follows:

$$RWS \, ¼ = (V_c * P_g) / ETp$$

Where, "P_g is gross precipitation, in mm". The target value of the RWS indicator was considered as 2.0 (Molden et al., 1998).

Depleted fraction (DF)

Depleted fraction shows changes in actual water use by crops and quantifies differences in the water balance of the areas under study. The depletion in an irrigation scheme is governed by ETa. This indicator is defined by (Molden, 1997):

$$DF \, ¼ = ET_a / (V_c * P_g)$$

Where "ET_a is actual evapotranspiration, in mm". DF should be considered as a function of time. For semi-arid and arid regions the critical value of the depleted fraction averages about 0.6 (Bos et al., 2005). The acceptable range of DF was considered as 0.6 to 1.1(Bastiaanssen et al., 2001).

Crop water deficit (CWD)

Crop water deficit over a period is defined as the difference between ETp and ETa of the cropping pattern within an area. A common period is 1 month, and an average CWD of 30 mm month is acceptable (Bastiaanssen et al., 2001). Crop water deficit is defined as follows:

$$CWD^{1/4} = ET_p - ET_a$$

Relative evapotranspiration (RE_T)

Relative evapotranspiration quantifies reduction in evapotranspiration and detects water-short areas. To evaluate the adequacy of irrigation water delivery to a selected command area as a function of time, the dimensionless ratio of ET_a over ET_p gives valuable information to the water manager.

Conclusions

There is only a small group of researchers working on remote sensing interpretations for irrigation performance assessment. Their work is summarized in this paper to make a larger audience aware of their progress. The accuracy of measuring individual parameter from remote sensing data ranges between 80 to 90%. Performance indicators are usually based on several parameters combined together, and the accuracy is therefore approximately 75 to 85%. It is apparent that scientific progress has been ahead of implementation because the images and processing equipment were a financial burden in the eighties. The current reduction in costs of raw satellite images, direct availability of National Oceanic and Atmospheric Administration (NOAA) satellite images through commercial ground receiving stations, cheaper PC based processing software, and the availability of more robust interpretation algorithms, make applications nowadays more attractive. Irrigation managers, consultants and policy makers are usually not aware of the opportunities remote sensing can offer. This is partially due to the overselling of the possibilities satellites could offer at the onset of the remote sensing era in the seventies. Some of the case studies shown in this paper reveal, however, that irrigation performance indicators can be calculated for

1. Diagnosing water management practices at small unit areas up to farm plot level.
2. Monitoring irrigation events at scales > 500 ha in a regular fashion.
3. Revealing the overall water resources utilization.

This is an appropriate extension of using flow rates in the conveyance network. Aspects of adequacy, productivity, equity, reliability and sustainability in irrigation management can be computed from remotely sensed data. Standardization of irrigation performance indicators is an important issue for the International Committee on Irrigation and Drainage (ICID), and satellite measurements can help in surveying the conditions of irrigated land in a consistent and objective manner. It will allow comparing conditions within and between different irrigation schemes in a normalized manner. Most of the reviewed literature needs a thermal-infrared band to infer surface temperature for crop water stress and surface energy balance modeling.

REFERENCES

Ahmad MD, Biggs T, Turral H, Scott CA (2006). Application of SEBAL approach and MODIS time-series to map vegetation water use patterns in the data scarce Krishna river basin of India. IWA J. Water Sci. Technol., 53(10): 83–90.

Akbari M, Toomanian N, Droogers P, Bastiaanssen W Gieske A (2007). Monitoring irrigation performance in Esfahan, Iran, using NOAA satellite imagery. Agric. Water Manage., 88: 99–109.

Bastiaanssen WGM, Ahmad MD, Chemin Y (2002). Satellite surveillance of evaporative depletion across the Indus. Water Resour. Res. Water Resour. Res., 38(12): 1-9 1273.

Bastiaanssen WGM, Menenti M, Feddes RA, Holtslag AAM (2001). A remote sensing surface energy balance algorithm for land (SEBAL). Part 1. Formulation. J. Hydrol., 212–213: 198–212.

Bastiaanssen WGM (2000). SEBAL-based sensible and latent heat fluxes in the irrigated Gediz Basin, Turkey. J. Hydrol., 229: 87–100.

Bastiaanssen WGM, Bos MG (1999). Irrigation performance indicators based on remotely sensed data: a review of literature. Irrigation Drainage Syst., 13: 291–311.

Clemmens AJ, Bos MG (1990). Statistical methods for irrigation system delivery performance evaluation. Irrigation and Drainage Systems 4.

Menenti M, Bastiaanssen WGM, VanEick D (1989).Determination of surface hemispherical reflectance with Thematic Mapper data. Remote sensing Environ., 28:327-33.

Molden D (1997). Accounting for Water Use and Productivity SWIM Paper 1. International Irrigation Management Institute (IIMI), Colombo, Sri Lanka.

Molden D, Sakthivadivel R (1999). Water accounting to assess use and productivity of water. Water Resour. Dev., 15: 55–71.

Moran MS (1994). Irrigation management in Arizona using satellites and air planes. Irrigation Sci., 15: 35-44.

Limnochemical characteristics of river Yamuna in Yamunanagar, Haryana, India

Girish Chopra, Anita Bhatnagar and Priyanka Malhotra

Department of Zoology, Kurukshetra University Kurukshetra-136119 India.

The present paper deals with the monthly variations of limnochemical characteristics of river Yamuna, Yamunanagar which is polluted with industrial effluents and domestic sewage via, maskara nala. Three sampling points that is, Station-Y1: Upstream of the river; Station-Y2: Point of influx of industrial effluents and domestic sewage; Station-Y3: About 5 km downstream from station Y2 were selected for the investigation. Studies revealed high values of conductivity, free CO_2, Biochemical oxygen demand (BOD), Chemical Oxygen Demand (COD), total alkalinity, calcium, hardness, chloride, nitrite and nitrate whereas, low values of Dissolved Oxygen (DO) at the point of getting effluent samples. The differences in various parameters were statistically significant ($P<0.05$) when compared from upstream and downstream stretches of the river. DO, BOD and COD were found to be important parameters which showed strong correlation with other parameters and hence can serve as good indices of river water quality. Water quality may be designated as 'medium' according to Brian Oram's water quality Index and 'severely polluted' according to Kaur's water quality index at station Y2.

Key words: Water quality index, Yamuna River, limnochemical characteristics, industrial effluent, sewage.

INTRODUCTION

Rivers have always been the most important fresh water resources, and most developmental activities are still dependent upon them. Rivers play a major role in assimilating or carrying industrial and municipal waste water, manure discharge and runoff water from agriculture fields, road ways and streets which are responsible for river pollution (Ward and Elliot, 1995). Host of workers have contributed to our knowledge of riverine pollution, notable studies are on Scotland's rivers (Benzie et al., 1999); St. Clair river, Canada (Griffiths et al., 1991); Chinese rivers (Zhang et al., 1995); Orogodo river, Agbor Nigeria (Rim-Rukeh et al., 2006); Huangpu river, China (Yang et al., 2007), Sava river, Slovenia (Toman, 2009), Indian rivers such as river Chambal (Kulshrestha et al., 1991), river Ghaggar (Bhatnagar and Garg, 1998), basins Kothaiyar and Pazhayar in Kanyakumari district (Raj and Jayasekher, 2007), river Kosi (Bhandari and Nayal, 2008), river Hindon (Dalal and Arora, 2008; Suthar et al., 2010), river Godawari (Deshmukh and Sonawane, 2008), river Damodar (Singh et al., 2009),river Yamuna (Kaushik et al., 2001; Khaiwal et al., 2003; Bhatnagar et al., 2009) and river Karmana (Sujitha et al., 2011).

According to Kaushik et al. (2009), in India, it is important to systematically study the status of pollution of the rivers in relation to various anthropogenic activities as river water has been used as drinking water, for agriculture and for fish culture throughout the history of mankind. The Yamuna is one of the most important rivers of the Indogangetic plains in India and outnumbers any other river in the number of industries on its bank. The industrial belt of Haryana state in India is mainly situated along the north-eastern part of the state along the river Yamuna. The waste from these industries, agricultural runoff and the drains carrying municipal sewage of the cities enter into the river and affect its water quality. Yamunanagar, one of the major industrial cities of Haryana, receives effluents from sugar mills from the district of Saharanpur through 'Maskara nala'. Therefore, this present study has been undertaken to evaluate the

*Corresponding author. E-mail: anitabhatnagar@gmail.com.

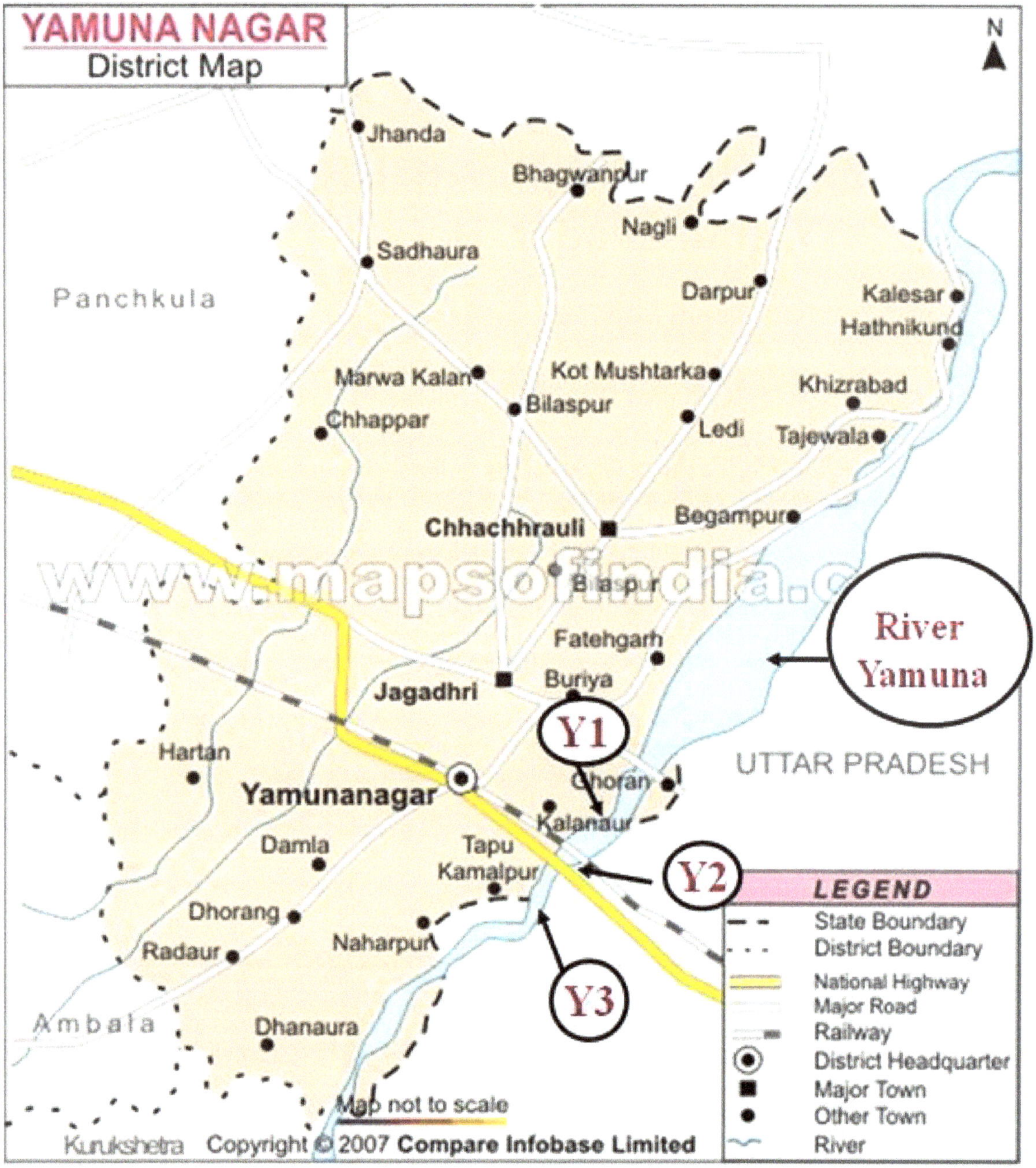

Figure 1. Map of Yamunanagar showing selected stations on river Yamuna.

water quality of river Yamuna in Yamunanagar in terms of limnochemical characteristics and water quality indices. The results obtained are useful to the authorities and the other stakeholders.

MATERIALS AND METHODS

As noticed in Figure 1, taking into cognizance the points of discharges into the river, three stations Y1, Y2 and Y3 were selected. Station-1 (Y1) lies in village Kalanaur, upstream of the river before the influx of discharges. Station-2 (Y2) lies 4 to 5 km downstream from station Y1 at middle reach of the river where the mill effluents joins the river. Station-3 (Y3) at 5 km downstream from station-2 after the influx of discharges (Figure 1). Water samples were collected monthly from August 2008- July 2009 in three replicates from all the sampling stations.

Limnochemical characteristics

Surface water samples were collected monthly from August, 2008 to July, 2009 from river Yamuna, from all the selected three stations, in triplicate, in polythene bottles. The limnochemical characteristics, viz., temperature, dissolved oxygen, free CO_2 and

and alkalinity were analyzed at the site itself as their values are liable to change soon, whereas, other parameters were analyzed in the laboratory according to standard procedures (Golterman et al., 1978; APHA, 1998) within the following 3 to 4 days, during which samples were kept in cold storage. Dissolved oxygen, pH, conductivity were analyzed using MultiSet F Line three Water analysis kit (E Merck). Free CO_2, alkalinity, chlorides, hardness, calcium, magnesium were anlysed by titrimitric method. BOD was determined following dilution and incubation method (APHA, 1998). COD was analyzed by reflux method and ammonia, nitrite, nitrate and orthophosphate were determined spectrophotometerically (APHA, 1998).

Water quality index calculation

Water quality index, a measure of overall water quality, was calculated by using Brian Oram's on line calculator (Oram, 2007) (WQI A) and Horton Water Quality Index (Horton, 1965) modified by (Kaur et al., 2001) (WQI B). Kaur's water quality index calculation was based on the weightage of the individual parameters and its rating scale.

Statistical analysis

The coefficient of correlation was calculated using SPSS packages.

RESULTS

Monthly and stational variations in different limnochemical characteristics are shown in Figures 2 and 3. Table 1 depicts the mean values of the water quality characteristics at different stations. The values of DO and pH decreased significantly from station Y1 to Y2. DO showed significant ($P<0.05$) negative correlation with free CO_2 ($r = -0.521$) and positive with alkalinity ($r = 0.529$). The values of BOD during present investigation ranged from 4.2 to 8.2 mg L^{-1}. BOD values showed a sharp and significant ($P<0.05$) increase at station Y2 and thereafter the values significantly decreased at Y3. BOD showed a statistically significant positive correlation (Table 2) with COD ($r = 0.499$).

The mean value of COD was recorded maximum at station Y2 (282 mg L^{-1}). Conductivity, free CO2, BOD, COD, total alkalinity, hardness, calcium, chloride, significantly ($P<0.05$) increased from station Y1 to Y2 and then decreased at Y3. Amongst nutrients the values of orthophosphate ranged between 0.1 mg L^{-1} to 0.2 mg L^{-1} at station Y1, 0.1 mg L^{-1} to 0.3 mg L^{-1} at station Y2 and 0.1 mg L^{-1} to 0.4 mg L^{-1} at station Y3. The mean values of orthophosphate showed an increasing trend from station Y1 (0.1±0 mg L^{-1}) to Y2 (0.2±0 mg L^{-1}) and Y3 (0.2±0 mg L^{-1}). The mean values of nitrite increased from station Y1 (0.2±0.1 mg L^{-1}) to Y2 (0.4±0.1 mg L^{-1}) and then further decrease from station Y2 to Y3 (0.3±0.1 mg L^{-1}). No significant variations were observed in the values of ammonia from station Y1 and Y2; however the values decreased at station Y3. The values of nitrate ranged between 0.1 mg L^{-1} to 0.7 mg L^{-1} at station Y1, 0.2 mg L^{-1} to 0.9 mg L^{-1} at station Y2 and 0.1 mg L^{-1} to 1.9 mg L^{-1} at station Y3. The mean values showed a gradual and significant ($P<0.05$) increase from station Y1 (0.3±0.1 mg L^{-1}) to Y3 (0.5±0.1 mg L^{-1}). No significant variations were observed in the values of ammonia from station Y1 and Y2, however, the values decreased at station Y3 Water quality may be designated as 'bad' according to Brian Oram's water quality Index and 'severely polluted' according to Kaur's water quality index at station Y2.

DISCUSSION

Conductivity, free CO2, BOD, COD, total alkalinity, hardness, calcium, chloride, nitrite and nitrate significantly ($P<0.05$) increased whereas the values of DO, pH and values of water quality index decreased significantly from station Y1 to Y2 (where the mill effluents joins the river) depicting that introduction of industrial waste and domestic sewage has changed the overall condition of the river's water.

DO in water is usually depleted, if high amount of organic matter undergoing biological degradation is present (Rim-Rukeh, 2006). The mean concentration of DO in the Yamuna river was below 5.0 mg L^{-1} at all the stations and, therefore, the river water appeared to be less conducive for aquatic organisms. ICMR guidelines also recommend DO >5.0 mg L^{-1}. Increase in BOD at Y2 may be attributed to influx of pollutants and decrease at Y3 may be the result of self purification process of the river. There occurs a natural process of self purification which extends over several kilometers of a river (Hawkes, 1978). This results in a regeneration of polluted river back into this normal original state. Maximum permissible limit for the discharge of effluent for BOD is 30 mg L^{-1} (Polestya et al., 2008). However, the BOD was higher at station Y2 and Y3 when compared with ICMR (1975) standards (<5 mg L^{-1}). BOD and COD are important index of organic pollution in the river. The values of these two parameters significantly ($P<0.05$) increased with influx of pollutants at station Y2. Statistically also COD showed a significant positive correlation with BOD ($r = 0.499$, $P<0.05$).

pH was alkaline through out the study period. The values were significantly ($P<0.05$) high at station Y1 and decreased at station Y2. The lowering of pH at station Y2 seems to be due to input of wastes (Bhatnagar and Sanghwan, 2009). At this station, pH was low but alkalinity was higher.

The high value of free carbon-dioxide at Y2 may be associated with low DO at this station. Statistically, free CO2 showed a significant negative correlation with DO ($r = -0.521$) $P<0.05$). In the present study, it has been observed that amount of total hardness was high during rainy season than during summer season, confirming that the rain water brings more amounts of Ca and Mg (Devis and DeDewiest, 1996; and Mathivanan et al., 2008). Nitrate, the most highly oxidized form of nitrogen compounds is commonly present in surface waters

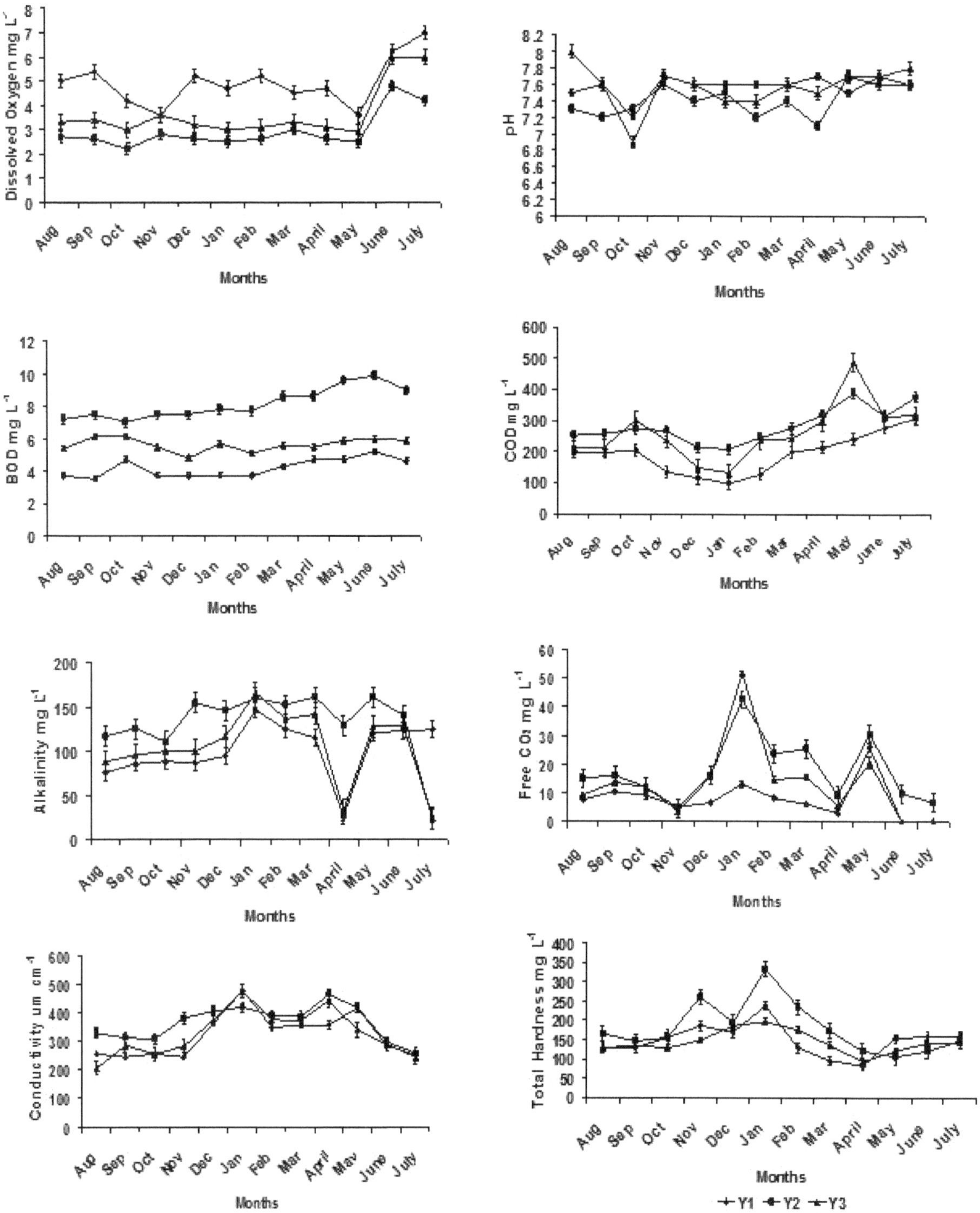

Figure 2. Monthly variations in DO, pH, BOD, COD, alkalinity, free CO_2, conductivity and total hardness on various stations.

because it is the end product of aerobic decomposition of organic nitrogenous matter (Polestya et al., 2008). The values of nitrate were in the range of 0.2 to 0.4 mg L^{-1}. Kaur et al. (1997) also measured nitrate in the same range, supporting the present results. Nitrite was observed in the range of 0.3 to 0.5 mg L^{-1}. A similar trend in nitrite-nitrogen was observed by Jaji et al. (2007) in Ogun River, Nigeria.

A comparison of the mean values of analyzed physico-chemical conditions with drinking water standards of

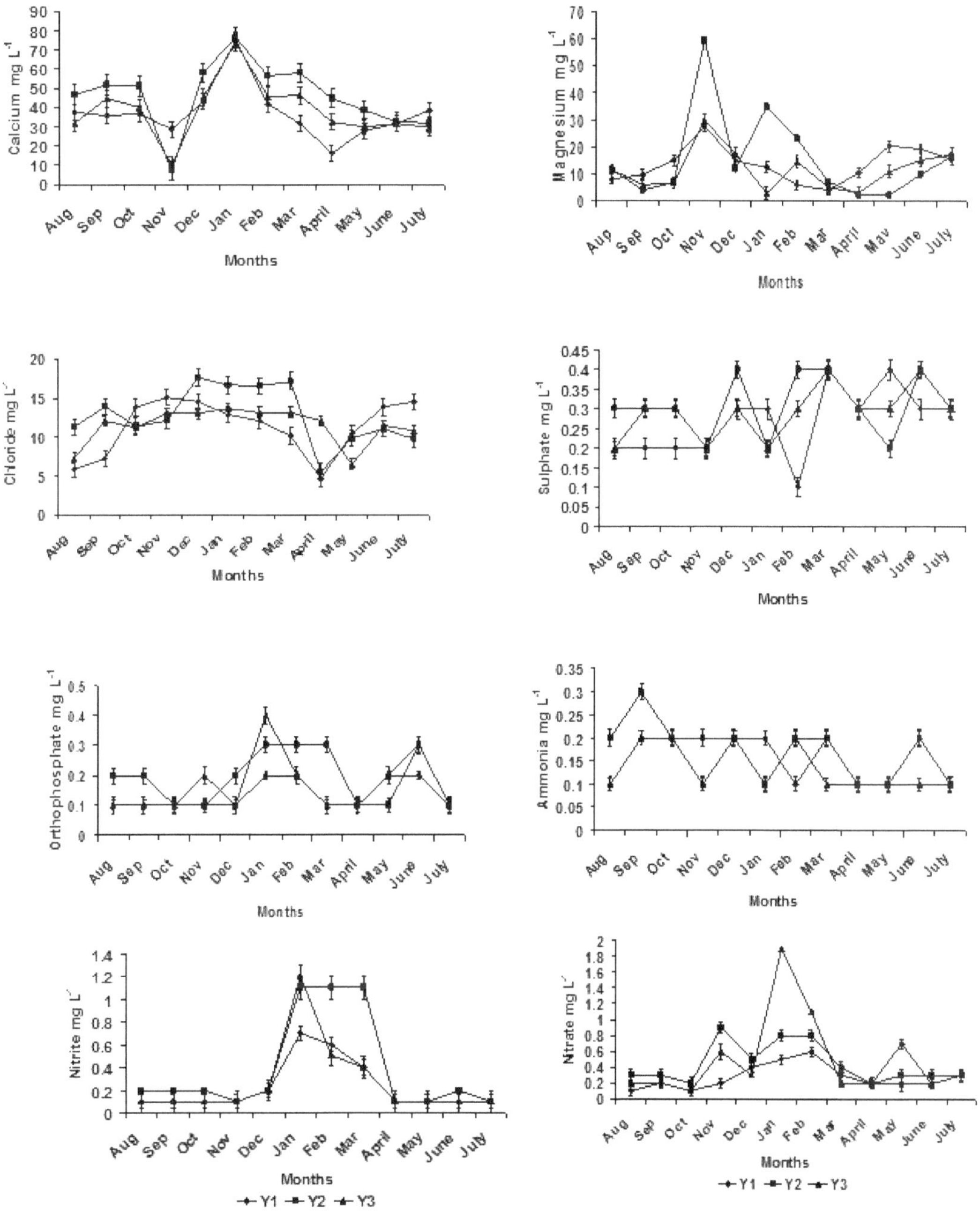

Figure 3. Monthly variations in Calcium, magnesium, chloride, sulphate orthophosphate, ammonia, nitrite and nitrate on various stations.

ICMR (1975), WHO (1984, 1993) reveal that DO is low in the present studies whereas BOD, COD, calcium and hardness are high. The values of other parameters are within permissible limits. It is necessary to treat the water of river Yamuna and to establish efficient treatment plants by the effluents generating industries for maintaining the important parameters within the permissible limit prescribed by world health organization/Indian council ofmedical Research.

To communicate the information on water quality depicting the overall quality of water at all the selected stations of river Yamuna, the mathematical values in terms of water quality index (WQI) according to Brian Oram's (on line calculator), Horton (1965) and Kaur et al. (2001) was calculated. According to Oram's water quality index (Oram, 2007) water quality legend (WQI A) in the range of 90-100 indicate excellent quality, 70-90 good, 50-70 medium, 25-50 bad and 0-25 very bad. It was found that the water was considered of 'medium quality' according to Brian Oram's water quality index at

Table 1. Mean values of water quality characteristics of river Yamuna on various stations

Water quality characteristics	Y1	Y2	Y3	ICMR, 1975 Standard	WHO(1984,1993) Standard for drinking Water
Temperature °C	22.7 ± 0.9^{A}	23.3 ± 0.9^{A}	22.7 ± 0.9^{A}	-	-
pH	7.6 ± 0^{A}	7.4 ± 0.1^{B}	7.6 ± 0.1^{A}	<7.0-8.5	6.5-8.5
Conductivity µm cm^{-1}	$320\text{-}\pm21.1^{B}$	363 ± 17^{A}	328 ± 22.5^{B}	-	0-300
Turbidity NTU	6 ± 1.7^{A}	5.6 ± 1.5^{A}	4.6 ± 1.5^{A}	>5	5-10
Free CO_2 mg L^{-1}	9.6 ± 1.9^{C}	17.8 ± 3.0^{A}	16.3 ± 4.0^{B}	-	
DO mg L^{-1}	4.9 ± 0.3^{A}	2.9 ± 0.2^{B}	3.7 ± 0.3^{B}	>5	
BOD mg L^{-1}	4.2 ± 0.2^{C}	8.2 ± 0.3^{A}	5.6 ± 0.1^{B}	<5	<6.0
COD mg L^{-1}	192 ± 17.7^{B}	282 ± 15.8^{A}	262 ± 25.6^{A}	<120	<10.0
Total Alkalinity mg L^{-1}	100 ± 8.6^{B}	131 ± 10.5^{A}	104 ± 11.4^{B}	600	13-246
Total Hardness mg L^{-1}	148 ± 11.2^{B}	179 ± 18.3^{A}	144 ± 7.9^{B}	<75	300-600
Calcium mg L^{-1}	37.1 ± 3.8^{B}	46.1 ± 4.8^{A}	38.5 ± 4.1^{B}	<50	28-48
Magnesium mg L^{-1}	13.6 ± 1.8^{A}	15.8 ± 4.5^{A}	11.7 ± 2.1^{A}	<200	9.23-26.24
Sulphate mg L^{-1}	0.3 ± 0^{AB}	0.3 ± 0^{A}	0.3 ± 0^{B}	<250	50-91
Chloride mg L^{-1}	11.3 ± 1.0^{B}	12.8 ± 1.0^{A}	11.5 ± 0.6^{B}	-	7-26
Orthophosphate mg L^{-1}	0.1 ± 0^{B}	0.2 ± 0^{A}	0.2 ± 0^{B}	-	
Ammonia mg L^{-1}	0.2 ± 0^{B}	0.2 ± 0^{A}	0.1 ± 0^{B}	-	
Nitrite mg L^{-1}	0.2 ± 0.1^{B}	0.4 ± 0.1^{A}	0.3 ± 0.1^{B}	-	
Nitrate mg L^{-1}	0.3 ± 0.1^{B}	0.4 ± 0.1^{A}	0.5 ± 0.1^{A}	-	0-1.77
WQI A	60 ± 0.50^{A}	56 ± 0.50^{B}	58 ± 0.52^{A}	-	
WQI B	67.4 ± 2.8^{A}	39.9 ± 4.3^{B}	47.5 ± 3.0^{B}	-	

Means with different letters in the same row are significantly ($P<0.05$) different. (Data were analyzed by Duncan's multiple range tests).

Table 2. Coefficient of correlation between various limnochemical characteristics of river Yamuna.

	pH	Conductivity	DO	BOD	COD	Free CO_2	Alkalinity	Calcium	Hardness	Magnesium	Turbidity	Chloride	Ammonia	Sulphate	O-PO_4	Nitrite	Nitrate
Temp	0.105	-0.514	0.363	0.101	0.519	-0.247	-0.222	-0.406	-0.376	-0.069	0.385	-0.354	-0.182	-0.006	-0.080	-0.559	-0.428
pH	-	-0.194	0.394	-0.348	-0.09	-0.257	0.236	-0.381	-0.135	0.144	-0.085	-0.200	-0.307	-0.080	-0.118	-0.222	-0.092
Conductivity	-	-	-0.492	0.250	0.012	0.554	0.486	0.469	0.309	-0.043	-0.487	0.190	0.020	0.148	0.367	0.485	0.505
DO	-	-	-	-0.278	-0.065	-0.521	-0.529	-0.349	-0.228	0.026	0.048	-0.125	-0.305	0.131	0.273	-0.289	-0.243
BOD	-	-	-	-	0.499	0.362	0.463	0.270	0.280	0.038	-0.043	0.259	0.180	0.092	0.310	0.269	0.104

Table 2. cont.

COD	-	-	-	-	-	-0.011	0.166	-0.231	-0.292	-0.112	-0.018	-0.191	-0.141	0.140	0.052	-0.228	-0.230
Free CO_2	-	-	-	-	-	-	0.558	0.650	0.415	-0.071	-0.067	0.247	0.047	-0.115	0.648	0.666	0.615
Alkalinity	-	-	-	-	-	-	-	0.485	0.483	0.084	-0.186	0.428	0.264	0.103	0.593	0.528	0.469
Calcium	-	-	-	-	-	-	-	-	0.489	-0.269	0.001	0.446	0.278	0.076	0.512	0.704	0.363
Hardness	-	-	-	-	-	-	-	-	-	0.558	-0.138	0.593	0.201	-0.202	0.404	0.558	0.529
Magnesium	-	-	-	-	-	-	-	-	-	-	-0.159	0.218	0.006	-0.260	-0.024	-0.023	0.205
Turbidity	-	-	-	-	-	-	-	-	-	-	-	-0.191	0.198	-0.252	-0.111	-0.226	-0.290
Chloride	-	-	-	-	-	-	-	-	-	-	-	-	0.301	0.107	0.334	0.468	0.340
Ammonia	-	-	-	-	-	-	-	-	-	-	-	-	-	0.131	0.073	0.095	0.028
Sulphate	-	-	-	-	-	-	-	-	-	-	-	-	-	-	0.113	0.133	-0.156
O-PO_4	-	-	-	-	-	-	-	-	-	-	-	-	-		-	0.642	0.618
Nitrite	-	-	-	-	-	-	-	-	-	-	-	-	-	-	-	-	0.627
Nitrate	-	-	-	-	-	-	-	-	-	-	-	-	-	-	-	-	-

the point of influx.

According to Kaur et al. (2001) (WQI B) based on extent of pollution, water has been designated as Absolutely clean when WQI is 100, Slightly to moderately polluted when WQI is between 60-80, Excessively polluted when WQI is between 40-60 and Severely polluted when WQI is between 0-40. Water quality was considered severely polluted according to Kaur's water quality index at station Y2. This may be due to influx of effluents through Maskara nala at this point. Moundiotiya et al. (2004) and Sisodia and Moundiotiya (2006) have also reported excessive pollution in Kalakho Lake and Jamwaramgarh wetland respectively in Rajasthan on the basis of WQI.

Thus, in the present study, the values of WQI indicated that water was being polluted at station Y2.

Conclusion

Study of different limnochemical parameters revealed that the intensity of pollution increased as the river was subjected to sewage and industrial wastes. In the growing awareness of relationships between human health and water pollution, it is essential to undertake regular monitoring and surveillance of such important aquatic ecosystems. In order to manage the pollution load of river Yamuna that pass nearby Yamunanagar, it is recommended that various methods of sewage/industrial wastes treatment should be used before the disposal of effluents.

REFERENCES

APHA, AWWA, WPCF (1998). Standard methods for the examination of water and waste water. American Public Health Association, (American Public Health Association, New York). DC 20005-2605.

Benzie JAH, Pugh KB, Davidson MB (1999). The rivers of North East Scotland: physicochemical characteristics. Hydrobiologia. 218(2): 93-106.

Bhandari N, Nayal K (2008). Correlation study on physicochemical parameters and quality assessment of Kosi river water, Uttarakhand. J. Chem., 5(2): 342-346.

Bhatnagar A, Chopra G, Malhotra P (2009). Water quality indices and abiotic characteristics of western Yamuna canal in Yamunanagar, Haryana. J. Appl. Nat. Sci. Haridwar 1(2): 149-154.

Bhatnagar A, Garg SK (1998). Environmental impact assessment in river Ghaggar in Haryana. J.Nat. Conservators, 10(2): 215-224.

Bhatnagar A, Sanghwan P (2009). Impact of mass bathing on water quality. Int. J. Environ. Res., 3(2): 247-252.

Dalal A, Arora MP (2008). Seasonal physicochemical fluctuations of water quality of river Hindon. J. Exp. Zool., 11(1): 125-131.

Deshmukh DR, Sonawane SR (2008). Studies on seasonal variation in abiotic factors of Godawari river at Paithan. J. Exp. Zool., 11(1): 143-145.

Devis SN, De Dewiest RJM (1996). Hydrogeology. John Wiley Inc., New York.

Golterman HL, Clymo RS, Ohnstad MAM (1978). Methods for chemical analysis of fresh water. IBP Handbook No. 8, Second Edition. Blackwell Scientific Publications: 178.

Griffiths RW, Thornley S, Edsall TA (1991). Limnological aspects of the St. Clair river. Hydrobiologia. 219: 97-123.

Hawkes HA (1978). Ecology of polluted water. Comprehensive Biotechnology IV.

Horton RK (1965). An index number for rating water quality.J.Water Pollut. Control Federation. 37(3): 300-306.

ICMR (1975). Manual of standards of quality of drinking water supplies. 2nd edn. Special series report No. 44.

Jaji MO, Bamgbose O, Odukoya OO, Arowolo TA (2007). Water quality assessment of Ogun River, South West Nigeria. Environment Monitoring Assessment, 133: 437-482.

Kaur H, Dhillon SS, Bath KS, Mander G (1997). Inter-relations between physicochemical factors at Harike Wetland (Punjab-India). J. Environ. Pollut., 4(3): 237-240.

Kaur H, Syal J, Dhillon SS (2001). Water quality index of the river Satluz. Pollut. Res., 20(2): 199-204.

Kaushik A, Jain S, Dawra J, Sahu R, Kaushik CP (2001). Heavy metal pollution of river Yamuna in the industrially developing state of Haryana. Indian J. Environ. Health. 43(4): 164-168.Kaushik A, Kansal A, Meena S, Kumari S, Kaushik CP (2009). Heavy metal contamination of river Yamuna, Haryana, India: Assessment by metal enrichment factor of the sediments. J. Hazard. Mater., 164: 265-270.

Khaiwal R, Ameena M, Monika R, Kaushik A (2003). Seasonal variations in physicochemical characteristics of River Yamuna in Haryana and its ecological best designated use. J. Environ. Monitor., 5: 419-426.

Kulshrestha SK, Adholia UN, Bhatnagar A (1991). Zooplankton community of river Chambal and its tributaries with special reference to industrial pollution. J. Environ. Stud., 40: 207-216.

Mathivanan V, Jeyachitra O, Selvisabhanayakam VP, Elanchezhiyan C (2008). Environmental monitoring studies on river Cauvery at Thanjavur district, Tamilnadu in relation to pollution. J. Exp. Zool., 11(1): 225-23

Moundiotiya C, Sisodia R, Kulshrestha M, Bhatia AL (2004). A case study of the Jamwaramgarh wetland with special reference to physicochemical properties of water and its environment. J. Environ. Hydrol., 12: Paper 24.

Oram B (2007). Wilkes university environmental engineering earth science. WQI Index- consumer support group online calculator. Retrieved September 2007 from http:// www.csg network.com/ h2oqualindexcuttemponlycalc html.

Polestya A, Sharma S, Reddy AS (2008). Assessment of water quality of Phagwara drain in Punjab. Environment Conservation J. 9(1&2):129-134.

Raj ADS, Jayasekher T (2007). Hydrogeochemistry of the river basins of Kanyakumari district. Int. J. Environ. Prot., 27(2): 145-152.

Rim-Rukeh A, Ikhifa OG, Okokoyo AP (2006). Effects of agricultural activities on the water quality of Orogodo river, Agbor Nigeria. J. Appl. Sci. Res., 2(5): 256-259.

Singh BK, Srivastava KK, Srivastava SK (2009). Water quality assessment: Five stations of river Damodar and one station of river Nalkari of south Karanpura coalfield. Int. J. Environ. Prot., 29(8): 673-680.

Sisodia R, Moundiotiya C (2006). Assessment of water quality index of wetland Kalakho lake, Rajasthan, India. J. Environ. Hydrol., 14: paper 23.

Sujitha PC, Mitra DD, Sowmya PK, Mini PR (2011). PhysicoChemical parameters of Karamana River Water in Trivandrum District, Kerala, India Int. J. Environ. Sci., 2 (2): 472-490.

Suthar S, Sharma J, Chabukdhara M, Nema AK (2010). Water quality assessment of river Hindon at Ghaziabad, India: impact of industrial and urban wastewater. Environ Monit Assess., 165(1-4):103-112.

Toman MJ (2009). Physicochemical characteristics and seasonal changes of plankton communities in a river reservoir. Lakes and reservoirs: research and management, 2(1&2): 71-76.

Ward AD, Elliot WJ (1995). Environmental Hydrology. Lewis Publishers. Boca Radittaton, Florida.

WHO (1984). World Health Organization Guidelines for Drinking Water, Geneva, Switzerland. Vol.1.

WHO (1993). World Health Organization, Guidelines for drinking water quality-I, Recommendations,2ndEd.GenevaWHO, 1993.www.lenntech.com/drinking-water-standards.htm.

Yang H, Shen Z, Zhang J, Wang W (2007). Water quality characteristics along the course of the Huangpu river (China). J. Environ. Sci., 19(10): 1193-1198.

Zhang J, Yan J, Zhang ZF (1995). National river chemistry trends in China: Huanghe and Chanjiang. Ambio- A J. Hum. Environ., 24(5): 275-279.

Assessment of water quality and impact of effluents from fertilizer factories to the Lakhya River

Mohammad Hafizul Islam[1]*, Md. Mafizur Rahman[2] and Fahmidah Ummul Ashraf[3]

[1]Public Works Department, Sub-Divisional Engineer, Bangladesh.
[2]Department of Civil Engineering, BUET, Bangladesh.
[3]Water Resources Engineering, BUET, Bangladesh.

A study was carried out in Polash and Ghorasal Urea fertilizer factories to assess the impact of effluent on water quality of the Lakhya River. Comprehensive waste water sampling by grab sampling method and flow measurement by float velocity method were carried out for five weeks (one sample per week) at five sampling stations at Polash and Ghorasal Urea fertilizer factories during June to July, 2007. Water quality samplings by grab sampling method were also carried out for five weeks (one sample per week) at four stations in the Lakhya River at the same time and River flows on the period of October 2006 to September 2007 were collected from Institute of Water Modelling. Effluents at both the factories and water sample from selected points in the river were analysed for pH, temperature, DO, BOD_5, COD, NH_3-N, NH_4-N, TS, TSS, and TDS during June to July, 2007 at the Environmental Engineering Laboratory of Bangladesh University of Engineering and Technology, Bangladesh. The results showed that the effluents were alkaline while the level of DO, BOD_5, COD, NH_3-N, NH_4-N, TS, TSS, and TDS relatively high and 17425-20012 kg/day Ammonia load discharged from fertilizer factories into the Lakhya River. The upstream water was near to neutral pH (average pH, 7.66 ± 0.102) with high dissolved oxygen but low in the levels of the other parameters. The river water after the effluent receiving points was basic (average pH, 8.16 ± 0.08) and the levels of other parameters were high due to heavy pollution load especially Ammonia discharged from fertilizer factories. The results suggested that water in the river was polluted and not good for human consumption. It is therefore recommended that the disposal of improperly treated or untreated wastes should be stopped to save the river water from further deterioration.

Key words: Industrial effluents, Impact, river water, pollution, water quality.

INTRODUCTION

Water is essential to all forms of life and makes up 50 to 97% of the weight of all plants and animals and about 70% of human body (Buchholz, 1998). Water is also indispensable for agriculture, manufacturing, transportation and many other human activities. Despite its importance, water is the most poorly managed resource in the world (Fakayode, 2005). Ground and surface waters can be contaminated by several sources. In farming areas, the routine application of agricultural fertilizers is the major source of contamination (Altman and Parizek, 1995; Emongor et al., 2005). In urban areas, the careless disposal of industrial effluents and other wastes may contribute greatly to the contamination and poor quality of the water (Chindah et al., 2004; Emongor et al., 2005; Furtado et al., 1998 and Ugochukwu, 2004). A study on the impact of industrial effluent on water quality of a river carried out in Nigeria (Fakayode, 2005) showed that the chemical parameters studied were above the allowable limits and also tended to accumulate downstream. Most of the rivers in the urban areas of the developing world are the end points of effluents discharged from the industries. Industrial effluents, if not treated and properly controlled can also pollute ground water (Olayinka, 2004). Therefore, both bore holes and rivers generally have poor quality

* Corresponding author. E-mail: hafizbuet@gmail.com.

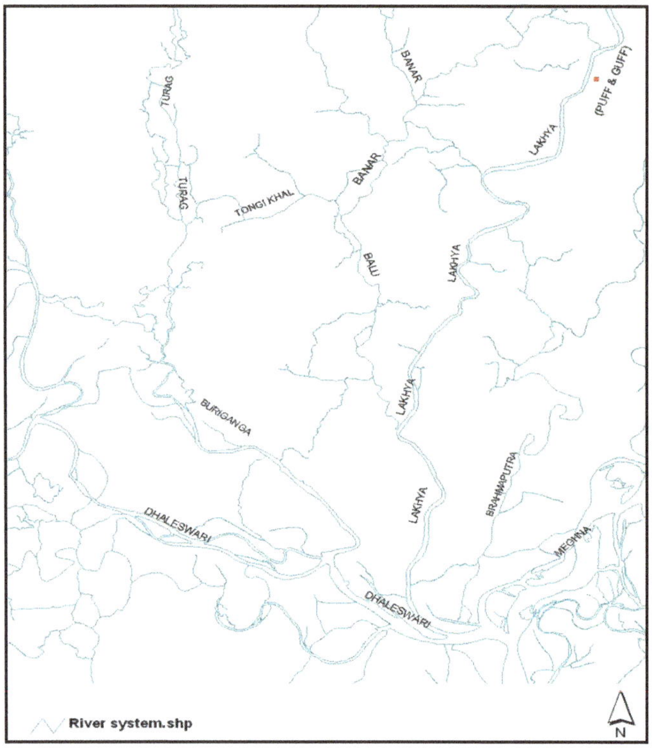

Figure 1. Location of fertilizer factories and perepheral river system around Dhaka City.

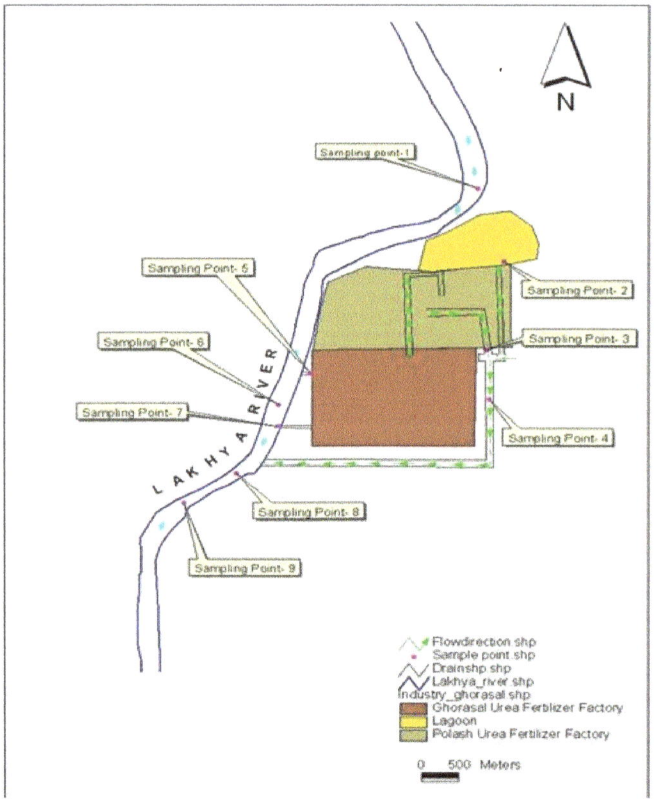

Figure 2. Sample and sampling.

water in the affected areas. Bangladesh is experiencing rapid industrial growth and this is making environmental conservation a difficult task (WB, 2007).

In Dhaka, the capital city of Bangladesh, there are a number of rivers that run through several industrial sites carrying large amount of discharged industrial effluents (Figure 1). People living in the river side use such contaminated water for domestic purposes. Unfortunately, there is no enough information about the quality of the effluent discharged into this river and also about the quality of the river water which is standard for human use.In this circumstances it is difficult for the authorities to take proper action in preventing pollution of water environment which is essential for surrounding living communities and Global Environment. The objective of this study was therefore to assess the extent of chemical pollution of the Lakhya river as it is increasingly polluted by industrial effluents.

MATERIALS AND METHODS

Study area

The study was conducted in the effluent channels from two industries and in the Lakhya River that runs through an industrial area in Dhaka, the capital city of Bangladesh. Water and wastewater sampling, flow measurement points are shown in Figure 2. The industries which discharge effluents into the Lakhya River are the Polash and Ghorasal urea fertilizer factories.

Sample and sampling

Samples of effluents and discharge data were collected for five weeks (one sample per week) at sampling points-2,3,4,5 and 7. Water samples in the river were also collected for five weeks (one sample per weeks) at sampling points-1,6,8 and 9. Discharge of river were collected from secondary sources. Sampling point-1: This point is located upstream of the fertilizer factories. Water sample collected from this point was unaffected by the effluent of the fertilizer factories (Polash and Ghorasal). Sampling point-2: This is the lagoon where effluent discharged from the fertilizer factories (Polash and Ghorasal) was retained. This point was very close to the pump. Ultimately this effluent was discharged into the Lakhya River through pump without treatment. Sampling point-3: This is the channel of the Polash urea fertilizer factory. The treated effluent of this factory was discharged through this channel. The performance of the treatment process of Polash urea fertilizer factory was evaluated from this point. Sampling point-4 is located in a channel where the discharged effluent was the combination of treated effluent discharged from the Polash urea fertilizer factory and untreated effluent discharged from Lagoon. This channel discharged effluent into the Lakhya River. Sampling point-5: Untreated effluent of Ghorasal urea fertilizer factory was discharged through this point. Sampling point-6: This point is 1000 m downstream from the sampling point-1. The water samples at this point were affected by the untreated effluent of Ghorasal urea fertilizer factory. Sampling point-7: This point is on the channel of the Ghorasal urea fertilizer factory. The treated effluent of this factory was discharged through this channel. This channel discharged effluent into the Lakhya River. The performance of the treatment process of Ghorasal urea fertilizer factory was evaluated from this point. Sampling point-8: This point is 538 m downstream from the sampling point-6. The water samples of this point were totally affected by both the fertilizer factories (Polash and Ghorasal). Combined effects of effluent from both the factories on the Lakhya were evaluated from the investigation at this point. Sampling point-9: This point is 514 m downstream from the sampling point-8. The water samples of this point were also totally affected by both the fertilizer factories (Polash and Ghorasal). This point was necessary to evaluate the impact of effluent from fertilizer factories along the length of the river. All samples were placed into thoroughly cleaned 1 liter polyethylene bottles and tightly closed. Each bottle was rinsed with the appropriate sample before the final sample collection. The samples were placed in a cooler box and then taken to the Environmental Engineering Laboratory, Bangladesh University of Engineering and Technology for analysis. Samples were carried out in the June to July, 2007.

Chemical analysis

pH, Temperature, DO: The parameters were analysed by electronic method by using pH meter (Hach), Temperature meter and DO meter respectively.
BOD_5: The parameter were analysed by using standard method for analysis of water and waste water, 20th Edition, APHA, AWWA, WEE, 1998 COD (Dichromate value): The parameter was analysed by Reactor Digestion Method by using Spectrophoto meter (Hach DR/4000U) .
NH_3-N, NH_4-N : The parameters were analysed by Nessler Method by using Spectrophoto meter (Hach DR/4000U).
TS, TSS, TDS : The parameters were analysed by Standard Method 2540C by using Oven (Despatch Co.USA) (Standard methods for Analysis of water and wastewater, 20'h Edition, APHA, AWWA, WEE, 1998)

RESULTS

Analysis of water quality along the Lakhya River to assess the impact of effluent from Polash and Ghorasal urea fertilizer factories

Biochemical oxygen demand (BOD)

The level of BOD_5 (Figure 3) was 8 to11 mg/l (9.2 ± 1.16 mg/l) in the sampling point-1. In the sampling point-6 it was 9 to 11 mg/l (10 ± 0.89 mg/l) in which 146.97 to 348.19 kg/day (226.18 ± 66.26 kg/day) BOD_5 discharged from sampling point-5. In the sampling point-8, it was 11 to 14 mg/l (12.4 ± 1.01 mg/l) in which 2518.3 to 3094.42 kg/day (2902.92 ± 211.31 kg/day) BOD_5 discharged from sampling point-4 and 7. For the dilution factor the BOD_5 reduced in the sampling point-9 and it was 10 to 13 mg/l (11.4 ± 1.01 mg/l).

Chemical oxygen demand (COD)

The level of COD (Figure 4) was 12 to 19 mg/l (15.8 ± 2.78 mg/l) in the sampling point-1. In the sampling point-6 it was 15 to 21 mg/l (18 ± 2.52 mg/l) in which 258.94 to 595.29 kg/day (369.49 ± 118.68 kg/day) COD discharged from sampling point-5. In the sampling point-8 it was 18 to 23 mg/l (20.6 ± 2.24 mg/l) in which 4455.64 to 4644.45 kg/day (4543.72 ± 67.82 kg/day) COD discharged from sampling point-4 and 7. For the dilution factor the COD reduced in the sampling point-9 and it was 17 to 22 mg/l (19.6 ± 2.24 mg/l).

Dissolved oxygen (DO)

The level of DO (Figure 5) was 4 to 4.8 mg/l (4.476 ± 0.26 mg/l) in the sampling point-1. In the sampling point-6 it was 3.9 to 4.4 mg/l (4.08 ± 0.19 mg/l) due to pollution load discharged from sampling point-5. In the sampling point-8 it was 3.78 to 4.1 mg/l (3.9 ± 0.12 mg/l) due to pollution load discharged from sampling point-4 and 7. For the dilution factor pollution load reduced, so DO increased in the sampling point-9 and it was 3.81 to 4.2 mg/l (3.948 ± 0.14 mg/l).

pH

The level of pH (Figure 6) was 7.5 to 7.8 (7.66 ± 0.102) in the sampling point-1. In the sampling point-6 it was 7.9 to 8.3 (8.1 ± 0.12) due to pollution load discharge from sampling point-5. In the sampling point-8 it was 8 to 8.4 (8.18 ± 0.13) due to pollution load discharge from sampling point-4 and 7. For the dilution factor pollution load decreased at the sampling point-9, so pH decreased at that point and it was 8.1-8.3 (8.16±0.08).

Temperature

The level of temperature (Figure 7) was 28 to 31.4°C (29.68 ± 1.13°C) in the sampling point-1. In the sampling point-6 it was 28 to 32°C (30.4 ± 1.43°C) due to pollution load in high temperature discharged from sampling point-5. In the sampling point-8 it was 28.2 to 31.7°C (30.14 ± 1.26°C) due to pollution load in high temperature discharged from sampling point-4 and 7. In the sampling point-9, it was 28.1 to 30.5°C (29.72 ± 1.04°C).

Total ammonia (NH_3-N + NH_4-N)

The concentration of total ammonia (Figure 8) was 0.685 to 0.71 mg/l (0.69 ± 0.009 mg/l) in the sampling point-1. In the sampling point-6 it was 0.97 to 1.19 mg/l (1.086 ± 0.079 mg/l) due to 7368 to 11203 kg/day (8624 ± 1344 kg/day) total ammonia discharged from sampling point-5. In the sampling point-8, it was 1.09 to 1.30 mg/l (1.2 ± 0.072 mg/l) due to 8808-10467 kg/day (9577 ± 533 kg/day) total ammonia discharged from sampling point-4 and 7. For the dilution factor the total ammonia reduced in the sampling point-9 and it was 0.95 to 1.15 mg/l (1.05 ± 0.07 mg/l). It was observed that fertilizer factories are the major sources of total ammonia discharged into the Lakhya River in the

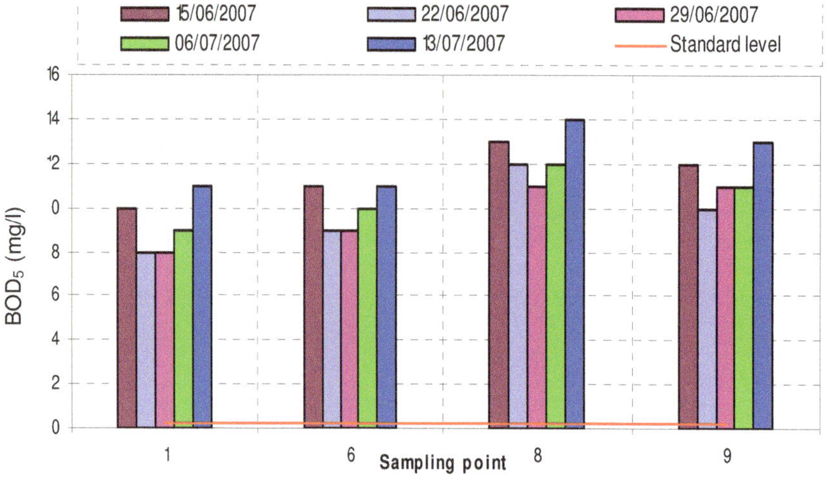

Figure 3. Variation of BOD_5 in mg/l along the Lakhya River.

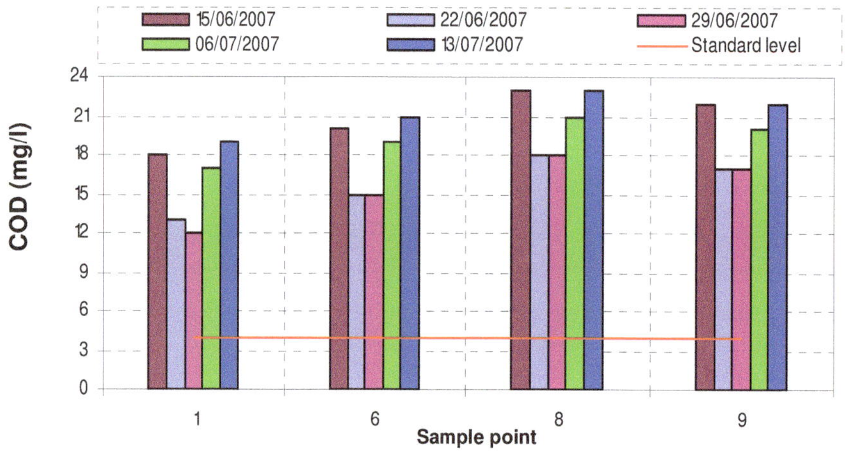

Figure 4. Variations of COD in mg/l along the Lakhya River.

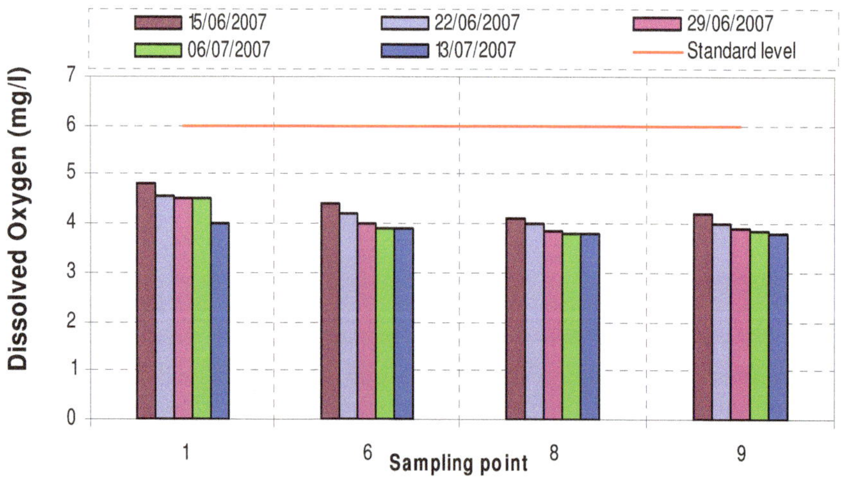

Figure 5. Variation of DO in mg/l along the Lakhya River.

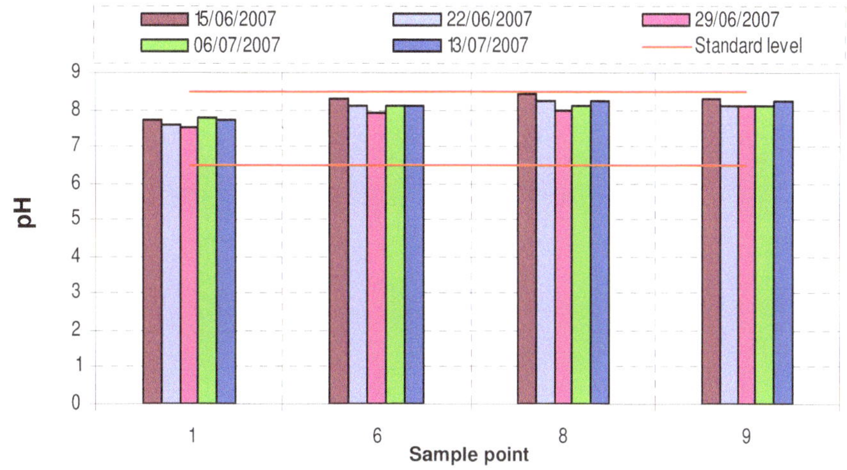

Figure 6. Variation of pH along the Lakhya River.

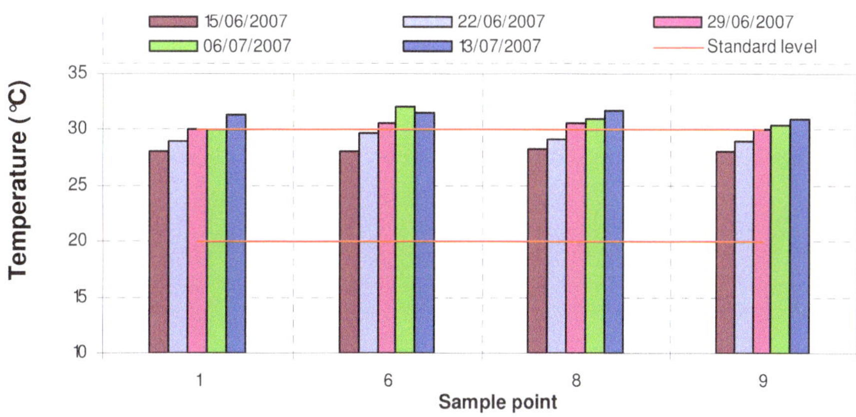

Figure 7. Variation of temperature in °C along the Lakhya River.

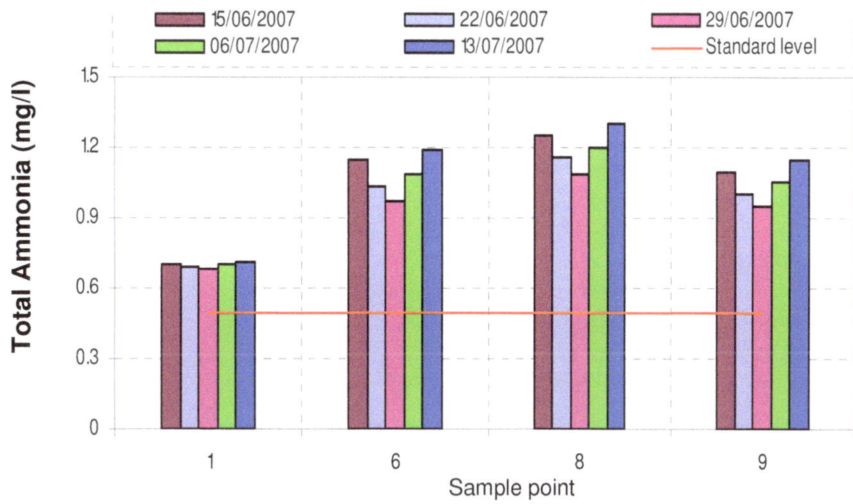

Figure 8. Variation of total Ammonia in mg/l along the Lakhya River.

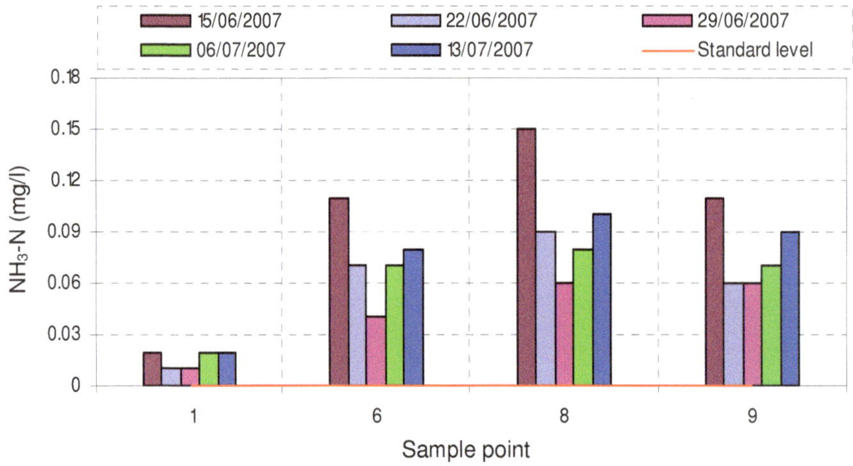

Figure 9. Variation of NH_3-N in mg/l along the Lakhya River.

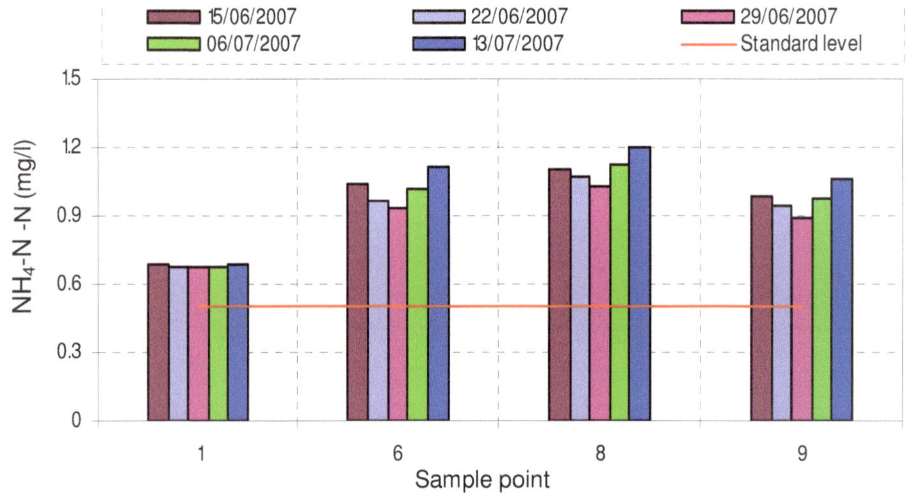

Figure 10. Variation of NH_4-N in mg/l along the Lakhya River.

study area. The concentration of total ammonia in the sampling point-9 is greater than the concentration of total ammonia in the sampling point-1. This increment was due to the ammonia load discharged from fertilizer factories. So the Ammonia load discharged from fertilizer factories has a large impact on water quality of the Lakhya River.

Ammonia as nitrogen (NH_3-N)

The concentration of Ammonia as Nitrogen (NH_3-N) (Figure 9) was 0.01 to 0.02 mg/l (0.016 ± 0.0004mg/l) in the sampling point-1. In the sampling point-6 it was 0.08-0.11 mg/l (0.074 ± 0.022 mg/l) due to 4975.23 to 6598.68 kg/day (5643.73 ± 590.45 kg/day) NH_3-N discharged from sampling point-5. In the sampling point-8 it was 0.06 to 0.15 mg/l (0.096 ± 0.03 mg/l) due to 2322.54 to 5265.65 kg/day (3542.03 ± 1080.86 kg/day) NH_3-N discharged from sampling point-4 and 7. For the dilution factor the NH_3-N reduced in the sampling point-9 and it was 0.06-0.11 mg/l (0.078 ± 0.01 mg/l). It was observed that fertilizer factories are the major sources of Ammonia (NH_3-N) discharged into the Lakhya River in the study area. The concentration of Ammonia in sampling point-9 is greater than the concentration of Ammonia in sampling point-1. This increment was due to the Ammonia discharged from fertilizer factories. So the Ammonia discharged from fertilizer factories has a large impact on water quality of the Lakhya River.

Ammonium as nitrogen (NH_4-N)

The concentration of Ammonium as Nitrogen (NH_4-N) (Figure 10) was 0.67 to 0.69 mg/l (0.68 ± 0.007 mg/l) in the sampling point-1. In the sampling point-6 it was 0.93 to

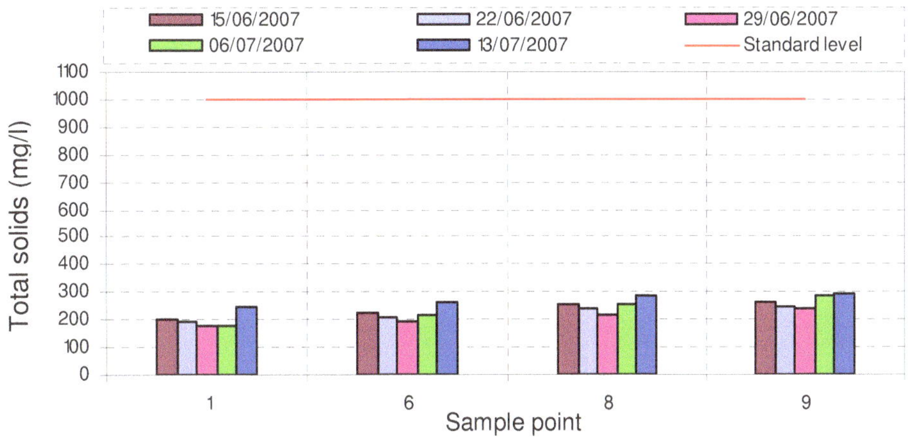

Figure 11. Variations of total solids in mg/l along the Lakhya River.

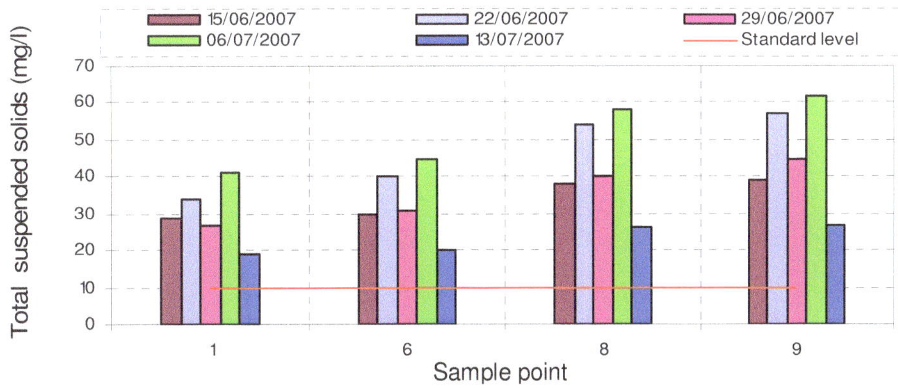

Figure 12. Variations of total suspended solids in mg/l along the Lakhya River.

1.11 mg/l (1.01 ± 0.06 mg/l) due to 1903.08 to 5988.22 kg/day (2980.78 ± 1541.86 kg/day) NH_4-N discharged from sampling point-5. In the sampling point-8 it was 1.03 to 1.20 mg/l (1.104 ± 0.05 mg/l) due to 5201.7 to 7064.64 kg/day (6035.72 ± 687.03 kg/day) NH_4-N discharged from sampling point-4 and 7. For the dilution factor the NH_4-N reduced in the sampling point-9 and it was 0.89-1.06 mg/l (0.97 ± 0.05 mg/l). It was observed that fertilizer factories are the major sources of Ammonium (NH_3-N) discharged into the Lakhya River in the study area. The concentration of Ammonium in sampling point-9 is greater than the concentration of Ammonium in sampling point-1. This increment was due to the Ammonium load discharged from fertilizer factories. So the Ammonium load discharged from fertilizer factories has a large impact on water quality of the Lakhya River.

Total solids (TS)

The concentration of TS (Figure 11) was 179 to 247 mg/l (200 ± 24.88 mg/l) in the sampling point-1. In the sampling point-6 it was 193-261 mg/l (220 ± 22.82 mg/l), in which 1523 to 3077 kg/day (1908 ± 594 kg/day) TS discharged from sampling point-5. In the sampling point-8 it was 220 to 286 mg/l (251 ± 21.6 mg/l) in which 5925 to 7302 kg/day (6816 ± 606.69 kg/day) TS discharged from sampling point-4 and 7. Total solids were accumulated in the downstream, for this, the value of TS were increased in the sampling point-9 and it were 237 to 298 mg/l (267.6 ± 22.08 mg/l).

Total suspended solids (TSS)

The concentration of TSS (Figure 12) was 19 to 41 mg/l (30 ± 7 mg/l) in the sampling point-1. In the sampling point-6 it was 20 to 45 mg/l (33.2 ± 8.65 mg/l) in which 1394 to 2623 kg/day (2060 ± 451 kg/day) TSS discharged from sampling point-5. In the sampling point-8 it was 26 to 58 mg/l (43.2 ± 11.56 mg/l) in which 118 to 752 kg/day (348 ± 213 kg/day) TSS discharged from sampling point-4 and 7. Total suspended solids were accumulated in the downstream, for this, the value of TSS were increased in the sampling point-9 and it were 27 to 62 mg/l (46 ± 12.55 mg/l).

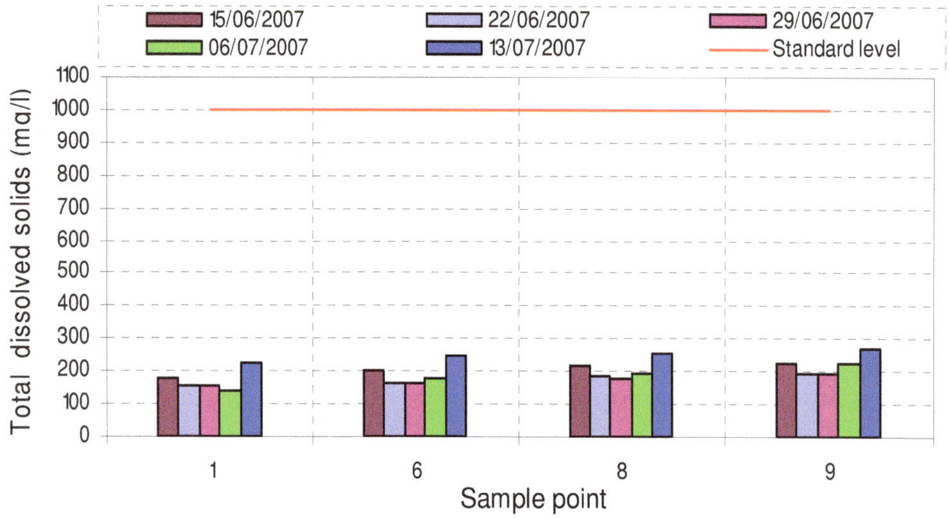

Figure 13. Variation of total dissolved solids in mg/l along the Lakhya River.

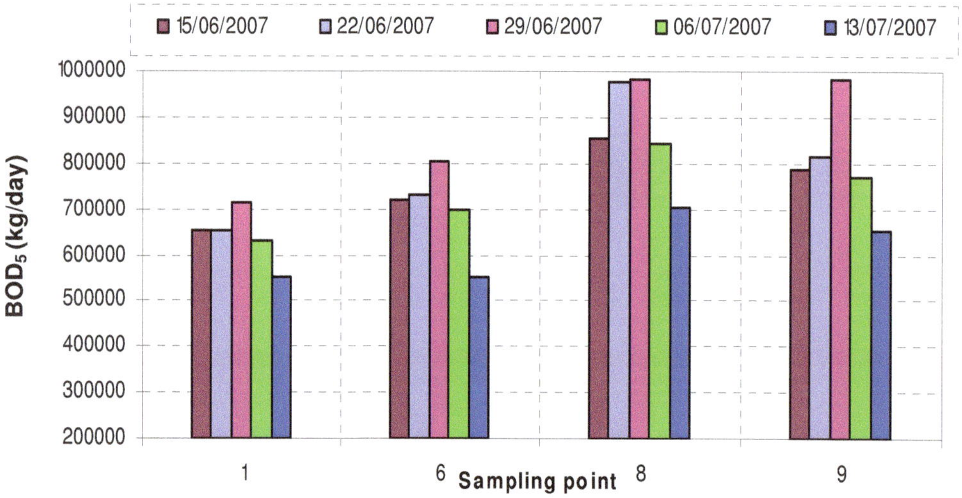

Figure 14. Variation of BOD_5 in kg/day along the Lakhya River.

Total dissolved solids (TDS)

The concentration of TDS (Figure 13) was 141 to 228 mg/l (170 ± 30 mg/l) in the sampling point-1. In the sampling point-6 it was 162 to 245 mg/l (190.2 ± 30.91 mg/l) in which 1394 to 2623 kg/day (2060 ± 451 kg/day) TDS discharged from sampling point-5. In the sampling point-8 it was 181 to 257 mg/l (207.4 ± 27.8 mg/l) in which 118 to 752 kg/day (348 ± 213 kg/day) TDS discharged from sampling point-4 and 7. Total suspended solids were accumulated in the downstream, for this, the value of TDS were increased in the sampling point-9 and it were 192 to 271 mg/l (221.6 ± 28.7 mg/l).

Analysis of pollution load along the Lakhya River to assess the impact of Polash and Ghorasal urea fertilizer factories effluent

Biochemical oxygen demand (BOD)

The amount of pollution load along the Lakhya River of BOD_5 (Figure 14) was 552878 to 713615 kg/day (640523 ± 51939 kg/day) in the sampling point-1. In the sampling point-6 it was 553607 to 803513 kg/day (702374 ± 82048 kg/day) in which 146.97 to 348.19 kg/day (226.18 ± 66.26 kg/day) BOD_5 discharged from sampling point-5. In the sampling point-8 it was 705054 to 982497 kg/day (871789

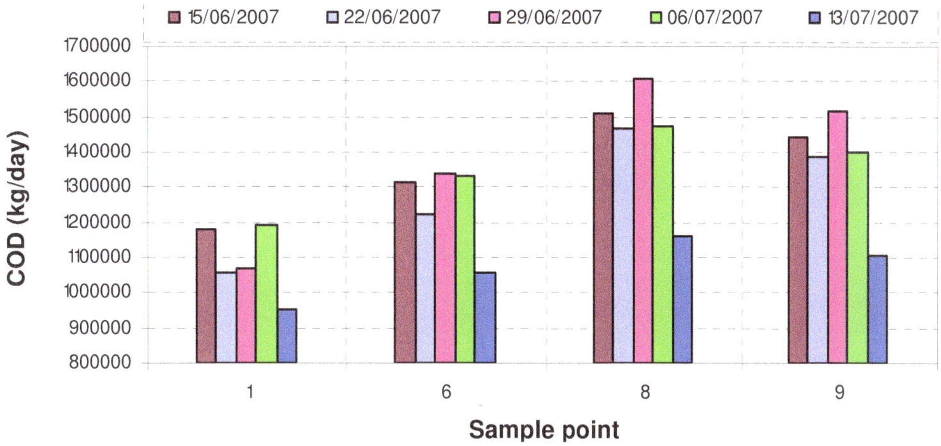

Figure 15. Variation of COD in kg/day along the Lakhya River.

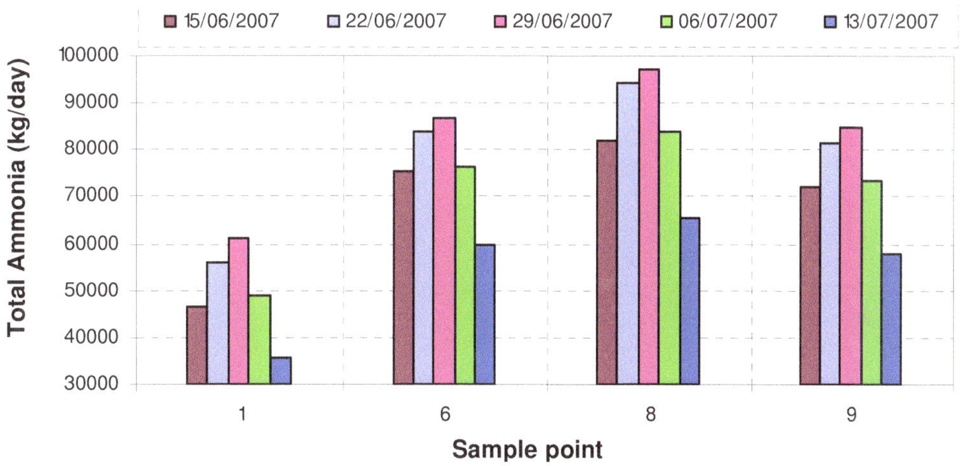

Figure 16. Variation of total ammonia in kg/day along the Lakhya River.

± 102518 kg/day) in which 2518.3 to 3094.42 kg/day (2902.92 ± 211.31 kg/day) BOD_5 discharged from sampling point-4 and 7. In the sampling point-9, it was 655124 to 982923 kg/day (802362 ± 105472 kg/day). It was shown that there was insignificant impact of pollution load of BOD_5 discharged from fertilizer factories.

Chemical oxygen demand (COD)

The amount of pollution load along the Lakhya River of COD (Figure 15) was 954971 to 1188933 kg/day (1090493 ± 86782 kg/day) in the sampling point-1. In the sampling point-6 it was 1056886 to 1339188 kg/day (1252013 ± 106198 kg/day) in which 258.94 to 595.29 kg/day (369.49 ± 118.68 kg/day) COD discharged from sampling point-5. In the sampling point-8 it was 1158303 - 1607722 kg/day (1442595 ± 150924 kg/day) in which 4455.64 to 4644.45 kg/day (4543.72 ± 67.82 kg/day) COD discharged from sampling point-4 and 7. In the sampling point-9, it was 1108672 to 1519062 kg/l (1371856 ± 139415 kg/day). It was shown that there was insignificant impact of pollution load of COD discharged from fertilizer factories.

Total ammonia (NH_3-N + NH_4-N)

The amount of pollution load along the Lakhya River of total ammonia (Figure 16) was 35686 to 61103 kg/day (49663 ± 8707 mg/l) in the sampling point-1. In the sampling point-6 it was 59890 to 86601 kg/day (76427 ± 8309 kg/day) due to 7368 to 11203 kg/day (8624 ± 1344 kg/day) total ammonia discharged from sampling point-5. In the sampling point-8 it was 65469 to 97357 kg/day

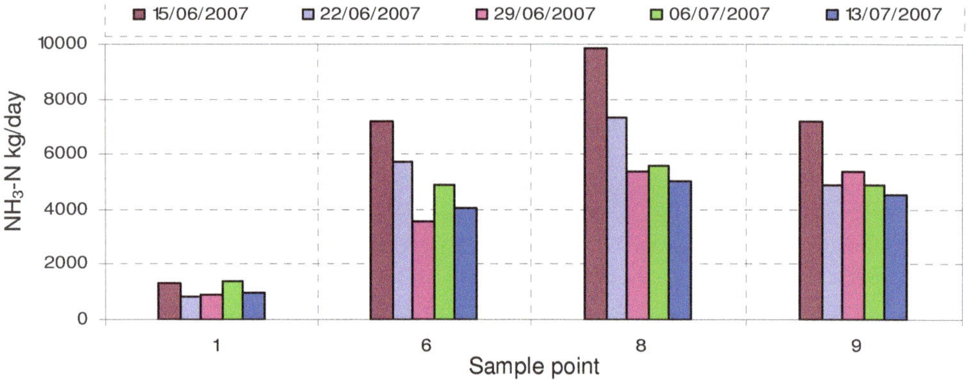

Figure 17. Variation of Ammonia as Nitrogen (NH_3-N) in kg/day along the Lakhya River.

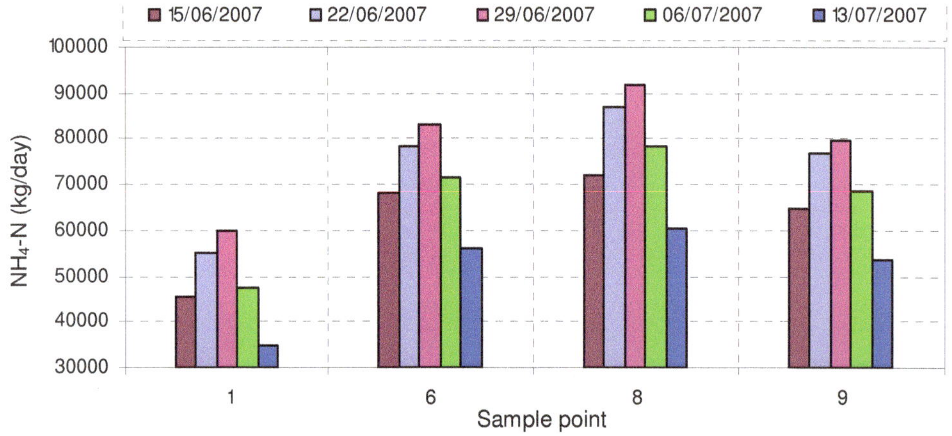

Figure 18. Variation of Ammonium as Nitrogen (NH_4-N) in kg/day along the Lakhya River.

(84685 ± 11257 kg/day) due to 8808 to 10467 kg/day (9577 ± 533 kg/day) total ammonia discharged from sampling point-4 and 7. For the dilution factor, the total ammonia reduced in the sampling point-9 and it was 57953 to 84889 kg/day (74030 ± 9336 kg/day). It was observed that fertilizer factories are the major sources of total ammonia load discharged in Lakhya River in the study area. The amount of total ammonia load in sampling point-9 is greater than the load of total ammonia in sampling point-1. This increment was due to the ammonia load discharged from fertilizer factories. So the ammonia load discharged from fertilizer factories has a large impact on the Lakhya River water quality.

Ammonia as nitrogen (NH_3-N)

The pollution load of Ammonia as Nitrogen (NH_3-N) (Figure 17) was 814 to 1399 kg/day (1084 ± 231 kg/day) in the sampling point-1. In the sampling point-6 it was 3571 to 7215 kg/day (5083 ± 1293 kg/day) due to 4975.23-6598.68 kg/day (5643.73 ± 590.45 kg/day) NH_3-N discharged from sampling point-5. In the sampling point-8 it was 5036-9840 kg/day (6635 ± 1788 kg/day) due to 2322.54 to 5265.65 kg/day (3542.03 ± 1080.86 kg/day) NH_3-N discharged from sampling point-4 and 7. For the dilution factor the NH_3-N reduced in the sampling point-9 and it was 4535 to 7217 kg/day (5383 ± 954 kg/day). It was observed that fertilizer factories are the major sources of ammonia load (NH_3-N) discharged into the Lakhya River in the study area. The amount of ammonia load in sampling point-9 is greater than the load of Ammonia in sampling point-1. This increment was due to the ammonia load discharged from fertilizer factories. So the ammonia load discharged from fertilizer factories has a large impact on water quality of the Lakhya River.

Ammonium as nitrogen (NH_4-N)

The pollution load of Ammonium as Nitrogen (NH_4-N) (Figure 18) was 34681 to 59765 kg/day (48515 ± 8670 kg/day) in the sampling point-1. In the sampling point-6 it was 55864 to 83030 kg/day (71344 ± 9300 kg/day) due to 1903.08 to 5988.22 kg/day (2980.78 ± 1541.86 kg/day)

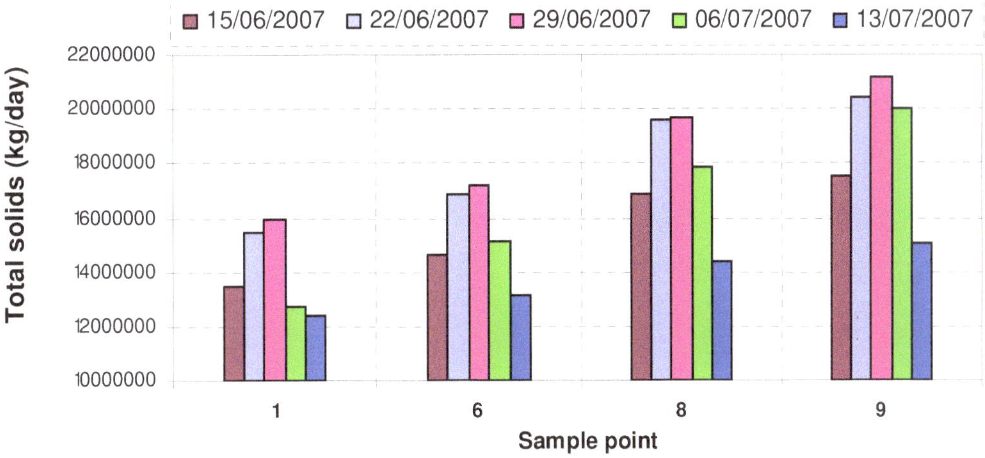

Figure 19. Variation of total solids in kg/day along the Lakhya River.

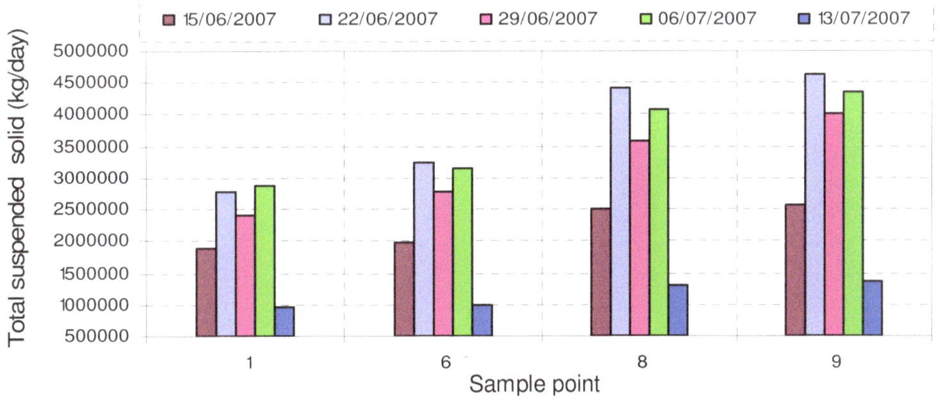

Figure 20. Variation of total suspended solids in kg/day along the Lakhya River.

NH_4-N discharged from sampling point-5. In the sampling point-8 it was 60433 to 91997 kg/day (78050 ± 11161 kg/day) due to 5201.7 to 7064.64 kg/day (6035.72 ± 687.03 kg/day) NH_4-N discharged from sampling point-4 and 7. For the dilution factor the NH_4-N reduced in the sampling point-9 and it was 53418 to 79527 kg/day (68648 ± 9245 kg/day). It was observed that fertilizer factories are the major sources of Ammonium load (NH_3-N) discharged into the Lakhya River in the study area. The amount of Ammonium load in sampling point-9 is greater than the load of Ammonium in sampling point-1. This increment was due to the Ammonium load discharged from fertilizer factories. So the Ammonium load discharged from fertilizer factories has a large impact on water quality of the Lakhya River.

Total solids (TS)

The amount of pollution load of TS (Figure 19) was 12414626 to 15967134 kg/day (14002508 ± 1444285 kg/day) in the sampling point-1. In the sampling point-6 it was 13135585 to 17230884 kg/day (15395465 ± 1501169 kg/day) in which 1523 to 3077 kg/day (1908 ± 594 kg/day) TS discharged from sampling point-5. In the sampling point-8 it was 14403247 to 19649938 kg/day (17683404 ± 1956566 kg/day) in which 5925 to 7302 kg/day (6816 ± 606.69 kg/day) TS discharged from sampling point-4 and 7. Total solids were accumulated in the downstream, for this, the load of TS were increased in the sampling point-9 and it were 15017466 to 21177516 kg/day (18831382 ± 2269732 kg/day). It was shown that the pollution load of TS discharged from sampling point 5, 4 and 7 were negligible as compared to the load discharged from residential area on the bank of the Lakhya river. So it can be concluded that there was insignificant impact of pollution load of TS discharged from fertilizer factories.

Total suspended solids (TSS)

The amount of TSS (Figure 20) was 954971 to2867428 kg/day (2179768±699440 kg/day) in the sampling point-1.

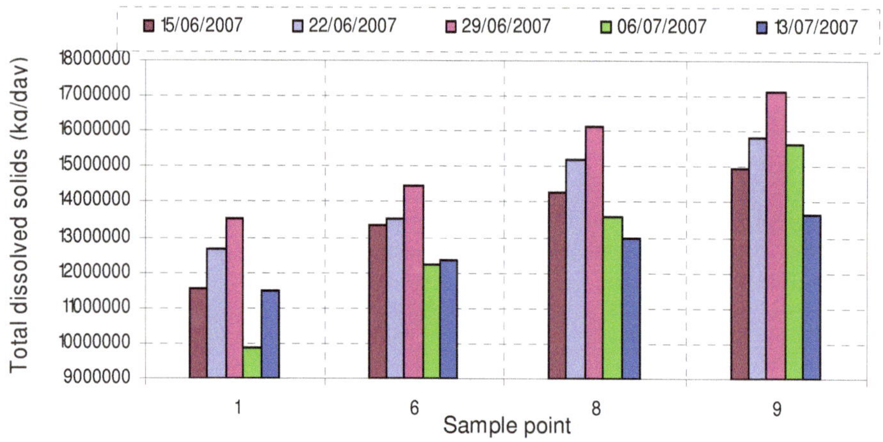

Figure 21. Variation of total dissolved solids in kg/day along the Lakhya River.

In the sampling point-6 it was 1006558 to3257035 kg/day (2430154±843632 kg/day) in which 1394 to 2623 kg/day (2060 ± 451 kg/day) TSS discharged from sampling point-5. In the sampling point-8 it was 1309386-4398998 kg/day (3167783 ± 1130850 kg/day) in which 118 to 752 kg/day (348 ± 213 kg/day) TSS discharged from sampling point-4 and 7. Total suspended solids were accumulated in the downstream, for this, the load of TSS were increased in the sampling point-9 and it were 1360643 to 4645510 kg/day (3386845 ± 1241162 kg/day). It was shown that the pollution load of TSS discharged from sampling point 5, 4 and 7 were negligible as compared to the load discharged from residential area on the bank of the Lakhya River. So it can be concluded that there was negligible impact of pollution load of TS discharged from fertilizer factories.

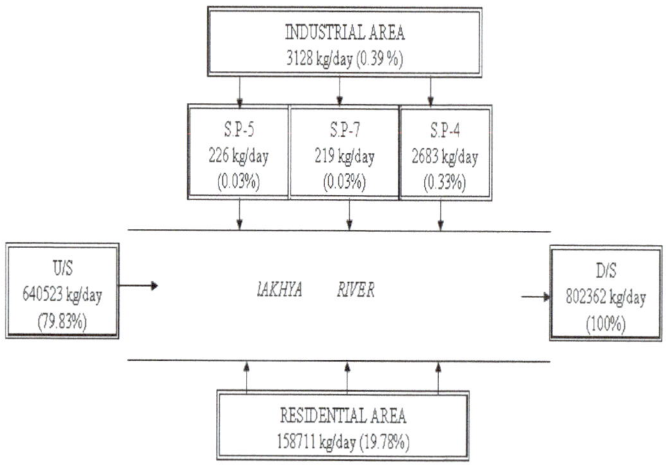

Figure 22. Different sources of BOD_5 along the Lakhya River.

Total dissolved solids (TDS)

The amount of pollution load of TDS (Figure 21) was 9861153 to 13558684 kg/day (11822740 ± 1251002 kg/day) in the sampling point-1. In the sampling point-6 it was 12256499 to 14463229 kg/day (13176508 ± 818925 kg/day) in which 1394 to 2623 kg/day (2060 ± 451 kg/day) TDS discharged from sampling point-5. In the sampling point-8 it was 12942778 to 16166540 kg/day (14448096 ± 1147385 kg/day) in which 118 to 752 kg/day (348 ± 213 kg/day) TDS discharged from sampling point-4 and 7. TDS were accumulated in the downstream, for this, the load of TDS were increased in the sampling point-9 and it were 13656823 to 17156469 kg/day (15444538 ± 1143304 kg/day). It was shown that the pollution load of TDS discharged from sampling point 5, 4 and 7 were negligible as compared to the load discharged from residential area on the bank of the Lakhya River. So it can be concluded that there was negligible impact of pollution load of TS discharged from fertilizer factories.

Schematic figures of water quality parameters for different sources along the Lakhya River are shown in Figures 22 to 27.

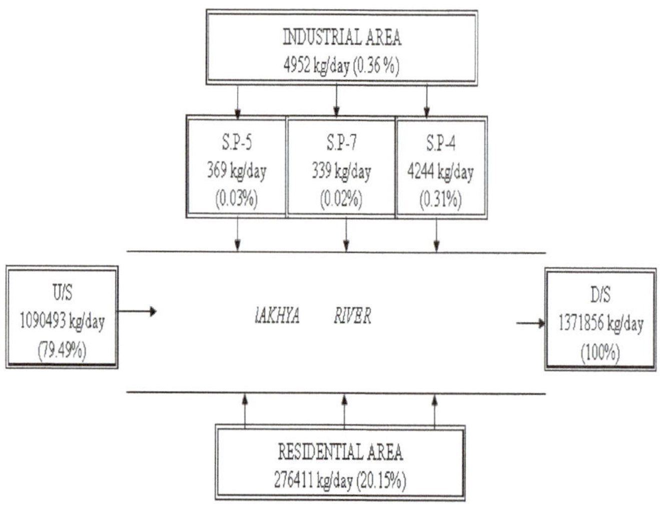

Figure 23. Different sources of COD along the Lakhya River.

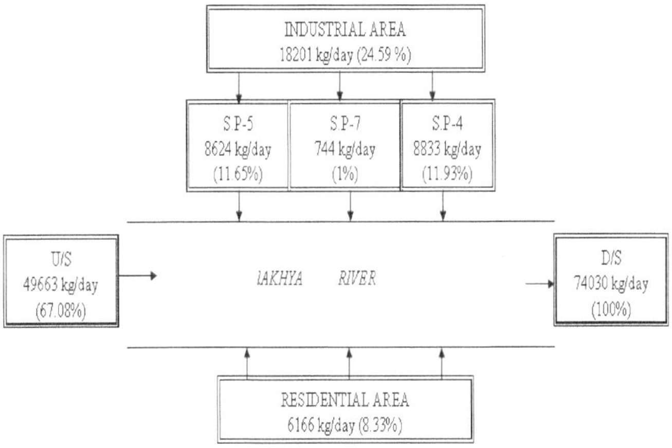

Figure 24. Different sources of Ammonia along the Lakhya River.

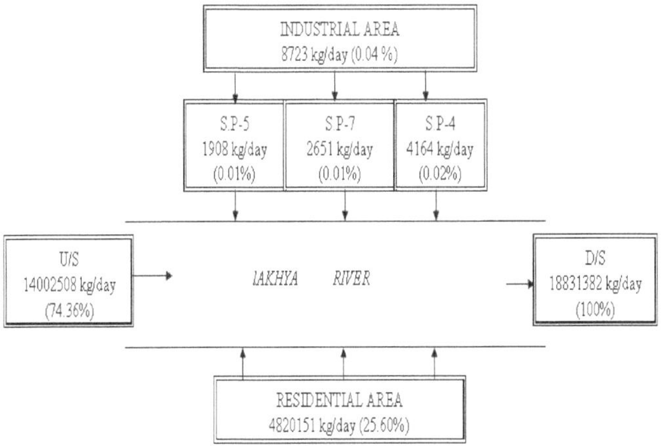

Figure 25. Different sources of total solids along the Lakhya River.

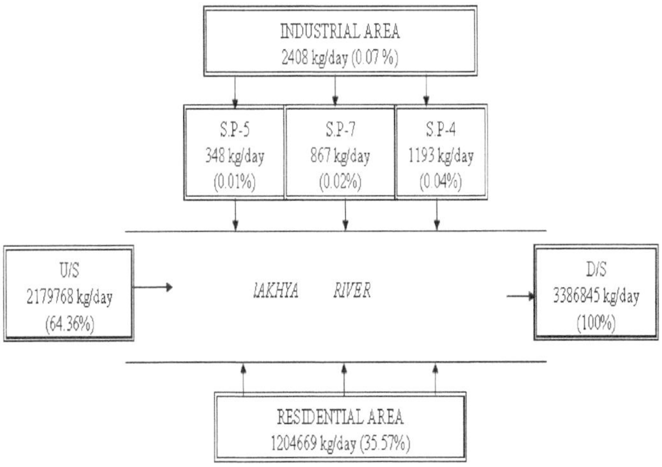

Figure 26. Different sources of total suspended solids along the Lakhya River.

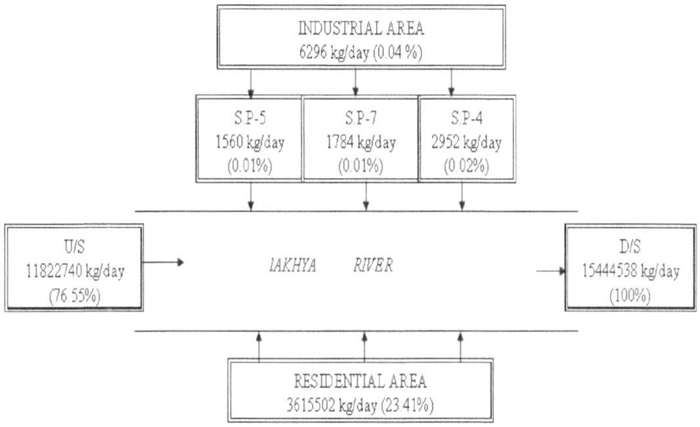

Figure 27. Different sources of total dissolved solids along the Lakhya River.

DISCUSSION

(1) The concentration of most of the water quality parameters in down stream (sampling point-9) is greater than the concentration in up stream (sampling point-1).
(2) In the sampling point-1, the level of BOD_5, COD, DO, pH, Temperature, Total Ammonia, Ammonia as Nitrogen (NH_3-N), Ammonium as Nitrogen (NH_4-N), TS, TSS, TDS was 9.2, 15.8 and 4.476 mg/l, 7.66, 29.68°C, 0.69, 0.016, 0.68, 200, 30 and 170 mg/l respectively. In the sampling point-9, these were 11.4, 19.6, 3.95 and 8.16, 29.72°C, 1.05, 0.078, 0.97, 267.6, 46 and 221.6 mg/l respectively.
(3) In the sampling point-1, the amount of BOD_5, COD, total ammonia, Ammonia as Nitrogen (NH_3-N), Ammonium as Nitrogen (NH_4-N), TS, TSS, TDS was 640523, 1090493, 49663, 1084, 48515, 14002508, 2179768 and 11822740 kg/day respectively. In the sampling point-9, these were 802362, 1371856, 74030, 5383, 53418, 1883138, 3386845 kg/day and 15444538 kg/day, respectively.
(4) The total amount of pollution load discharged from fertilizer factories into the Lakhya River were BOD_5 were 2665 to 3369 kg/day, COD were 4714-5221 kg/day, Total Ammonia (NH_3-N) were 17425 to 20012 kg/day, Ammonia as Nitrogen (NH_3-N) were 7850.68 to 10671.21 kg/day, Ammonium as Nitrogen (NH_4-N) were 7164.34-12161.79 kg/day, TS were 7449 to 10380 kg/day, TSS were 1513 to 3375 kg/day, TDS were 5467 to 7004 kg/day
(5) It was observed that fertilizer factories are the major sources of Total ammonia load discharged into the Lakhya River in the study area. The amount of total ammonia load in the sampling point-9 is greater than the load of Total Ammonia in the sampling point-1. This increment was due to the Ammonia load discharged from fertilizer factories. So the Ammonia load discharged from fertilizer factories has a large impact on the Lakhya River water quality. It was also shown that the pollution load of BOD, COD and TS discharged from sampling point 5, 4 and 7 were negligible as compared to the load discharged from residential

area on the right bank of the Lakhya River. So it can be concluded that there was insignificant impact of BOD, COD and TS load discharged from fertilizer factories.

(6) Contribution of different sources are for total Ammonia: u/s 67.08%, fertilizer factories 24.59%, residential area 8.33% ; BOD_5: u/s 79.83%, fertilizer factories 0.39%, residential area 19.78%; COD: u/s 79.49%, fertilizer factories 0.36%, residential area 20.15%; TS: u/s 74.36%, fertilizer factories 0.04%, residential area 25.60%; TSS: u/s 64.36%, fertilizer factories 0.07%, residential area 35.57%; TDS: u/s 76.55%, fertilizer factories 0.04%, residential area 23.41%;

Conclusion

The study has shown that the effluents from the industries have a big impact on the water quality of the Lakhya River. Although, the values of some water quality parameters in some cases were lower than the allowable limits, the continued discharge of the effluents in the river may result in severe accumulation of the contaminants and unless the authorities implement the laws governing the disposal of wastes this may affect the lives of the people. The river water after the effluent receiving points was alkaline (average pH, 8.16 $\pm$ 0.08) and the levels of other parameters were high due to heavy pollution load especially Ammonia discharged from fertilizer factories. The results suggested that water in the river was polluted and not good for human consumption. It is therefore recommended that the disposal of improperly treated or untreated wastes should be stopped to save the river water from further deterioration.

RECOMMENDATIONS

(1) Treatment process must be updated as high level of Ammonia was found on the test result and untreated effluent discharged from fertilizer factories through the sampling point-4, 5 must be stopped.

(2) The retaining period of effluent in lagoon was not sufficient as the effluent enters in the lagoon on one side and exit on the other side. So it is necessary to increase retaining period. It is also noted that the capacity of lagoon should be increased as in the rainy season flooding were occurred in the lagoon.

(3) Leakage from pipeline, valves or ammonia storage tank should be prevented through regular maintenance.

(4) The ground water quality in and around the factory should be checked periodically in order to ensure no adverse effect on the ground water quality.

(6) Smooth and effective implementation of laws and regulations should be ensured.

REFERENCES

Altman SJ, Parizek RR (1995). Dilution of non-point source nitrate in ground water. J. Environ. Quality, 24: 707-717.

Buchholz RA (1998). Principles of environmental management. The Greening of Business, 2nd. Prentice-Hall, London, UK.

Chindah AC, Braide AS, Sibeudu OC (2004). Distribution of hydrocarbons and heavy metals in sediment and a crustacean (shrimps-Penaeus notialis) from the bonny/new calabar river estuary. Niger Delta, Ajeam-Ragee, 9: 1-14.

Department of Environment (1997). Environmental quality standards for Bangladesh. Government of the People's Republic of Bangladesh.

Emongor V, Kealotswe E, Koorapetse I, Sankwasa S, Keikanetswe S (2005). Pollution indicators in Gaberone effluent. J. Appl. Sci., 5: 147-150.

Fakayode SO (2005). Impact of industrial effluents on water quality of the receiving Alaro River in Ibadan. Nigeria, Ajeam-Ragee, 10: 1-13.

Furtado AAL, Albuquerque RT, Leite SGF, Pecanha RP (1998). Effect of hydraulicretention time on nitrification in an airlift biological reactor. Br. J. Chem. Eng., 15: 1-7.

Olayinka KO (2004). Studies on industrial pollution in Nigeria: The effect of textile effluents on the quality of groundwater in some parts of Lagos. Niger. J. Health Biomed. Sci., 3: 44-50.

Ugochukwu CNC (2004). Effluent monitoring of oil servicing company and its impact on the environment. Ajeam-Ragee, 8: 27-30, 244.

World Bank (2007). Industrial Environmental Compliance and Pollution Control in Greater Dhaka. Final Report; Dhaka Water Supply and Sewerage Authority.

Making a case for systems thinking approach to integrated water resources management (IWRM)

Bloodless Dzwairo[1]*, Fred A. O. Otieno[2] and George M. Ochieng[1]

[1]Tshwane University of Technology, Faculty of Engineering and the Built Environment, Civil Engineering Department, Private Bag X680, Pretoria, 0001, South Africa.
[2]Durban University of Technology, Deputy Vice Chancellor for Technology, Innovation and Partnerships, P. O. Box 1334, 4000, South Africa.

While definite benefits could be gained from integrating the management of natural and human systems, taking into account the effects of human activities on the environment, the task of operationalising and implementing this approach in practice represents an enormous challenge. Some limitations highlighted include the definition of IWRM itself, where it appears to be broad, all-encompassing and impressive, but contains "lofty phrases" that have little practical importance for either present or future water management practices. This paper makes recommendations to uptake and incorporate the systems thinking approach (STA) in IWRM decision making processes, an approach envisaged to be more participatory and also process-oriented. Incorporation of socio-economic dynamics is also recommended, with a suggestion that the biophysical relationships like interactions in a basin between water, the human population and settlements should be analysed to the extent possible.

Key words: operationalising IWRM, socio-economic dynamics, South Africa, systems thinking approach.

INTRODUCTION

IWRM and the economic interdependence

The last two decades have seen a call for increased integration in water management, driven by recognition of the limits of a fragmented organisational approach and an increased understanding of the interconnected nature of water related issues. The following documents: Agenda 21 chapter 18 (United Nations, 2004b), summary report of the Bellagio report (WSSCC, 2000), Rio Declaration principles 15 and 16 (United Nations, 1992), MDGs (United Nations, 2008) and the Johannesburg Declaration 3 (United Nations, 2004a) are examples.

At a global scale the Brundtland report (United Nations, 1987) warns of many forms of development which erode the environmental resources upon which they must be based. For example too much water and non-renewable resources of high quality are taken from the eco-system and returned to the eco-sphere as pollution (Gumbo, 2005). Compatibility of environmental and economic objectives has thus been lost in the pursuit of individual or group gains, with very limited regard for the impacts on other sectors. There is pursuance of blind faith in science's ability to find solutions and an ignorance of the not too distant consequences of today's decisions. Institutional rigidities are alleged to add to this myopia. One such important rigidity which is echoed in the Brundtland report (United Nations, 1987) is the tendency to deal with one industry or sector in isolation, failing to

*Corresponding author. E-mail: dzwairob@tut.ac.za, ig445578@gmail.com, lessyd@iwsd.co.zw.

Abbreviations: DWA, Department of water affairs; **DWAF,** department of water affairs and forestry; **IWRM,** integrated water resources management; **K9,** monitoring point on Klip River; **L1,** monitoring point on Leeuspruit River; **MDG,** millennium development goals; **MWB,** Midvaal water board; **M3,** monitoring point on Mooi River; **NWA,** national water act; **PPP,** polluter-pays principle; **RWB,** Rand water board; **SWB,** Sedibeng water board; **S4,** monitoring point on Suikerbosrant River; **T1,** monitoring point on a stream from Webb's farm; **VR,** Vaal River; **Vs3,** monitoring point on Vals River; **V1 - V19,** monitoring points along the Vaal River; **WRC,** water research commission; **WSSCC,** water supply and sanitation collaborative council.

recognize that the strategy for sustainable development is to integrate economic and ecological considerations in decision making.

For river basins, their sub-basins or the recharge areas of groundwater, various water uses may compete or even be in conflict with each other, thus creating management problems, particularly if water is scarce and/or its quality is deteriorating. Knowledge about economic development patterns becomes important for understanding how water management problems may worsen in the future (United Nations, 2007). A few of these problems identified at a catchment level include the effects of excess nutrients, suspended and bottom sediments, heavy metals and other chemicals in surface/groundwater on drinking-water purification and use; as well as the impacts of mining activities on water resources.

Yet in conventional economic appraisals the water industry has concentrated on economic efficiency defined almost entirely in allocative terms, the key issues being the efficiency with which available water resources are allocated between users and with which factors of production are invested in new facilities. In the spirit of IWRM the link with economic operations has not been so obvious, highlighting the economic and social good of water as a resource (Liu et al., 2003).

The United Nations status report by Young et al. (2008) tracks progress towards full implementation of IWRM and highlights thorny issues where developed countries have advanced on almost all major issues, but that there is still much room for further improvement. For developing countries there has been some improvement in the IWRM planning processes at national level but more needs to be done to implement the plans. Insufficient alignment and cooperation between the policies of different government departments and the practices of different water use sectors that impact on water are pointed out by Funke et al. (2007), as huge impediments to the limited successes in IWRM, so far. Van der Zaag (2005), though, argues that the implementation of IWRM is a 'must' because systematically pursuing IWRM constitutes a path of short-term risk that leads to long-term security, which supports the current and future environmental and developmental needs to manage water resources in a holistic manner.

While definite benefits could be gained from integrating the management of natural and human systems, and taking into account the effects of human activities on the environment, the task of operationalising and implementting this approach in practice represents an enormous challenge (Funke et al., 2007). The same report further argues that while the Global Water Partnership's definition of IWRM appears to be broad, all-encompassing and impressive, it however contains "lofty phrases" that have little practical importance for either present or future water management practices. The report further notes that the phrases do not help water planners or managers to solve problems in the water sector. Ultimately, Biswas (2004) in Funke et al. (2007) considers the IWRM concept to be "unimplementable" because of the difficulty in integrating the actions of different sectors such as water and energy.

For operational reasons in South Africa, notwithstanding many other IWRM related activities, the Department of Water Affairs (DWA) had moved from broadly IWRM to primarily domestic water supply and sanitation (Jonker, 2007). The report assets that this move was due to a lack of conceptual clarity of IWRM, although in South Africa, the concept is an important aspect of the 1998 National Water Act (NWA). This is evidenced by the establishment of Water Management Areas and other supporting institutions described in DWAF (2004). Jonker (2007) thus proposes a re-worded definition for IWRM that allows for measurement of the concept's critical components, namely, people's activities, improvement of their livelihoods and disruption of the water cycle.

South Africa is water-scarce and the provision of water should always be considered in terms of the socio and ecological economic benefits accruing from contributions by specific users, where successful implementation of the NWA 1998 is dependent on having the right tools available to support water resources decision-making and competent people to apply these tools. Futuristic systems thinking (ST) and a visionary approach are useful tools in dealing with the complex institutional environment and assessment of future reform impacts. A STA could also provide an assessment platform for the measurable activities.

The STA to integrated catchment management

Global environmental change and sustainability are characterised as the challenge of managing change in dynamic systems riddled with uncertainty (Dovers and Handmer, 1992) and for economic exchanges to become beneficial for all involved, two conditions must be met. Firstly, the sustainability of that ecosystem must be guaranteed. Secondly, the partners to the economic exchanges must be satisfied that the basis of exchange is equitable. Unfortunately for the water and economy sectors neither condition is met, hence the unsustainable situation that many developing countries find themselves in, where fresh water resources are deteriorating at an alarming rate.

Integration of disciplines

An increasing awareness of the complexity of environmental problems has triggered the development of new management approaches. Pahl-Wostl (2007) discusses the importance of focusing on the transition to new management paradigms based on the insight that the systems to be managed are complex and adaptive. The paper provides arguments for the role of social learning processes and the need to develop methods combining approaches from hard and soft systems analysis. Many of the traditional approaches tend to strive for objectivity and are thus classified as hard systems. The STA has been proposed as a useful method of integrating disciplines based on `reductionism' and the degree of `subjectivity'. A reductionist approach like IWRM tends towards breaking down complex systems into simple constituents, based on the view that they can be fully understood in terms of their isolated parts. In contrast, the holistic approach recognises the tendency in nature to form `wholes' that are more than the sum of the parts by ordered grouping, *a STA*. Although some systems can be studied in a reductionist manner, a truly holistic approach is, by virtue of its breadth and complexity, difficult to adopt thus in practice there is a continuum of approaches incorporating degrees of reductionism. Systems based research, as a paradigm, tends towards the holism where sustainable planning is influenced by the concept of holism, a central tenet of ST.

Examples of practical application of the STA

By way of moving towards operationalising IWRM, Moore (2004) provides an approach to ecosystem toxicity using ST, even though practical challenges are also highlighted. Stakeholder cooperation is emphasized as key to implementation of the STA towards greener production in the textile industry, a major industry contributing towards environmental toxicity. One of the approaches in this case was applying limits to success for toxicity reduction by product substitution as well as making follow-ups on a flowchart depicting an inclusive process for aquatic toxicity resolution. A bench-marked dialogue acceptable to both legal and regulatory authorities was thus created, without which the whole process could have been sidetracked by the diverse goals of multiple stakeholder groups (Moore, 2004).

From Figure 1, Chan and Huang (2004) argues that the viewpoint of sustainable development dictates that a sound growth pattern of a community should be a logistic S-curve type. Because the cybernetic thinking tends towards the idle, environments in crisis currently due to pollution and over allocation of natural resources, are already signifying signs of collapsed.

The 4 interrelated challenges

In the context of decision-making and decision support systems, targets pose 4 inter-related challenges:

- Going beyond 'end-of-pipe' solutions.
- Enhancing integration.
- Encouraging meaningful participation.
- Linking actions and measures across multiple scales.

In the process of holistic planning, stakeholders are encouraged to step outside of the boundaries of their disciplines, and to have a 'life cycle basis' view so that the implications of a project over its total time span are considered (Tippett, 2005).

The importance of participation

In a reflection on lessons learnt in the twentieth century, the need for broad participation in planning could not be over emphasised, especially as participatory elements could increase the quality of public decision making by improving information management and the legitimacy of the decision process (Messner et al., 2006). Participation is thus a crucial element in IWRM, and du Toit and Pollard (2008) note that in South Africa 'participation fatigue' might eventually lead to growing frustration with the implementation of the content of the National Water Act (1998).

The STA to integrated catchment management views the interaction between parts of the catchment, including human activities and population dynamics in such a way as to avoid problems like optimising one part of the system at the expense of the whole. Recognition of equity in handling upstream-downstream impacts promotes a foaminess of collaboration, allowing for individual ideas to share the burden of the entire system. This assures excellent stability and adaptability – characteristics lacking in systems which insist on homogeneity and absolute efficiency. Even as the European Union-Wide Water Framework Directive enters into public policy; researchers like Collins et al. (2007) are calling for a systemic approach to managing multiple perspectives and stake holding in water catchments.

Upstream - downstream relationships

River basin management is currently based on a narrow evidence base in which science is used to inform policy principally through modelling of catchment functioning. Limitations arise because many issues cannot be understood or defined from one particular perspective alone, nor resolved by unilateral action on the part of a single stakeholder. Even when the importance of context and multiple perspectives is recognised by policy makers, they may struggle to develop a more holistic or systemic view of the interdependency between individual actions and catchment (Collins et al., 2007). Scientific understanding is essential and informative but, where sustainable management is the goal; it is most effectively seen as only part of how stakeholders understand catchments. This situation is unlikely to change unless more systemic

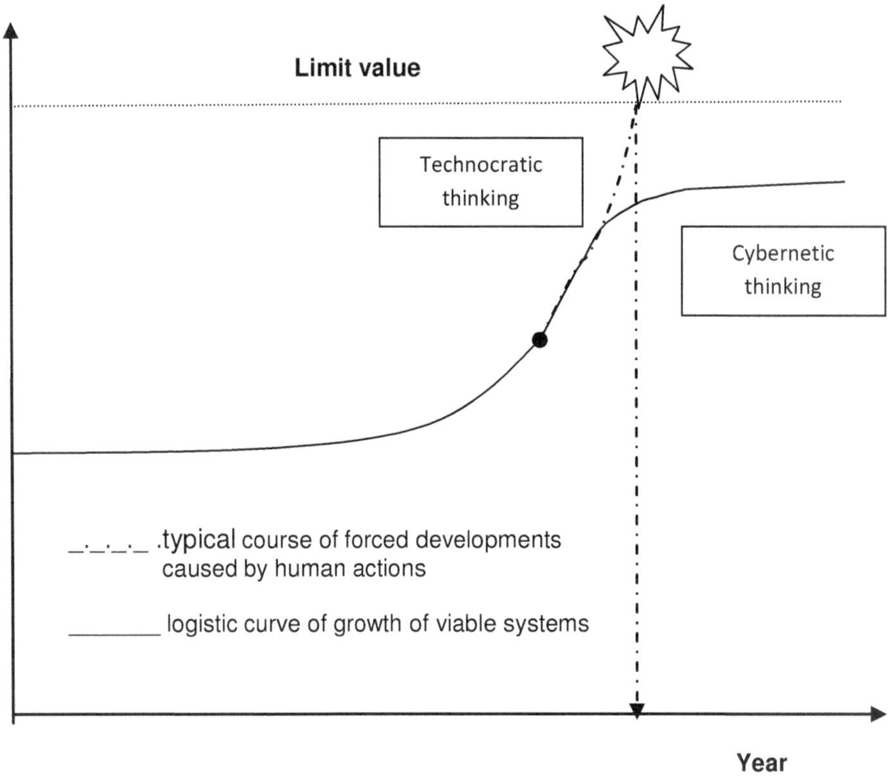

Figure 1. Logistic S-curve of sustainable development (Adapted from Chan and Huang, 2004).

approaches to managing multiple stake holding become more prevalent in policy and practice (Collins et al., 2007). Sustainable management of water requires integration, recognising the interconnections between upstream and downstream systems operating at different levels of scale. This is an endeavour in which ST could provide useful tools. Systems oriented models could enhance dialogue and facilitate work across these scales so as to increase focus on the 'whole picture' (Tippett, 2005).

Recognising the spatial placements of various stakeholders along waterways and within river basins, environmental policy principles such as the polluter-pays principle (PPP) and the precautionary principle are meant to equitably internalise the pollution externalities for basin sustainability. Their actual implementation however, depends on factors like different pollution assimilative capacities of the environment, different social objectives and priorities attached to environmental protection and different degrees of industrialisation and population density, which justify differing national environmental policies (Correljé et al., 2007).

The polluter-pays and the user-pays principle are both related to who should bear the costs of environmental degradation. According to the PPP, those who cause pollution should meet the costs to which it gives rise (Correljé et al., 2007). This is because environmental risks arising from upstream developmental decisions and actions impinge on downstream areas that have little or no influence on those decisions. In order to take account of downstream interests, the two principles may be implemented so at to set an economically and socially acceptable scenario within the basin.

Waterborne effluents as outputs from a range of activities are disposed onto land and into streams and oceans via various sewerage and drainage systems. This tends to present a scenario where an operator is also responsible for polluting the downstream outlets, compromising environmental sustainability principles. Even where these services are handled by different players like in the case of South Africa, accountability is greatly compromised by the power of economics. This approach is currently running into the law of diminishing returns in the face of complex environmental challenges as stressed by various reports regarding the critical pollution problems for South African water resources, these being externalities to water treatment processes.

Pretty et al. (2001) cautions that if such externalities are not internalised, they distort the market by encouraging activities that are costly to society even if the private benefits are substantial. Such is the case when polluted water is treated for drinking water purposes. The tradition is to internalise the cost of treatment due to pollution load by setting the tariff structures, thereby offsetting the cost of production. This costing model is structured without any regard to the other factors like the cost of depleting the water resources itself, a cost which should ideally be borne by the user of the natural resource.

Application of the sensitivity model (SM) to study area

For South Africa examples of deteriorating ecosystems are highlighted in DWAF (2003), DWAF (2006) and Winde and Stoch (2010). The Vaal basin, which is the focus of current on-going research, is one such example. It is highly impacted by mining, industrial and domestic waste (Cloot and Roux, 1997; DWAF, 2007; Gouws and Coetzee, 1997; Naicker et al., 2003; Pieterse et al., 1987; Stevn and Toerien, 1976; Winde and Jacobus van der Walt, 2004).

Various ecological problems have ensured over more than 100 years (Winde and Stoch, 2010), of which the most acute ones are growing water deficit; pollution of open and underground water; enormous over-norm water losses and exacerbation of the problem of quality drinking water supply to the population (Leendertse et al., 2008). The report further highlights that addressing these and other water environmental needs in South Africa has been quite different where the focus has been strongly on organisational, institutional and legal aspects and not so much on implementation. In the meanwhile pollution continues to be a huge dilemma which calls for workable solutions, even if those solutions are slow. A ST tool called the sensitivity model (SM) (Chan and Huang, 2004) was proposed for implementing a STA in the study area, the Upper and Middle Vaal water management areas in order to systematically tackle the huge water pollution problems currently threatening the Vaal basin. The layout as adapted from Chan and Huang (2004) is shown in Figure 2. System description to understand the major issues as well as to identify the key players was done (Dzwairo et al., 2009) and a variable set established. Dzwairo and Otieno (2010) traced some pollution pathways within the study area that were critical to understanding the impact matrix (Figure 3) as well as the available data from key stakeholders, these being DWA and three Water Boards (utilities that treat water for potable use) falling within the study area . Implementation of the tool is on-going, hopefully to be taken up by the regulator and custodian of water resources in the country.

DISCUSSION AND CONCLUSION

A holistic approach to water management is beneficial for the environment only if it takes cognizance of environmental concerns. The benefits and implications of an integrated management system for the environment cannot be over-emphasized. Several practical examples

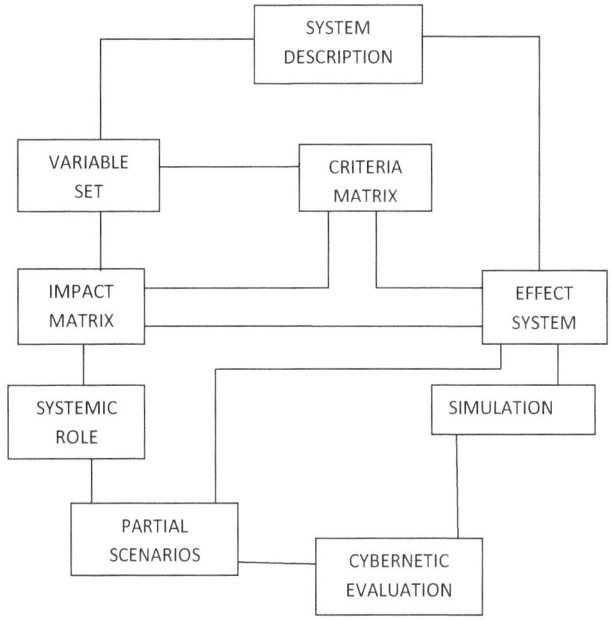

Figure 2. The procedure of sensitivity model (Source: Chan and Huang, 2004).

from Southern Africa and elsewhere make a strong case for IWRM to be an effective approach for sustainable management at river basin level. Operationalising IWRM is the huge challenge currently.

Stake holding is noted to be essential as the IWRM process should be driven by local interests and should address real needs without which IWRM cannot work. Although the immediate aim is not to be over-formulaic, it is imperative to align people in the common vision of IWRM, including sustainable environmental management. This requires strong institutional support if it is to be successful. The SM as one of many ST tools is envisaged to be adopted systematically and in phases beyond the current research in order to fill the gaps which are impediments to IWRM operationalisation in the Vaal basin. The model could incorporate a payment mechanism for natural resources through Simulation, reciprocal to pollution load as a trade-off for upstream-downstream pollution within a river basin.

It is thus recommended to uptake and incorporate the STA in IWRM decision making processes, an approach envisaged to be more participatory and also process-oriented. Incorporation of socio-economic dynamics is also recommended, with a suggestion that the bio-physical relationships like interactions in a basin between water, the human population and settlements should be analysed to the extent possible.

ACKNOWLEDGEMENT

The financial assistance of the Department of Science Technology (DST) is hereby acknowledged. Opinions expressed and conclusions arrived at, are those of the authors and are not necessarily to be attributed to the

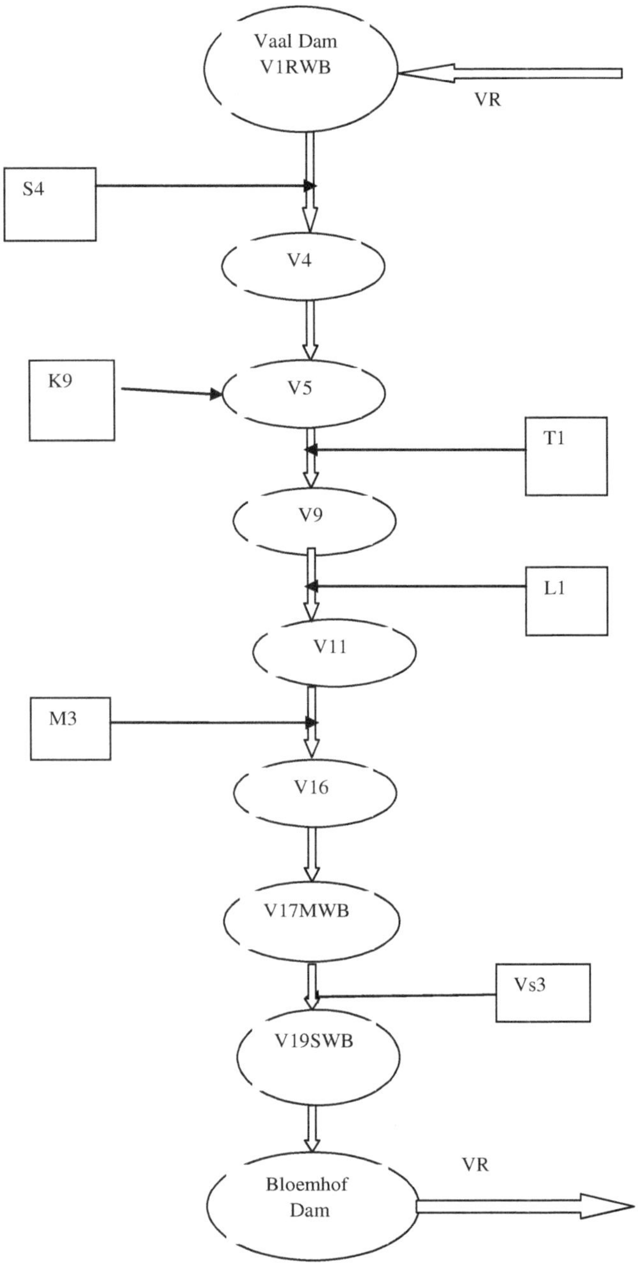

Figure 3. Layout of Vaal River (VR) monitoring points in relation to its tributaries (Source: Dzwairo and Otieno, 2010)

DST. The authors would also like to thank Tshwane University of Technology for hosting and also funding this research. Rand Water, Midvaal Water and Sedibeng Water Boards, DWA and the WRC are also acknowledged.

REFERENCES

Biswas AK (2004). Integrated water resources management: A reassessment. Water Int., 29: 248-256.

Chan SI. Huang SL (2004). A systems approach for the development of a sustainable community - the application of the sensitivity model (SM). J. Environ. Manage., 72: 133-147.

Cloot A, Roux GL (1997). Modelling algal blooms in the middle Vaal River: a site specific approach. Water. Res., 31: 271-279.

Collins K, Blackmore C, Morris D, Watson D (2007). A systemic approach to managing multiple perspectives and stakeholding in water catchments: some findings from three UK case studies. Environ. Sci. Policy., 10: 564-574.

Correljé A, François D, Verbeke T (2007). Integrating water management and principles of policy: towards an EU framework, J. Cleaner. Prod., 15: 1499-1506.

Dovers SR, Handmer JW (1992). Uncertainty, sustainability and change. Global Environ. Change., 2: 262-276.

du Toit D, Pollard S (2008). Updating public participation in IWRM: A proposal for a focused and structured engagement with Catchment Management Strategies. Water SA., 34: 707-714.

DWAF (2003). River Health Programme: state-of-rivers report: Free State. Pretoria, South Africa: Department of Water Affairs and Forestry.

DWAF (2004). National water resources strategy: our blue print for survival, Pretoria, South Africa: Department of Water Affairs and Forestry.

DWAF (2006). Vaal river system: large bulk water supply reconciliation strategy: first stage reconciliation strategy. DMM Development Consultants, Golder Associates Africa, SRK, WRP Consulting Engineers and Zitholele Consulting. Pretoria, South Africa: Department of Water Affairs and Forestry.

DWAF (2007). Integrated water quality management plan for the Vaal River system. Pretoria, South Africa: Department of Water Affairs and Forestry.

Dzwairo B, Otieno FAO, Ochieng' GM (2009). Incorporating Surface Raw Water Quality into the Cost Chain for Water Services: Vaal Catchment, South Africa. Paper presented at the 10th WaterNet/WARFSA/GWP-SA Symposium in association with the International Commission on Water Resources Systems of the IAHS. IWRM: Environmental Sustainability, Climate Change and Livelihoods, Imperial Resort Beach Hotel, Entebbe, Uganda, 28-30 October.

Dzwairo B, Otieno FAO (2010). Integrating quality and cost of surface raw water: Upper and Middle Vaal Water Management Areas South Africa. Water Sci. Technol: Water Supply. 10: 201-207.

Funke N, Oelofse SHH, Hattingh J, Ashton PJ, Turton AR (2007). IWRM in developing countries: Lessons from the Mhlatuze Catchment in South Africa. Phys. Chem. Earth, Parts A/B/C. 32: 1237-1245.

Gouws K, Coetzee PP (1997). Determination and partitioning of heavy metals in sediments of the Vaal Dam System by sequential extraction. Water. SA. 23: 217-226.

Gumbo B (2005). Short-Cutting the Phosphorus Cycle in Urban Ecosystems. Delft University of Technology, Delft, Netherlands.

Jonker L (2007). Integrated water resources management: The theory-praxis-nexus, a South African perspective. Phys. Chem. Earth, Parts A/B/C. 32: 1257-1263.

Leendertse K, Mitchell S, Harlin J (2008). IWRM and the environment: A view on their interaction and examples where IWRM led to better environmental management in developing countries. Water. SA. 34: 691-698.

Liu J, Savenije HHG, Xu J (2003). Water as an economic good and water tariff design: Comparison between IBT-con and IRT-cap. Phys. Chem. Earth, Parts A/B/C. 28: 209-217.

Messner F, Zwirner O, Karkuschke M (2006). Participation in multi-criteria decision support for the resolution of a water allocation problem in the Spree River basin. Land Use Policy., 23: 63-75.

Moore SB, Ausley LW (2004). Systems thinking and green chemistry in the textile industry: concepts, technologies and benefits, J. Cleaner. Prod., 12: 585-601.

Naicker K, Cukrowska E, Mccarthy TS (2003). Acid mine drainage arising from gold mining activity in Johannesburg, South Africa and environs. Environ. Pollut., 122: 29-40.

Pahl-Wostl C (2007). The implications of complexity for integrated resources management. Environ. Modell. Softw. 22: 561-569.

Pieterse A, Roos J, Roos K, Pienaar C (1987). Preliminary observations on cross-channel and vertical heterogeneity in environmental and algological parameters in the Vaal River at Balkfontein, South Africa. Water SA. 12: 173-184.
Pretty J, Brett C, Gee D, Rachel H, Mason C, Morison JIL, Raven H, Rayment M and van der Bijl G (S.a). Policy Challenges and Priorities for internalising the externalities of modern agriculture. J. Environ. Plann. Manage., 44: 263-283.
Stevn DJ, Toerien DF (1976). Eutrophication levels of some South African impoundments. IV. Vaal dam. Water SA. 2: 53-57.
Tippett J (2005). The value of combining a systems view of sustainability with a participatory protocol for ecologically informed design in river basins. Environ. Modell. Softw. 20: 119-139.
United Nations (1987). United Nations development and international economic co-operation: environment: our common future: the Brundtland report (No. 25). Oslo.
United Nations (1992). United Nations report of the conference on environment and development: Annex 1: Rio declaration.
United Nations (2004a). Johannesburg declaration on sustainable development: from our origins to the future Johannesburg, South Africa.
United Nations (2004b). United Nations Agenda 21: United Nations department for economic and social affairs. New York.
United Nations (2007). United Nations convention on the protection and use of transboundary watercourses: recommendations on payments for ecosystem services in integrated water resources management and international lakes. New York.
United Nations (2008). The Millennium Development Goals report. New York.
Van D, Zaag P (2005). Integrated Water Resources Management: Relevant concept or irrelevant buzzword? A capacity building and research agenda for Southern Africa. Phys. Chem. Earth, Parts A/B/C. 30: 867-871.
Water Supply and Sanitation Collaborative Council (WSSCC) (2000). Summary report of Bellagio expert consultation on environmental sanitation in the 21st century. EAWAG, SANDEC, WSSC. Switzerland.
Winde F, Jacobus Van Der Wl (2004). The significance of groundwater-stream interactions and fluctuating stream chemistry on waterborne uranium contamination of streams-a case study from a gold mining site in South Africa. J. Hydrol., 287: 178-196.
Winde F, Stoch E (2010). Threats and opportunities for post-closure development in dolomitic gold mining areas of the West Rand and Far West Rand (South Africa) – a hydraulic view Part 1: Mining legacy and future threats. Water SA. 36: 69-74.
Young G, Shah B, Kimaite F (2008). Status report on IWRM and water efficiency plans. Prepared for the 16th session of the Commission on Sustainable Development - May .

Improvement of water treatment plants capacity and effluent quality by introducing "NINIVITE" as new filtration media

Sati M. Al-Rawi* and Kahtan A. Al-Najjar

Mosul University Technical College/Kirkuke Iraq.

This work compares the performance of four types of filters, three of them are dual media/capping filters and the fourth is a single medium sand filter. The filters were made to operate with the same effective size and the same depth. These filter were subjected to the same operating conditions of filtration rates and influent turbidity. A new locally available filter medium called ninivite (recently discovered rock) was introduced to act as an upper layer of one of the selected dual media filters. To verify the goals of this work a pilot plant similar to the classical water treatment plant was constructed. Filtration system consisted of four glass column filters, acting parallel and simultaneously. The first three columns contained 20 cm of granular ninivite rock GNR, granular activated carbon GAC, and anthracite coal respectively over 40 cm of sand. The fourth column was kept with 60 cm sand to act as a single –medium filter. The effective size of each media grains was 0.82 mm with a uniformity coefficient 1.6. A total of 18 runs for natural raw water turbidity were conducted. Run times of these experiments amounted to 12-27 h. The filters were operated at rates of 4.9, 7.3, and 9.8 m/h; which were 1, 1.5 and 2 times the current rate at treatment plants. The results were encouraging .Using filters consisting of GNR and sand operated at mentioned rates appeared to have the best performance for the studied parameters. Efficiency of turbidity removal amounted to 91.31% for sample having 9.9 NTU at filtration rate 4.9 m/h. Such performances were also recorded for other dual – media filters for the considered rates. GNR filter also showed considerable efficiency in bacterial removal.

Key words: Water treatment, dual- media filters, capping filtration, ninivite/porcelanite, drinking water filtration.

INTRODUCTION AND REVIEW OF LITERATURE

Municipal water treatment plants in Iraq adopt the rapid sand filtration technique for clarification of water for drinking and other uses. Filters of such kind operate at rates 1-2 gpm/sq.ft amounting to 2.5-5.0 m/h. Most of Iraqi plants are employed at 4 m/h as an average and employ a sand layer supported by gravel as a porous medium (Al-Rawi, 1995).

These filters suffer some problems that may affect its performance. Examples are: removal of suspended matter occurs in the top 1-4 inches (2.5-10 cm) and the whole remaining depth of the filter is not utilized. Besides, these filters are incapable of satisfying growing per capita demand and consumption. Furthermore, a deficiency of such filters to produce high quality furnished water is often noted.

The construction of upstream impoundment has greatly affected raw water quality received by treatment plants leaving the existing plants act merely as a passing through units.

Dual-media beds have gradually replaced sand alone in filters of many water treatment plants all over the world. Dual-media filters possess several distinct advantages compared to those of conventional rapid sand filters. Higher filtration rates, longer runtimes, flexibility to tackle variable turbidity loading and better performance are some examples (Amirtharaga, 1988; Al- Rawi, 1987; Degremont, 1991; Al- Ani et al., 1988; Twort et al., 2001; Peavey et al., 1987; Montgomery, 1985)

Usually the depth of upper lighter layers equals twice that of the bottom heavier layers. However, from view of cost, the developing countries –and Iraq is no exception- have used sand filter capping. Here the depth of the up-

*Corresponding author. E-mail: alrawism@yahoo.com.

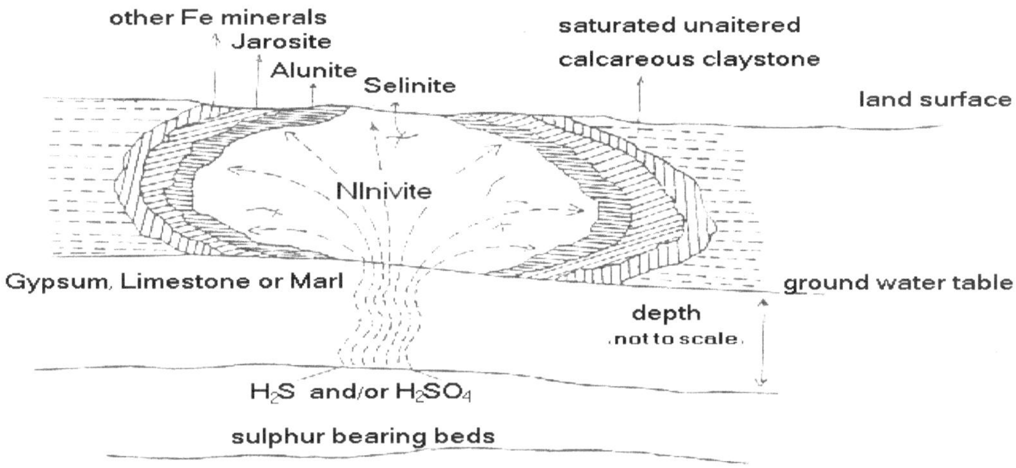

Figure 1. Model for the genesis of ninivite.

Table 1. Some physical characteristics of ninivite.

Sample No.	Porosity %	Density(gm/cm^3)	Absorption ratio %
1	72.0	0.787	91.43
2	72.7	0.743	97.82
3	68.0	0.753	89.51
4	76.2	0.720	104.5
5	72.2	0.740	96.37
6	71.1	0.750	95.17
7	70.5	0.726	97.20
8	69.6	0.745	93.46
9	62.1	0.760	81.60
10	70.3	0.670	104.70

Permeability = 6.9×10^{-5} cm/s. Surface area = 739-765.5 m^2/gm, Acid loss = zero color: white-grey, white when dry, grey when wet.

Table 2. some chemical constituents of ninivite

Oxides	SiO_2	Al_2O_3	Fe_2O_3	CaO	MgO	SO_3	TiO_2	L.O.I
%	95.7	0.22	0.08	0.98	0.02	0.07	0.59	2.2

per layers is often half that of the sand layer.

This work aims at improving the performance of existing water treatment filters through capping sand layers with GNR (granular ninivite rock, newly introduced filter medium), GAC(granular activated carbon), and anthracite coal, thus acting as a dual media/capping filters. Besides, filter with sand and the introduced ninivite is to be compared with different types of filters that employs sand alone (single layer filter) and sand with activated carbon and sand with anthracite (as dual media/capping filters).

The comparison will be based on using different levels of influent turbidity, and variable filtration rates and intends to add another savings to water treatment processes.

Background of ninivite

Ninivite is not widely known to researchers. It was discovered in 1987 during detailed geological mapping of some areas in Mosul suburbs. Ninivite was then defined as a new form of Porecelanite (Al- Naqib and Al-Dabbagh, 1990).

Ninivite is derived from the name of "Nineveh", the ancient capital of Assyrian civilization. Mosul, a city in the north of Iraq, now embraces the site. Figure 1 illustrates ninivite composition (Aswad et al., 1995).

This ninivite possesses the following physical and chemical process that enable this rock – as thought- to act as the upper layer of dual/capping media filters Tables 1 and 2 (Al- Naqib and Al- Dabbagh, 1990).

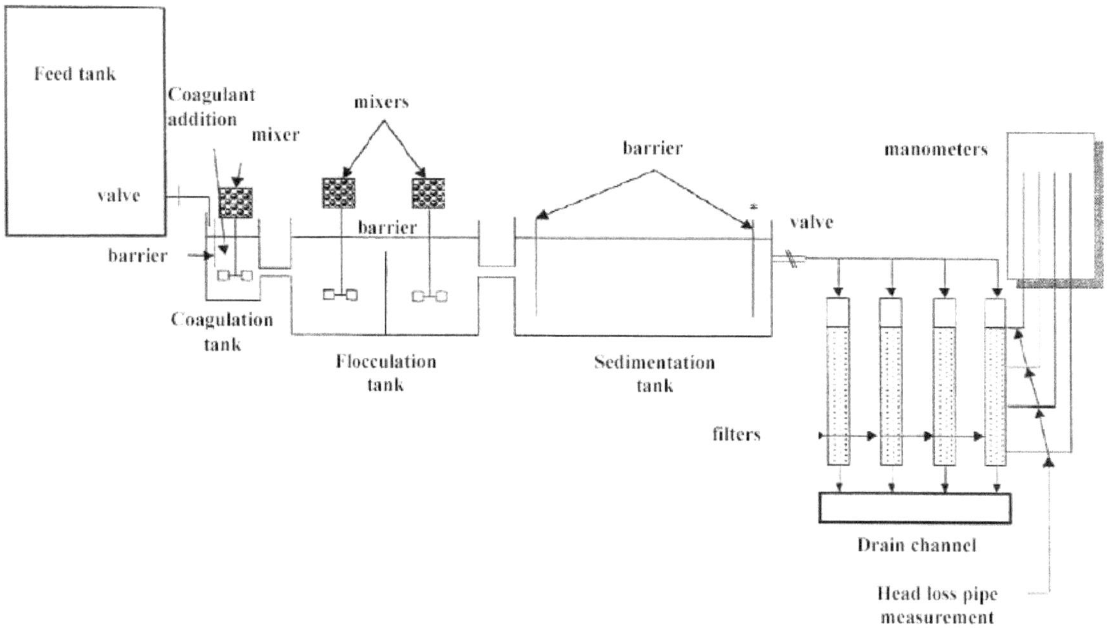

Figure 2. Pilot plant components.

Table 3. Components of pilot plant.

Item	Remarks
Feed tank	Three tanks 1 x 1 x 1 m each are used to provide a continuous supply of raw water .
Rapid mix	10.5 x 10.5 x 10.5 cm. Detention time is 1 min.
Flocculation	50 x 30 x 22 cm. Detention time is 30 min.
Sedimentation	Two tanks, 54 x 34 x 28 cm each. Detention time is 2 h.
Filter unit	4 Glass columns 5.3 cm diameter each connected to piezometers as shown in Figure 3

Table 4. Process variables and conditions

Parameters	Remarks
Raw water	River water having raw turbidity ranging from 16 56- NTU.
Temperature	11 – 19 °C
Coagulant	Alum is dosed as complied by raw water turbidity.
Filter media	Three dual/capping media filters consisting of 20 cm of ninivite, or anthracite, or activated carbon over 40 cm of sand .the 4th filter is kept 60 cm of sand as a single layer filter.
Treatment	Conventional processes.
Filtration rate	4.9, 7.3, and 9.8 m/h.
pH	It ranges from 7.3 to 8.6
Headloss	It is measured through water levels in 4 tubes connected at different heights. Filter cycle terminates when head loss reaches 80 cm.

MATERIALS AND METHODS

A pilot plant constructed similarly to the conventional water treatment plant is employed for the purposes of this work. Figure 2 illustrates the units of the used pilot plant. Tables 3 and 4 explain the units and process variables (Al-Najjar, 2000).

Ninivite is tested at ministry of health laboratories to check its suitability from health point of view as a filter medium. Besides, tests required to verify it as a filter medium such as acid loss, effective size and uniformity ... are also carried out.

Filtration system consists of four glass column filters, acting parallel and simultaneously. The first three columns contain 20 cm of granular ninivite rock GNR, granular activated carbon GAC, and anthracite coal respectively over 40 cm of sand. The fourth column is kept with 60 cm sand to act as a single -medium filter. The effective size of each medium grains is 0.82 mm (the current

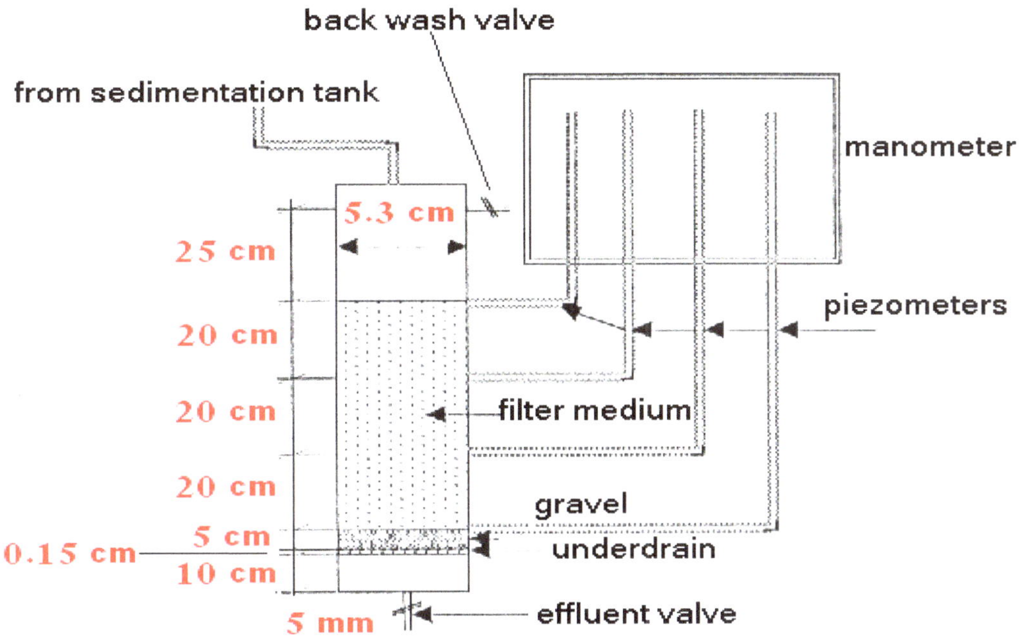

Figure 3. Filter column design.

Table 5. Achieved effluent turbidity under variable raw turbidity and variable filtration rates.

Average effluent turbidity NTU types of filters				Filtration rate m/h	Influent turbidity NTU
Sand alone	Sand + anthracite	Sand + Activated carbon	Sand + Ninivite		
1.10	0.93	0.87	0.86	4.9	9.9
0.91	0.82	0.76	0.75		5.8
1.10	0.93	0.88	0.86		9.8
0.91	0.83	0.75	0.73		5.6
0.59	0.53	0.47	0.47		3.6
0.60	0.52	0.46	0.46		3.3
1.13	0.94	0.89	0.87	7.3	9.8
0.94	0.84	0.76	0.74		5.6
1.14	0.95	0.88	0.87		9.8
0.95	0.83	0.77	0.75		5.5
0.62	0.55	0.50	0.48		3.5
0.63	0.55	0.51	0.49		3.2
1.25	0.97	0.94	0.92	9.8	9.9
0.98	0.85	0.77	0.76		5.7
1.24	0.97	0.94	0.93		9.8
0.99	0.85	0.77	0.76		5.5
0.68	0.55	0.52	0.51		3.6
0.66	0.56	0.52	0.50		3.5

medium size in the city treatment plants) with a uniformity coefficient 1.6. Figure 3 shows the details of filter unit.

A total of 18 runs for natural raw water turbidity are conducted. The filters were operated at rates of 4.9, 7.3, and 9.8 m/h; which is to 1 . 8 and 2.5 times the current prevailing filtration rate in the city water treatment plants. This practice may add to the economy of water treatment processes.

RESULTS AND DISCUSSION

Table 5 summarizes influent raw turbidity levels received by all filters. It also shows effluent turbidity for the employed filtration rates .It is worthy to note here that none of the resulting effluent turbidity for used dual/capping fil-

Table 6. Filter performance as filtration rate varies.

Influent turbidity NTU	Filtration rate m/h	Turbidity removal %			
		Sand + ninivite	Sand + activated carbon	Sand + anthracite	Sand alone
	4.9	100	100	100	100
3.6	7.3	95.92	92.16	94.44	93.65
	9.8	92.16	90.39	92.86	86.77
	4.9	100	100	100	100
5.6	7.3	97.3	97.40	97.62	95.79
	9.8	96.05	97.40	96.47	91.92
	4.9	100	100	100	100
9.9	7.3	98.85	97.75	97.90	94.93
	9.8	92.47	92.55	95.88	88.00

ters exceeds (1) NTU. This highly confirms the excellent performance of the used filters.

In practice and for single medium filters in treatment plant works, the listed results might not be achieved due-mainly- to improper alum doses, improper operation of units etc. Generally, as far as this work is concerned, effluent turbidity complies well with the WHO (1984), EPA (Hendricks, 2006), and Iraqi specifications (Ministry of health, 1988).

As shown, –for example- the influent turbidity for filtration rate of 4.9 m/h was 9.9 NTU. The resulting effluent turbidity for the used filters was 0.86, 0.87, 0.93 and 1.1. This means that removal achieved was 91.3, 91.1, 90.6, and 88.8% respectively.

Also, it is seen that filter consisting of 20 cm of ninivite over 40 cm of sand is superior in removal of turbidity compared to other filters.

It is worthy to display the role of porosity in achieving good filter performance. The porosity of ninivite, activated carbon, anthracite and sand were 0.77, 0.74, 0.62 and 0.40 respectively. This means that as porosity increases, the spaces among grains increase. Consequently a more ability for accumulation of flocs to be removed by various removal mechanisms. This ultimately will reduce the breakthrough of impurities inside the filters which may come out in the effluent and degrades filtrate quality.

From the other hand, the used filtration rates were 4.9, 7.3 and 9.9 m/h. This is equivalent to 1:1.5:2 times the current rates in country treatment works. This is very important from economic point of view. Authorities need not to construct a new plant to satisfy the demands on water. Water output in this case will be doubled with minimum cost and minor plant modifications.

Table 6 also shows the effect of changing filtration rates upon effluent turbidity. This effect appears to be minor when changing the rate from 4.9 to 7.3 m/h. This bears an economic point of view. Such change encourages the operation of the existing Iraqi water treatment filters at higher rates above the current rate of 3-5 m/h. Consequently, more furnished water will be produced, and a thinking of future extension of water works can be done at minimum costs.

However, when filtration rate becomes 9.8 m/h, some decrease in performance is noted, and water quality declines for all used filters. The reasons behind water deterioration can be accounted as follows:

Increasing filtration rates means an increase of collisions between suspended matter and the media. At the same time, it increases the hydraulic shear force. The latter tends to push the particulate matter deeper into the filter, where ultimately it emerges in the effluent and thus deteriorates its quality (Al-Najjar, 2000). However, the turbidity of the effluent for all rates remained below 1.0 NTU (except for sand filter).

Generally speaking, the competition among tried dual filters is so great, however, from an economical point of view, the focus will be made on GNR filter as it is locally available.

Runtimes

The elapsed time after backwash to the termination of a filter cycle is called length of run of that filter, or simply runtime. This parameter is dependent on many factors such as size and depth of the media, raw water characteristics, filtration rates, and other factors. Since this paper keeps the size and depth of the media constant for all runs, the effect of filtration rate on runtime will be discussed.

The fluctuation of filter runtimes at various rates is listed in Table 7. It is clearly evident that the lengths of filter runs are almost inversely proportional to filtration rates. This does not mean that doubling the rate from 4.9 to 9.8 m/h will necessarily reduce the runtime of filter cycle to half its original length of run. This is clearly shown in Figure 3. It is important to mention here that this runtimes are achieved after termination of filter cycles. The latter occurs as the head loss attains 80 cm, which is considered maximum value due to the used filter column.

Here GNR filter shoes longer runtime compared to other filters.

Table 7. Role of influent turbidity and filtration rate on filter runtimes.

Filtration rate m/h	Influent turbidity NTU	Filter runtime(h)			
		Sand + ninivite	Sand + activated carbon	Sand + anthracite	Sand alone
4.9	3.6	27	27	25	22
	5.6	26	26	24	20
	9.9	24	24	22	18
7.3	3.6	27	26	24	21
	5.6	26	24	22	19
	9.9	23	23	21	17
9.8	3.6	26	25	23	17
	5.6	24	23	20	15
	9.9	20	19	17	12

Table 8. Bacterial count of treated water*.

Filtration rate m/h	Bacterial count In the Filtrate influent (no/ml)	Bacterial count in the effluent and its percentage removal			
		Sand + ninivite	Sand +activated carbon	Sand +anthracite	Sand only
4.9	1300	60(95.4%)	50(96.2%)	70(94.6%)	120(90.8%)
7.3	1600	80(95%)	80(95%)	100(93.8%)	170(89.4%)
9.8	1400	90(92.6%)	80(94.3%)	120(91.4%)	170(87.9%)

* ( ): denotes percentage removal.

Bacterial removal

Table 8 lists bacterial count estimated at influent and effluent of the filter at different filtration rates. The removal efficiency ranges from 92 to 95.4% which is fairly high. Production of such water will provide complete satisfaction related to the biological quality of the treated water.

Conclusions

This work comes out with the following findings:

1. Ninivite - a locally available introduced material - has acquired good characteristics that enable it to be used as an upper layer in dual media filter in water treatment works.
2. A very clear furnished effluent that is palatable to consumers is produced. This is very vital from health point of view.
3. Great savings can be achieved represented by avoiding construction of new plants to meet increasing demands on water. Use of higher rates of filtration can yield greater plant output of water.
4. Further considerable economic consequences are expected when using ninivite. This includes avoiding importing upper coal or carbon layer material, as well as duplicating plant water productivity (compared to current city water treatment plants) and producing high quality water.
5. Turbidity as low as 0.50 NTU or less can be obtained, and a performance efficiency as high as 92 % can be achieved using ninivite.
6. Ninivite considerably minimizes bacterial count in the effluent water.
7. Dual - media filters consisting of 20 cm of ninivite over 40 cm of sand appear to fairly work under various conditions of turbidity loadings, runtime length, and filtration rates.
8. The study highly recommends using this Ninivite in water treatment works for its many merits.

REFERENCES

Al- Ani M, Al -Layla M, Al- Rawi SM (1988)." Selection of the size & depth of dual media filters in treating Tigris river water" UK. AQUA 6: 328-337

Al-Naqib SQ, Al- Dabbagh TH (1990). "Some physical and geotechnical properties of the new rock type Ninivite" Proc. of the 26th annual UK 9- 13 September.

Al- Rawi SM (1987). Turbidity removal of drinking water by dual - media filtration Mosul Univ., Iraq. MSc. Thesis.

Al-Najjar KA (2000): An Investigation of using locally Ninivite rocks in water treatment" Mosul University, M.Sc thesis.

Al-Rawi SM (1995). Water filtration units; existing and required status" Arabic J. Sci. Tunisia (text in Arabic). 26: 48-53

Amirtharaga A (1988). "Some theoretical & conceptual views of filtration "USA AWWA80:12 AWWA conference, June 19, NO(20028)

Aswad K, Amin M, Al -Naqib SQ (1995). "Marl- H2S interactions under surfacial oxidizing conditions" Dirasat J. Jordan. 22B: 6.

Degremont (1991). Water treatment handbook "France, 6th ed. 1. (chapter 5): 177-196.

Hendricks D (2006)." water treatment unit processes, physical and chemical" CRC publishing Book, USA. pp1266

Ministry of health (1988). Iraqi legislation for environmental protection, Iraq (text in Arabic).
Montgomery JM, John Wiley & Sons (1985). Consulting Engineers, Water treatment Principles & Design", USA.
Peavey HS, Rowe RR, Tchobanoglous G (1987). Environmental Engineering" McGraw Hill Book, YSA, 2nd ed.
Twort AC, Don D, Ratnayaka M, Brundt J (2001). "Water Supply" 5th edition, Chapter 7. Storage, Clarification, and Filtration of Water Butterworth Book Co. UK. pp. 276-316.
WHO Geneva (1984)." Guidelines for drinking water quality "

12

An economic assessment of the impact of information and communications technology (ICT) on performance indicators of water resource management in West Africa: A suggested strategy for avoiding the eminent international water wars

Sule Magaji and Eke, Chukwuemeka Ifegwu

Department of Economics, Faculty of Social Sciences, University of Abuja, Abuja Nigeria.

Using data from 16 West African countries, this paper examines the links between Per Capita Income, Trade and Financial indicators, Education and Freedom indicators. Others are internet users, broadband and mobile cell phone subscribers. Meanwhile fresh water supply (which is assumed as a bench mark public sector-led water resource management performance indicators) and access to safe drinking water (a bench mark private sector-led water resource management performance indicators) represents indicators of water resources management. The results show that income, information and communications technology (ICT) and government trade policies influence the efficient management of cross-country water resource. Freedom indicators strongly affect water resource management performance indicators (WRMPI). Internet Users, Broadband Subscribers, and Mobile cell phones subscribers have a positive association with WRMPI. However, contrary to wide spread expectations, education does not influence WRMPI. In areas where water resource management performance indicators of safe drinking water exhibited strong correlation are: Secondary school enrollment rate (0.57), fresh water supply with consumer price inflation (0.78) and a fair correlation of safe drinking water with corruption index.

Key words: Information and communications technology (ICT), resource management, safe drinking water.

INTRODUCTION

The diffusion of information communication technology has led to greater integration of economies around the world. An assessment of cross-country data however, reveals potential danger, as countries will definitely engage each other in water wars in the nearest future (Sule, 2003). It is common sense that with an increase in population, there will be a corresponding rise in the demand for fresh and safe drinking water.

Consequently, this water related crisis will fuel the tendency for governments of countries to strategise on ways to control water sources. There is a growing body of literature focusing on ICT tools, water resource management and its impact on the economy. The arguments presented by Swinford et al. (2007), Krantz and Kifferstein (2009), Shirley (2006) and Ribando et al. (1999) reveals that indicators of efficient water resource management has a significant positive impact on the gross domestic product (GDP) growth in varying degrees across the countries on the continent. Using simple descriptive statistics based on questionnaires, numerous studies have shown that ICT tools have greatly contributed to greater strategic management of the nations' water resources without necessarily escalating cross boarder tension (Van et al., 2003; Komsky et al., 2001; Swinford et al., 2007).

Based on these findings, it is plausible that in the era of the New Economy, new channels for assessing and improving water resource management is possible. Such a window may allow West African countries to fast track good governance in terms of cross country water resource and ensure more sustainable development. Thus it is important to conduct an economic assessment of the structural changes ushered in by the "New Economy" – described by Jorgenson (2007) as a mantra for faster, better and cheaper alternative systems and machines.

There is obviously a greater number of works on water economics, ICT and water resources and quality. Relatively, smaller works has been undertaken to understand in statistical terms, the impact of ICT and certain socio-economic indicators on water resources management (WRM), particularly in West African countries. For instance, Valipour (2014a) attempted to profile agriculture water management and identified platforms for this purpose. In focusing solely on West African economies, this work can add to the body of literature by addressing this salient issue. First, is an econometric assessment of the certain ICT and socio economic variables on WRM. Secondly, the direction of causality will be formally ascertained. Thirdly, other variables such as policy, civic society, and freedom can be crucial in this context.

Many papers have pointed out that water is plausible due to increasing demand on earth's water resources. More so, some other researchers have emphasized the catalysing role of citizen participation in the increased deployment of ICT (Heintz, 2004; Lagerskog, 2008; Moorhouse and Ellif, 2002; Jamieson and Fedra, 1996; Guimares et al., 2003; European Union, 2000; CET, 2003; Alder and Jacobs, 2000; Arnstein, 1969; CEC, 2001a). They argue that the increased participation of the public (through extensive use and deployment of ICT leads to more effective management of water resources as it broadens and guarantee speed of access to information and knowledge that can result in greater participation of people in decision making processes.

In Baliamoune-Lutz (2003)'s words "...a reduction in information asymmetry that enhances efficiency and access to knowledge for all would prevent one party from monopolizing opportunities ... and at the same time allow participation of previously excluded groups." However, the role of ICT in effectively combating cross-border challenges is yet to be empirically tested using data from West African countries. There are ambiguous conclusions concerning the link between some economic variables, ICT and water resources challenges in West African countries. Another challenging concern is the issue of causality. Can efficiently deployed ICT resources coupled with effective policy engineering on government's part and an engaging civil society forestall the emergence of cross country water wars? This is so because Valipour (2013) argued that water management is one of the most effective parameters towards achieving sustainable development in the world.

This paper examines the relationship between water resources management indicators, ICT and a set of time tested macroeconomic and policy variables for a sample of 53 African countries. These countries are listed in Appendix 1. Specifically, the paper explores the nature and direction of the links between WRM indicators and per capita income, trade and financial liberalization, literacy and education and freedom indicators including economic freedom, liberties and political rights. The fresh water supply, and safe drinking water are used as indicators of WRM bench marks in a cross country these two basis performance indicators as well as the set of macroeconomic and policy variables mentioned above.

DATA AND METHODOLOGY

In the words of Jorgenson and Stiroh (2002) "a new economy" is the mantra for technological and structural changes as individuals (once excluded from mainstream events) capitalize on new technologies, new opportunities and national investments in computing, information and communication technologies. Quah (2001) includes intellectual asset, electronic libraries and databases and biotechnology (Carbon – based libraries and database). On the other hand, Swinford et al. (2007) identified broadly some ICT-oriented performance indicators for water resource management: measures of openness, civil participation, accountability and trust. This paper focuses only on: WRM performance indicators (WRMPI) such as fresh water supply measured in KM^3/Yr across countries and the availability of safe drinking water, ICT indicators such as mobile phone subscribers per 100 inhabitants, broadband subscription per 10,000 inhabitants and Internet users per 10,000 inhabitant. The data on these variables are for the periods of 2007 unless otherwise stated, are taken from the website of World Water.Org. Fresh water supply may be viewed as an indicator of the state to control fresh water resources within its political boundary, while safe drinking water, the second performance indicator measures the access to bottle and sachet water usually private controlled. The model proposed is:

$$WRMPI_i = \alpha + g\prod + \varepsilon_i \qquad (1)$$

Where the vector $\prod$ contains all indicators. That is: $WRMPI_i$ = Fresh Water Supply/ Safe Drinking Water.

Table 1. Correlation among ICT-water resource management performance indicators.

Variables	SDW/POP	In PCI	BS	IPI	FWS(KM^3/YR)	SER	PR	HDI	CPI	CI	POP	SDW
SDW/POP	1.000											
In PCI	0.212	1.000										
BS	0.382**	0,238	1.000									
IPI	0.656**	0.311*	0.408**	1.000								
FWS(KM^3/YR)	-0.138	-0.052	0.101	-0.221	1.000							
SER	0.306*	0.446**	0.461**	0.669**	-0.304*	1.000						
PR	0.215	0.120	0.059	-0.166	0.113	-0.170	1.000					
HDI	0.314*	0.072	-0.120	-0.456**	0.225	-0.362**	-0.057	1.000				
CPI	-0.076	-0.182	0.086	-0.148	0.776**	-0.175	0.033	0.164	1.000			
CI	0.210	0.386**	0.306*	0.470**	0.301*	0.496**	-0.387	0.015	-0.163	1.000		
POP	-0.211	-0.053	0.323*	0.010	0.301*	0.085	0.076	0.167	-0.057	0.057	1.000	
SDW	0.524**	0.305*	0.382***	0.524**	-0.322*	0.569**	0.275*	-0.362**	-0.177	0.473**	-0.127	1.000

*indicates significance at 0.1, ** indicates significance at 0.05,*** indicates significance at 0.01. Sources: Author's Computation, ICT indicators are from International Telecommunication Union Website (2008); indexes for Civil Liberties and Political Rights are Freedom House website (2008); Economic Freedom index is from the Heritage Foundation website (2008); Education Index is from UNDP – Human Development Report 2008; Standard and Poor's Global Water Index website (2008); all other variables are from World Development Indicators CD-ROM (World Bank, 2008).

I = per capita income, F = Index of Economic Freedom, R = Political Rights, L = Civil Liberties, SR = Secondary School Enrollment Rate, HI = Human Development Index, CI = Corruption Index, CPI = Consumer Price Index, BS=Broadband Subscribers, IU=Internet Users, MCS=Mobile Cell Phone Subscribers.

Model (1) represents an equation to be estimated and the assumption of FWS and SDW as performance assessment indicators is fairly standard and plausible on both theoretical and empirical grounds (Swinford et al., 2007; Krantz and Kifferstein, 2009; Shirley, 2006; Ribando et al.,1999). In general, the association between WRMP indicators and income is expected to be very weak. This seems to be the case given the significant correlation of 0.31 between access to safe drinking water (Sachet/Bottle Water) and natural logarithm of per capita income. It is quite expected that income is not statistically significant with FWS as this is a natural resource endowment that has no link to income status of an economy. It is still not significant to the ratio of Fresh Water Supply and Safe Drinking Water to population (Table 1). However, in areas where water resource management performance indicators of safe drinking water are exhibited, strong correlation are: Secondary school enrollment rate (0.57), fresh water supply with consumer price inflation (0.78) and a fair correlation of safe drinking water with corruption index. WRM performance indicator safe drinking water corrects strongly with internet penetration index (0.52) but had a weak one though statistically too with broadband subscriber (0.38).

This paper also uses freedom indicator namely Index of Economic Freedom published by the Heritage Foundation. It scores an average score of 10 indexes measured on a one to five scale with 5 indicating the lowest level of economic freedom. These variables also assess trade policy, monetary policy, capital flows and foreign investment, wage and price control, banking and bureaucracy, government intervention in the economy and the fiscal burden of the government (taxes and expenditure). Others are political rights and civil liberties. This paper is adopting approaches by Norris (2000) and Baliamoune-Lutz (2003). These two indexes are published by Freedom House and measured on a one to one seven scale with 7 indicating lowest degree of freedom. The correlation safe drinking water with Liberties and political rights are (SDW: PR = -0.28, and SDW: CL = -0.36). The signs on the coefficient are negative, implying that a fall in the index (an improvement) is associated with an increase in access to SDW. In discussing other policy variables such as financial liberalization and international trade variables, Baliamoune-Lutz (2003) defined financial deepening as the ratio of broad money (M2) to GDP is used as a proxy for financial liberalization. Eke (2007) and Baliamoune-Lutz (2003) argued that with increased financial deepening banks strive to make information available to their customer and generate additional income from service charge. Therefore increased deepening would empower consumers more. Many economic opportunities would be explored. This implies that financial deepening should spur establishment of more bottling water firms, civic action initiatives on better water management policy and strategies.

The correlation coefficient in Table 1 shows that there a negative (though not statistically significant correlation between financial liberation core WRM performance indicators. (FWS: FL = -0.042; SDW: FL = -0.11). This discovery could be that since (M2) and GDP were measured in dollars terms, most West African countries have experience stiff decline (in dollar terms) over the years due to weakening exchange rates. The sensitivity of these countries water industry to trends in the international market was assessed using the net financial flows (NFF),

Table 2. An Economic assessment of the impact of ICT on performance indicators of water resource management in west africa dependent variable: FWS KM^3/Yr.

Equations	1	2	3	4
Number of observations	53	53	53	53
Adjusted R^2	0.651	0.160	0.129	0.661
Constant	35.366	23.008	-24.260	52.252
PCI	0.0005(0.009)	0.0003(0.014)	0.0009(0.013)	0.0007(0.009)
IEF	0.331(0.290)	-0.0004(0.435)	-0.0002(0.432)	0.349(0.301)
PR	10.665(23.643)			0.588(25.055)
CL	-8.000(31.381)	18.669(21.895)	23.364(19.707)	-4.781(32.682)
SER	-1.257(0.911)	-1.208(1.396)	-1.481(1.239)	-1.642(1.072)
HDI	0.009(0.510)	0.706(0.776)	0.575(0.720)	0.0002(0.542)
CI	-3.839(28.718)	-19.157(44.073)		-4.780(31.906)
CPI	2.207(0.278)			2.157(0.295)
NFF			0.0005(0.012)	0.0005(0.009)
BS				0.0002(0.292)
IU				-0.0004(0.018)
MCS 2007				0.0002(0.003)

Standard Errors in parentheses, Source: ICT indicators are from International Telecommunication Union Website (2008); indexes for Civil Liberties and Political Rights are Freedom House website (2008); Economic Freedom index is from the Heritage Foundation website (2008); Education Index is from UNDP – Human Development Report 2008; Standard and Poor's Global Water Index website (2008); all other variables are from World Development Indicators CD-ROM (World Bank, 2008).

however, this is not statistically significant.

However, House (1999) observed the role of ICTs in enabling greater access to water related issues/information. More so, Baliamoune-Lutz (2003) argued that ICT also foster the development of NGOs and information and Knowledge based communities that are more capable (relative individual citizen) to cause institutional changes. Consequently we expected efficient management of our cross border water resources as our stock of ICT increases in depth and breath. ICT diffusion may also affect the degree of effectiveness of the civil society and structures of the water industry across West African countries. For example, because water consumers have access to indexes and statistics of water coupled with market prices on World Wide Web, they can quickly organize around a problem, mount pressure and initiate a change process. The monopoly, strict control/limited access to knowledge hoisted on the populace by most West African government would become a thing of the past. Competition in the safe drinking water industry would amount to efficient use of resource and civic society watchful eyes through an active online community would act as a deterrent to waste and government excesses. The coefficient of correlation between Fresh water supply, safe Drinking Water is weak and highly significant (Table 1) this may reflect increased integration of these indicators.

DISCUSSION OF EMPIRICAL RESULTS

Equation (1) results from the model exploring the factors that influence WRM performance indicators are represented in Table 2 (Equations 1 to 4). To test the robustness of the model, four equations were estimated. Table 2 displays the statistical results from estimating the model with Fresh Water as the relevant WRMPI variable. Equation 1 uses consumer price inflation, political rights, index of economic freedom, per capita income, human development index, corruption index, secondary school enrolment rate and civil liberties in Equation (2) we excluded consumer price inflation and political rights. In Equation (3) we excluded corruption index and brought in net financial flows while in Equation (4), ICT indicators – Internet Users, broadband subscribers and mobile cell phone subscribers were brought in. Only variable that exhibits high significance will be discussed and estimate of all models is estimated using SPSS (Version 11.00).

In Table 2, Equation (1) displays the estimates using fresh water supply as the WRMP indicators. The empirical results shows these variables are strongly significant policy variables, per capita income, political rights, and liabilities, human development index and corruption index. More so, fresh water supply is assumed to be in the government's exclusive control. The variables that returned as not significant are index of economic freedom and secondary school enrolment rates. Effective management of our water is done by experts and politicians probably. Secondary school enrollment rate, theoretically seen as crucial part, does not really add the needed value statistically. This is so because in most West African Countries you do not need much education to get involve politics. However, finding of a negative effect may suggest there is a link.

Interestingly as expected all ICT indicators reported very significant (Table 2, Equation 4) from the theoretical point view increased broadband subscribers should boost WRMPI. Having a negative effect could suggest that some members of the elite that have laptops and wireless

Table 3. An Economic Assessment of the Impact of ICT on Performance Indicators of Water Resource Management in West Africa Dependent Variable: SDW.

Equations	1	2	3	4
Number of observations	53	53	53	53
Adjusted R^2	0.347	0.355	0.344	0.308
Constant	69.579	68.754	68.433	73.216
PCI	0.0006(0.001)	-0.0006(0.001)	-0.0006(0.001)	0.0004(0.001)
IEF	0.002(0.031)	-0.0002(0.031)	-0.020(0.031)	0.020(0.032)
SER	0.212(0.099)	0.204(0.097)	0.199(0.098)	0.129(0.117)
HDI	0.008(0.055)	-0.084(0.055)	-0.086(0.055)	-0.090(0.058)
CI	4.729(3.107)	4.901(3.076)	5.217(3.159)	4.959(3.438)
CL	-4.088(3.396)	-2.094(1.536)	-2.075(1.549)	-3.616(3.521)
PR	1.687(2.558)			0.768(2.714)
CPI	-0.0002(0.030)	0.007(0.029)	-0.007(0.029)	-0.003(0.032)
NFF			-0.0005(0.001)	0.0001(0.001)
MCS 2002				0.0018(0.002)
MCS 2007				-0.0014(0.000)
BS				-0.046(0.047)
IU				0.0001(0.002)

Standard Errors in parentheses, Source: ICT indicators are from International Telecommunication Union Website (2008); indexes for Civil Liberties and Political Rights are Freedom House website (2008); Economic Freedom index is from the Heritage Foundation website (2008); Education Index is from UNDP – Human Development Report 2008; Standard and Poor's Global Water Index website (2008); all other variables are from World Development Indicators CD-ROM (World Bank, 2008).

web access are yet not actively involved in water issues. If there is an increase in number of subscribers in the future then there could be a point where safe drinking water would be solely private sector controlled. The segment of the industry would be comprised of bottle and sachet water companies, private water distributors, and bore hole drilling firms.

In Table 3, the R-squared is fairly low compared to Fresh Water which is very low. Access to safe drinking water appears to be most influenced by internet Users. In addition, results indicate that factor such as secondary school enrolment rates, corruption index, civil liberties; human development index and political rights are not statistically significant. This is not surprising as corruption in the water sector is negligible. Pricing data and orders placed on the net or otherwise distributed are basically the essential ingredients that is expanding this industry in Africa. Competition is the order of the day in countries such as Ghana and Egypt. Since this is private sector led and statistics seems to solidly support the idea; market oriented variable such as per capita income index of economic freedom, consumer price inflation, net financial flows, political rights, mobile phone subscribers, internet users were strongly significant.

If you compare the relevant variables in Table 3 (private sector) to Table 2 (public sector) you will notice that the difference in policy is statistically relevant. For instance, net financial flows and index of economic freedom were not significant variables in influencing fresh water supply but were highly significant in influencing access to safe drinking water which within the domain of the private sector as against the former assumed to be exclusively controlled by the various West African states. Also, comparing Equations 1 to 4 in Table 3, you notice that equation that has all policy (irrespective of the sector) provides the best results in terms of R-square, 0.481.

The impact of population dynamics on Fresh water supply informed the use of the ratio – FWS/POP. This indicates the tendency of nation states to go war on water issues due to pressures arising from population dynamics. Both equations capture the process and show corruption index the most influential. Others are index of economic freedom, human development index and civil liberties. But human development index has a wrong sign (though statistically significant). It simply shows that improvement in human capital in West African countries does not necessarily translated into improving their water situation. This finding is similar to Broadband Subscribers notwithstanding, human capital and broadband subscribers must have positive roles in the effective management of fresh water supply but may be one possible reason for this anomaly. There could have been serious data measurement error.

In summary, the empirical results provide strong support for the role of ICT indicator as a major determinant of effective water resources management. This is consistent with the conclusion the studies of Van et al. (2003), Koinsky et al. (2001) and Swinford et al. (2007). Similarly all freedom indicators namely civil

Table 4. An Economic assessment of the impact of ICT on performance indicators of water resource management in West Africa Dependent Variable: FWS/POP.

Equations	1	2	3	4
Number of observations	53	53	53	53
Adjusted R^2	0.135	0.153	0.153	0.078
Constant	28.821	35.095	34.002	38.063
PCI	0.006(0.002)	-0.006(0.002)	-0.006(0.002)	0.006(0.002)
IEF	0.03(0.069)	-0.03(0.067)	-0.03(0.067)	0.030(0.070)
SER	-0.662(0.217)	-0.663(0.215)	-0.663(0.215)	-0.711(0.255)
HDI	0.06(0.122)	-0.07(0.119)	-0.07(0.119)	-0.070(0.128)
CI	0.661(6.639)	-0.150(6.776)	-0.150(6.776)	-0.328(7.418)
CL		-0.655(3.366)	-0.655(3.366)	-1.0256(3.615)
PR	0.371(2.583)			
CPI	-0.008(0.064)			
MCS 2002				-0.0008(0.004)
MCS 2007				-0.00009(0.001)
BS				-0.03(0.089)
IU				0.0002(0.004)

Standard Errors in parentheses, Source: ICT indicators are from International Telecommunication Union Website (2008); indexes for Civil Liberties and Political Rights are Freedom House website (2008); Economic Freedom index is from the Heritage Foundation website (2008); Education Index is from UNDP – Human Development Report 2008; Standard and Poor's Global Water Index website (2008); all other variables are from World Development Indicators CD-ROM (World Bank, 2008).

liberties, political rights and index of economic freedom influence the performance of WRM (Table 4).

Impact of selected economic and social development indicators on west water resource management

$$WRMPIi = \alpha + \delta' Z_i + \varepsilon_i \quad (2)$$

Where the vector Z contains economic and social indicators and ε_i is white noise. The only significant variable (though moderate) is fresh water supply followed by internet penetration index that showed p value. Meanwhile Equation (2) shows access to safe drinking water and internet penetration index returns significant (Table 5). In Table 6, Equation (1) report estimates of the impact of selected economic and social indicators as well as WRMPI on income (log transformation of per capita income). ICS, Inter Country scale is a dummy variable for representing UNDP's classification based on HDI as high, medium and low. In our case, 1 stands for countries grouped as medium and 0 for those that were grouped low.

These results indicates that an effective management of cross border and intra border water resource in both sectors of the economy increase in the subsector of cyber and broadband application, would lead to higher per capita income. From our selection the variable that seems to influence income are secondary school enrolment rate, human development index and inter country scale. The is plausible as secondary school enrolment in most West African economies is low. This will directly affect West Africa's Human Development Index which as a result is grossly low.

In theory education is expected to have a positive influence on management of resources. As consumer of sachet and bottle water, for instance, have access to more learning (post primary) it should in turn have a huge impact on politicians who majorly are in charge and monopolize decision making processes. However, the empirical literature presents another picture. In most West African countries school enrolment is quite low. More so, those that enrol do graduate. On the other hand having successful private sector led water firms may not necessarily mean that they are run by secondary school graduates.

In many developing countries, most powerful politicians are ex-military men with little or no formal education. The results in Table 7 seem to support this view. There is no empirical evidence in support of the influence of education on water resource management. Because economic freedom diffusion of ICT and efficient macroeconomic policies improves the changes of managing our water better thereby averting water wars, it is expected to foster economic development in an unprecedented scale as water permeates all facets of the society and economy. This is also supported in Table 8.

Concluding comments

This paper examined the relationship between selected social and economic development, information

Table 5. An Economic Assessment of the Impact of ICT on Performance Indicators of Water Resource Management in West Africa Dependent Variable: SDW/POP.

Equations	**1**	**2**
Number of observations	53	53
Adjusted R^2	0.089	0.093
Constant	138.061	35.095
PCI	0.02(0.009)	-0.006(0.002)
IEF	-0.258(0.301)	-0.03(0.067)
SER	1.062(1.085)	-0.663(0.215)
HDI	-0.361(0.540)	-0.07(0.119)
CI	5.404(31.816)	-0.150(6.776)
CL	-18.235(32.584)	-0.655(3.366)
PR	0.08(25.118)	
CPI	-0.04(0.296)	
NFF	0.006(0.010)	
MCS 2002	0.003(0.020)	
MCS 2007	-0.004(0.004)	
BS	-0.036(0.433)	
IU	0.01(0.020)	

Standard Errors in parentheses, Source: ICT indicators are from International Telecommunication Union Website (2008); indexes for Civil Liberties and Political Rights are Freedom House website (2008); Economic Freedom index is from the Heritage Foundation website (2008); Education Index is from UNDP – Human Development Report 2008; Standard and Poor's Global Water Index website (2008); all other variables are from World Development Indicators CD-ROM (World Bank, 2008).

Table 6. An Economic Assessment of the Impact of ICT on Performance Indicators of Water Resource Management in West Africa. Dependent Variable: PCI.

Equations	**1**	**2**
Number of observations	53	53
Adjusted R^2	0.374	0.377
Constant	3.038	2.504
IEF	0.008(0.001)	0.008(0.003)
SER	0.031(0.010)	0.027(0.010)
HDI	0.013(0.005)	0.013(0.005)
ICS	1.513(0.889)	1.413(0.900)
IPI	0.037(0.039)	0.031(0.040)
SDW		0.099(0.014)

Standard Errors in parentheses, Source: ICT indicators are from International Telecommunication Union Website (2008); indexes for Civil Liberties and Political Rights are Freedom House website (2008); Economic Freedom index is from the Heritage Foundation website (2008); Education Index is from UNDP – Human Development Report 2008; Standard and Poor's Global Water Index website (2008); all other variables are from World Development Indicators CD-ROM (World Bank, 2008).

communication technology and water resource management performance indicators, ICT are a major determinant of water resource management performance indicator. It influences civil societies providing a veritable platform for actively sharing information on an international scale and engaging government agents to further entrench the culture of excellence and corporate governance. These platforms enable West African economies imbibe international best practices that evolve over time thereby forcing respective governments to be accountable and transparent which are hallmarks of the ongoing anti water war campaigns. Secondly, there is a strong influence of macroeconomic policies on the sector. Third, freedom improves the water resource management. Fourthly, empirical results support the notion that effective management of our water resources

Table 7. An Economic Assessment of the Impact of ICT on Performance Indicators of Water Resource Management in Africa. Dependent Variable: SER.

Equations	1	2
Number of observations	53	53
Adjusted R^2	0.468	0.093
Constant	35.422	35.095
IEF	0.048(0.045)	-0.03(0.067)
SER	1.062(1.085)	-0.663(0.215)
HDI	-0.136(0.072)	-0.07(0.119)
ICS	4.419(13.410)	-0.150(6.776)
IPI	2.279(0.486)	
FWS KM^3/yr	-0.019(0.015)	

Standard Errors in parentheses, Source: ICT indicators are from International Telecommunication Union Website (2008); indexes for Civil Liberties and Political Rights are Freedom House website (2008); Economic Freedom index is from the Heritage Foundation website (2008); Education Index is from UNDP – Human Development Report 2008; Standard and Poor's Global Water Index website (2008); all other variables are from World Development Indicators CD-ROM (World Bank, 2008).

Table 8. An Economic Assessment of the Impact of ICT on Performance Indicators of Water Resource Management in West Africa Dependent Variable: CL.

Equations	1	2
Number of observations	53	53
Adjusted R^2	0.119	0.033
Constant	7.385	5.066
IEF	-0.012(0.003)	-0.0009(0.003)
HDI	-0.007(0.005)	-0.006(0.005)
ICS	1.247(0.894)	0.851(0.033)
IPI	-0.034(0.035)	-0.655(3.366)
SDW	0.034(0.013)	
FWS KM^3/yr		0.0013(0.001)

Standard Errors in parentheses, Source: ICT indicators are from International Telecommunication Union Website (2008); indexes for Civil Liberties and Political Rights are Freedom House website (2008); Economic Freedom index is from the Heritage Foundation website (2008); Education Index is from UNDP – Human Development Report 2008; Standard and Poor's Global Water Index website (2008); all other variables are from World Development Indicators CD-ROM (World Bank, 2008).

will help to develop West Africa's economy. Also, enhance political rights and civil liberties. It must be pointed out here that in West African countries, based on statistical analysis, education is not associated with effective water resources management.

The present finding seems to offer some new strategies embedded in an ICT – oriented culture for managing our nation's water resources. It would also compliment concerted efforts to boost irrigation farming in West Africa as Valipour (2014b) observed that the respective government should encourage irrigation farming. There is ample evidence that managing our water resources effectively can provide an additional source of economic growth. On the other hand, this paper provides crucial insights about those ICT, economic and social development variables that are important parts of WRMPI.

Conflict of Interest

The authors have not declared any conflict of interest.

REFERENCES

Alder J, Jacobs M (2000). Citizens and Wetlands: Evaluating the Ely citizens' jury. Ecol. Econ. (34):217-232.

Arnstein SJ (1969). A ladder of citizen participation. J. Environ. Insitute of Planners (35):216-224.

Baliamoune-Lutz M (2003). "An Analysis of the determinants and effects of ICT diffusion." Information Technology for Development 10:151 – 169. IOS press.

CEC (Commission of the European Communities, 2001a) European Governance: A White Paper; Com (2001) 428, Brussels, 27.7.2001

Eke CI (2007). "Electronic Commerce, Cyber Space & Mega Profits: Generating Profit Opportunities On the Internet through Banks using our Culture and Language." The First Annual Conference of the Department of Finance, University of Lagos. 3rd – 4th April, 2007.

European Union (2000). Water European Union Guidance on the Water Framework Directive: Active involvement, Consultation and Public access to information. EC WED Common Implementation Strategy. Working Group on Public Participation. Framework Directive. 2000/06/EC

Guimares P, Rinaudo A, Jeffrey JD, Blasques P, Corral QJS, Petit V (2003). ICT tools to support public participation in water resources governance and planning: Experiences from the design and testing of a multi media platform. J. Environ. Policy, Assessment and Management. 5(3):395-420.

Heintz TH (2004). 'Applying the Concept of sustainability to Water Resources Management'. Water Resources Update. 127:6-10, February

House M (1999) Citizen Participation in Water Management. Water, Sci. Technol. 40(10):125 -130.

Lagerskog A (2008). 'Water and Conflict in the Middle East'. *The Middle East Institute View Points.* No.7 June. Also available at www.mideeasti.org.

Jamieson DG, Fedra K (1996). The 'Water Ware' decision-support systems for river basin planning. Conceptual design. J. Hydrology 177:163 – 175.

Jorgenson DW, Stiroh KJ (2002). "Raising the Speed Limits: Economic Growth in the Information Age." Measuring and Sustaining the New Economy. Washington D.C: National Academy Press.

Jorgenson DW (2007). "The Emergence of the New Economy." In Enhancing Productivity Growth in the Information Age: Measuring and Sustaining the New Economy.

Krantz D, Kifferstein B (2009). 'Water pollution and society.

Moorhouse M, Ellif S (2002). Planning Process for Public participation in regional water resource planning. J. America Water Resources Associat.38(2):531.

Norris P (2000). 'The Global Divide: Information Poverty and Internet Access worldwide'

Shirley MM (2006). 'Urban Water Reform: What we know, What we need to know'.

Sule B (2003). Water Security: Now and the Future. Being the 65th Inaugural Lecture delivered at the University of Ilorin, Ilorin, Nigeria.

Swinford A, McIntosh B, Jeffrey P (2007). 'Supporting the Strategic Objectives of Participative Water Resources Management; an Evaluation of the Performance of Four ICT Tools.'

Quah D (2001). 'Technology and Economic Growth: Some lessons for the New Economy'.

Standard and Poor's Global Water Index website (2008). https://us.spindices.com/indices/equity/sp-global-water-index

Valipour M (2013). Evolution of Irrigation-Equipped Areas as Shared of Cultivated Areas.

Valipour M (2014a). Land Use policy and agricultural water management of the previous half century in Africa. Available online

Valipour M (2014b). Future of Agricultural Water Management in Africa.

Water resources utilization in Yenagoa, Central Niger Delta: Environmental and health implications

Koinyan A. A.[1], Nwankwoala H. O.[2] and Eludoyin O. S.[1]

[1]Department of Geography and Environmental Management, University of Port Harcourt, Nigeria.
[2]Department of Geology, College of Natural and Applied Sciences, University of Port Harcourt, Nigeria.

This study examined the groundwater in relation to the surface water quality analysis in Yenagoa Local Government Area (LGA), Bayelsa State, Central Niger Delta. Groundwater samples were collected from the boreholes in the randomly selected communities namely Igbogene, Akenfa, Edepie, Swali, Biogbolo and Amarata and surface water samples from a point in Epie Creek that flows through the selected communities. These water samples were subjected to laboratory analyses for pH, conductivity, colour, total dissolve solids (TDS), total hardness, alkalinity, nitrate, sulphate, calcium, lead, copper, Total hydrocarbon content (THC), total coliform and faecal coliform. Descriptive and inferential statistics were used to analyze and explain the results obtained. The result showed that the pH in the study area was slightly alkaline with a mean of 7.3. The conductivity level of groundwater in the area was not significantly varied but it was very high in Igbogene and Biogbolo communities with 35 and 31 μS/cm, respectively. The mean values of considered parameters in groundwater were generally lower than that of surface water but the values in both were lower than the standard of World Health Organization (WHO). The results also showed that 78.3% agreed that water from both groundwater and surface water is used for domestic purpose while 18.3% for agricultural purpose. The study revealed that there is a positively direct and strong relationship between water treatment and occurrence of diseases. It is recommended that residents be educated more on groundwater and surface water management and laws should be promulgated as well as adoption of participatory management approach.

Key words: Water resources, water quality, utilization, health, Yenagoa.

INTRODUCTION

The principal goal of water management globally is to assess and manage the water resources that are available and this is not in exemption of the developing countries. Where groundwater is the main resource, management requires information on both its quantity and quality (Mogheir and Singh, 2002).

Groundwater continues to serve as a reliable source of water for a variety of purposes, including industrial and domestic uses and irrigation. The use of generally high quality groundwater for irrigation (which is largely indifferent to water quality) dwarfs all other uses (Burke, 2002). In many developing country settings, reliance has turned to dependency and the establishment of perceptions of access and use that are intensely 'private', irrespective of the legal status of the groundwater. However, groundwater and the aquifers that host it are inherently vulnerable to a wide range of human impacts.

Thus, it is understood that ground water has its standard quality and has a tendency to be altered or polluted. The pollution of groundwater can occur naturally, from environmental changes or from certain circumstances and activities carried out by man in relation to the location of the ground water. Christopherson (2002), stated that pollution can enter ground water from industrial injection wells(wastes pumped into the ground), septic tanks outflow, seepage from hazardous-waste disposal sites, industrial toxic wastes sites, agricultural residues (pesticides, herbicides, fertilizers) and urban waste landfills. All of the sources mentioned above are basically man-made. In groundwater pollution, the main contaminants of concern include petroleum hydrocarbon, chlorinated organic compounds such as trichloroethylene, heavy metals such as lead, zinc and chromium, and inorganic salts (Philip et al., 1994). Many of these industrial operations release various toxic heavy metals in their effluents. Heavy metals, arsenic and arsenical compounds, barium, chromium, lead, cadmium, manganese, zinc and other harmful materials are known to cause serious health effects (Nwankwoala et al., 2011). Such health problems includes but not limited to acute and chronic respiratory problems, gastro-intestinal tract infection, cardiovascular system disorder, nervous system and blood forming organisms malformation cases of skin and lung cancer (Horsfall et al., 2001). These harmful heavy metals and compounds have been found in traces as well as substantial amount in ground water. The presence of these substances is due to industrial activities and solid waste disposal sites located within such areas. Various households and industrial waste material are found in many urban wastes open dumps and garbage sites. Usually they are not properly located and managed, and as such the liquid residue (leachates) percolates into the soil, after heavy rainfall or long-term decomposition resulting in the pollution of ground water.

The effect of the contact and consumption of contaminated and polluted groundwater has become a cause for concern. This resultant effects of serious health issues and diseases to man, plant and animal developmental/growth problem is on the increase. Therefore, this has prompted the need for a study on the spatial analysis of quality of groundwater in Yenagoa LGA of Bayelsa State, Nigeria. The study therefore focused on the use of certain water variables to evaluate the spatial variations in the quality of groundwater in the study area. Nevertheless, the study suggested participatory ways to improve and maintain good quality of water in the study area.

Study area description

The study area is Yenagoa Local Government Area (LGA) in Bayelsa State of Nigeria. Yenagoa LGA is geographically located within latitudes 4°49'N and 5°23'N and also within longitudes 6°10' E and 6°33'E (Figure 1). Yenagoa City is the capital of Bayelsa State and Yenagoa LGA. The city is located on the banks of Ekole Creek and Nun River; the latter being one of the major river courses making up the Niger Delta's river. Yenagoa is the northernmost city of the state's significant population centre.

The study area is located within the lower delta plain believed to have been formed during the Holocene of the Quaternary period by the accumulation of sedimentary deposits. The major geological characteristic of the state is sedimentary alluvium. The study area is situated within the lower floodplain of the Niger Delta. The terrain is poorly drained with a gentle syncline to the Gulf of Guinea in a southwestern direction. The majority of the communities within the local government area are completely surrounded by wetlands, generally less than 5 m above sea level. The main features of the drainage in the study area include the meandering of the distributaries of the River Niger which makes the place to be prone to flooding.

METHODS OF STUDY

This study involved the collection of water samples from boreholes and surface water within Yenagoa metropolis. Water samples were collected from across the study area (Figure 1) into a well-labeled polyethylene bottles. Meanwhile, the bottles were properly rinsed with the borehole water to be sampled before the main water sample was collected in order to avoid contamination. Water samples were then subjected to laboratory analyses on both the cations and anions. The considered cations include Calcium (Ca^{2+}), while the anions included Cholride (Cl^-), Nitrate (NO_3^-) and Sulphate (SO_4^{2-}). Properties such as electrical conductivity (EC) pH, temperature and total dissolve solids (TDS) were also considered for the analysis. Analyses of groundwater samples were carried out using standard methods as suggested by the American Public Health Association (1989). Cations were analysed using an Atomic Absorption Spectrophotometer (Perkin – Elemer AAS 3110) and the anions using the Colorimetric method with the UV- Visible Spectrophotometer WPAS 110. Standard solutions and blanks were commonly run to check for possible errors in the analytical procedures. Generally, the processes controlling the chemistry of the groundwater were identified by the systematic study of hydro-chemical data. Global Positioning Systems (GPS) was used to record the latitudes and longitudes of each sampled borehole. The location readings enabled the mapping of the boreholes in their respective locations.

RESULTS

Table 1 present the physico-chemical characteristics of groundwater in Yenagoa in relation to the World Health Organization (WHO) standard. It was discovered that pH of the ground water was slightly acidic in Akenfa and Edepie while it was slightly alkaline in the remaining sampled communities. The pH fell within the standard of WHO (2006). The conductivity level of ground water in the area was not significantly varied but it was higher in Igbogene and Biogbolo communities with 35 and 31 µS/cm conductivity level respectively than in other areas. Total dissolved solid (TDS) was highest in Igbogene with 66 mg/L while the least (48 mg/L) was experienced in Akenfa. The TDS in Yenagoa was generally low. These

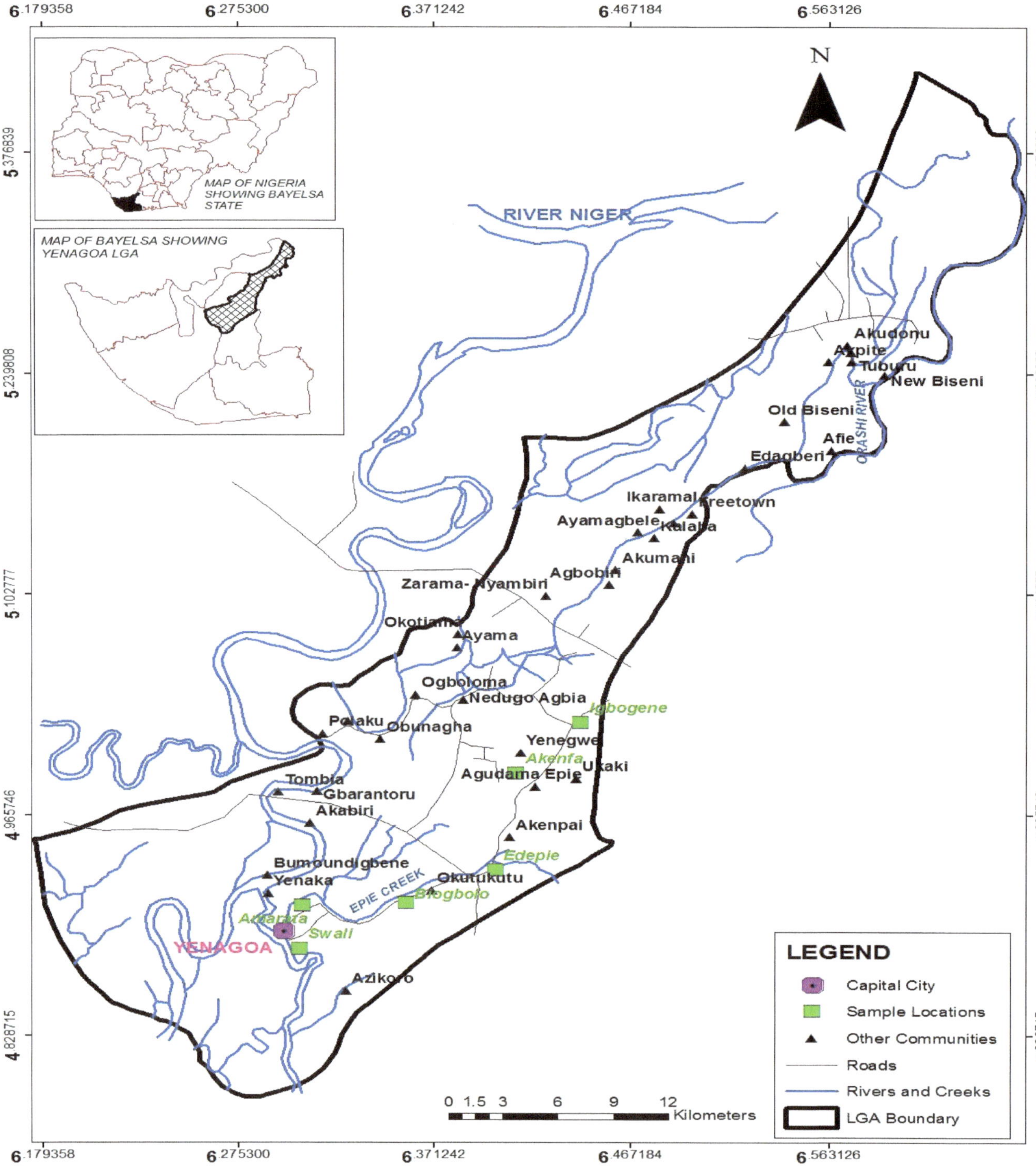

Figure 1. Map of Yenagoa showing sample locations (Inset: Map of Niger Delta showing the study area).

values are indications of the inland nature of the aquifers and their remoteness from the influence of any saline intrusion. The hardness of the borehole water samples ranged between 34 and 154 mg/L. The highest was observed in Amarata with 154 mg/L and the least in Edepie with 34 mg/L. The hardness was 137 and 133 mg/L in Igbogene and Akenfa, respectively. The hardness still generally was lower than the WHO standard. The levels observed could be attributed to the leaching of hardness enhancing species like magnesium and calcium, which abound in the soil systems around the study area. The alkalinity of the water samples

Table 1. Physico-chemical and microbiological quality of groundwater samples.

S/N	Parameter	Units	Igbogene	Akenfa	Edepie	Swali	Biogbolo	Amarata	WHO limits
1.	pH		7.8	6.8	6.9	7.5	7.6	7.7	6.0 – 8.5
2	Conductivity	µS/cm	35	23	26	24	31	23	-
3	TDS	mg/L	66	48	55	50	61	63	500
4	Colour	Pt-Co	0.2	0.3	0.8	0.4	0.6	0.5	5.0
5	Total hardness	mg/L	137	133	34	68	51	154	500
6	Alkalinity	mg/L	25.3	22.4	28.0	20.4	21.6	30.2	128-136
7	Chloride (Cl)	mg/L	4.3	3.2	5.6	1.5	1.8	2.4	250
8	Nitrate (NO_3)	mg/L	0.03	0.005	0.01	0.02	0.08	0.05	10
9	Sulphate (SO_4)	mg/L	0.002	0.009	0.004	0.006	0.050	0.005	25
10	Lead (Pb)	mg/L	0.001	0.013	0.013	0.00	0.01	0.02	0.05
11	Iron (Fe)	mg/L	0.003	0.073	0.005	0.035	0.055	0.001	0.3
12	Copper (Cu)	mg/L	0.016	0.092	0.002	0.105	0.077	0.006	1.0
13	Calcium (Ca)	mg/L	0.023	0.011	1.22	0.54	0.056	0.95	25
14	THC	cfu/ml	15000	30000	400000	10000	200000	30000	400
15	Total coliforms	cfu/100 ml	200	100	500	400	800	100	<10
16	Faecal coliforms	cfu/100 ml	30	27	65	30	85	70	0

Table 2. Sources of water.

Source	Igbogene		Akenfa		Edepie		Swali		Biogbolo		Amarata		Total	
	F	%	F	%	F	%	F	%	F	%	F	%	F	%
Ground water	16	80.0	18	90.0	18	90.0	16	80.0	20	100.0	19	95.0	107	89.2
Rainfall	1	5.0	1	5.0	1	5.0	0	0.0	0	0	1	5.0	4	3.3
Surface water	3	15.0	1	5.0	1	5.0	4	20.0	0	0	0	0	9	7.5
Total	20	100.0	16	100.0	20	100.0	20	100.0	20	100.0	20	100.0	120	100

N.B: F- Frequency, % - Percent.

ranged between 20.4 and 30.2 mg/L and the highest was observed in Edepie. Similarly, the alkalinity level of ground water in the study area was lower to WHO standard (2006). The range for the chloride concentration in the study area was between 1.5 and 5.6 mg/L. The concentration was 4.3 mg/L in Igbogene while it was 3.2, 5.6 and 1.5 in Akenfa, Edepie and Swali, respectively. The nitrate concentration was generally low in Yenagoa. The concentration ranged between 0.005 and 0.08 mg/L. The sulphate concentration level in Igbogene was 0.002 mg/L, Akenfa 0.009 mg/L, Edepie, 0.01 mg/L and Amarata with 0.005 mg/L. These values were generally less than WHO (2006) which is 10 mg/L. Calcium was also generally low in the water samples in Yenagoa. It is therefore observed that Igbogene borehole had a value of 0.023 mg/L, 1.22 mg/L in Edepie and 0.56 mg/L in Amarata.

The concentration of heavy metals (Lead, Iron, and Copper) in groundwater samples revealed that they were generally moderate. Lead ranged between 0.00 and 0.02 mg/L while Iron ranged between 0.001 and 0.073 mg/L and Copper which ranged between 0.0016 and 0.105. Total Hydrocarbon (THC) was quite high in the study area whereby the highest was discovered in Edepie with 400,000 cfu/100 ml. The concentration of THC in Yenagoa was generally higher than the WHO standard. The total coliform was highest in Biogbolo with 800 cfu/100 ml and the least in both Akenfa and Amarata with the concentration of 100 cfu/100 ml. Mean while, these values were higher than the WHO standard. The fecal coliform in the water samples in the study area were generally high.

Sources of water in the study area

The result of the analysis from the respondents in the study area reveals the sources of water that people depended upon. Table 2 and Figure 2 reveal that 80% of total respondents depended on ground water in Igbogene while 5 and 15% depended on rainfall and surface water respectively in the same community. In Akenfa and

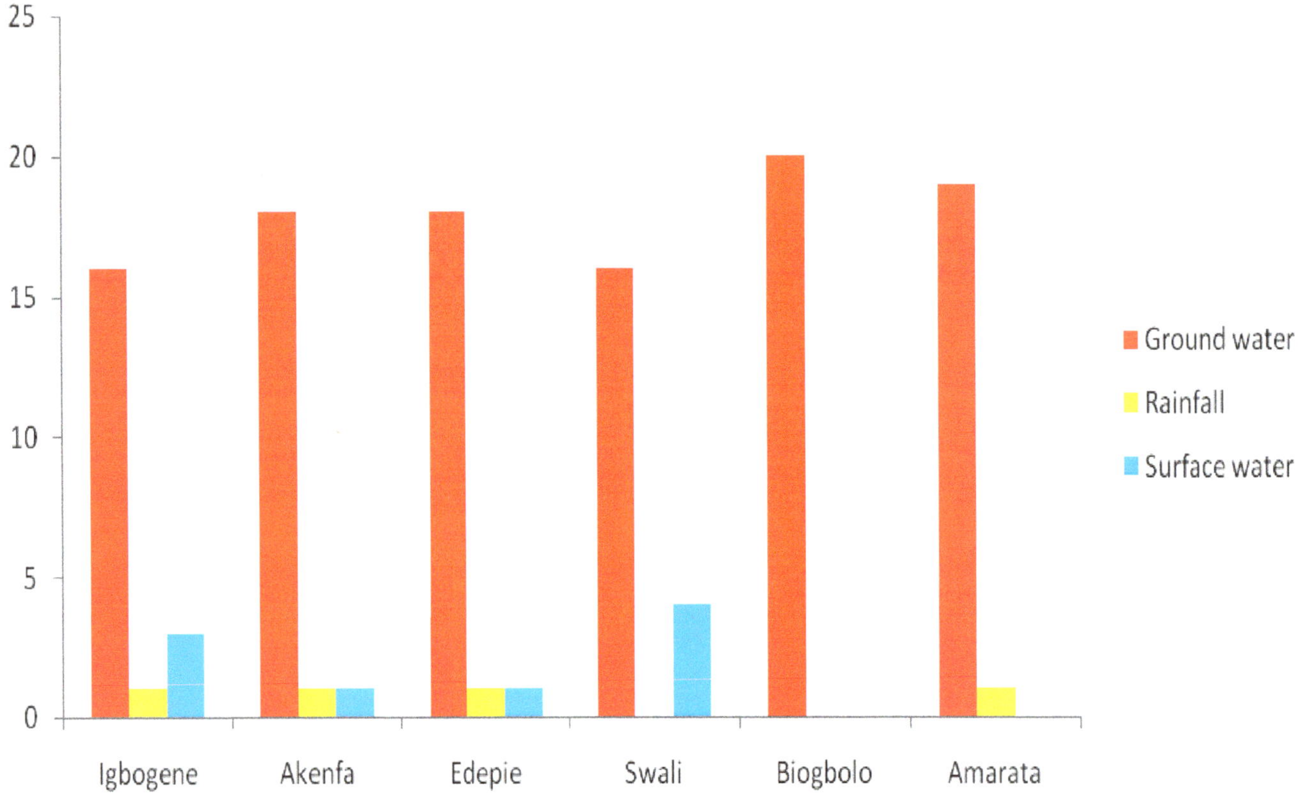

Figure 2. Sources of water in Yenagoa.

Edepie communities, 90% used borehole as the source of water while 5% each made use of rainfall and surface water. However, in Swali, 80% made use of groundwater while the remaining 20% depended on surface water. Similarly in Biogbolo and Amarata, 100 and 95% of total respondents depended on ground water. Generally, 89.2% of total respondents in the study area made use of ground water while 7.5 and 3.3% depended on surface water and rainfall. This shows that majority in the study area made use of borehole as their main source of their water.

Groundwater and surface water use in the area

Table 3, Figures 3 and 4 explains the different uses of ground and surface water in Yenagoa. In Igbogene, 65% of total respondents used water for domestic purposes while 25 and 10% used for agricultural and industrial purposes respectively. In Akenfa community, 80% used water for domestic purposes while 20% used it for agriculture. In Edepie, 70% used water for domestic purposes while 30% made use of water for agricultural purposes. In Biogbolo, 75% of total respondents used groundwater for domestic purposes while 15 and 10% used it for agricultural and industrial purposes. 85% of total respondents in Amarata community used water for domestic purpose while 15% used it for agricultural purpose. In all, 78.3% of total respondents used water for domestic purposes while 18.3 and 3.4% made use of water for agricultural and industrial purposes.

Water treatment methods

It is observed from Table 4 that people adopted three methods of treating water in the area namely boiling, filtering and addition of chemicals. In Igbogene, 10% of respondents agreed that boiling method is used to treat water while 15 and 5% believed in filtering and addition of chemicals. 70% of respondents in Igbogene do not treat water before use. In Akenfa, 5 and 5% treat water using boiling and filtering methods respectively while 90% do not treat water. In Edepie and Swali communities, 10 and 10% of respondent treat water using boiling and filtering methods respectively while 80% do not treat water before use. In Biogbolo, 15, 15 and 10% used boiling, filtering and addition of chemicals to treat water respectively while 60% do not treat water before use. In Amarata community, 20, 30 and 10% agreed to the use of boiling, filtering and chemical addition methods respectively for treating water while 40% do not treat water. In total, 11.7,

Table 3. Use of water in the study area.

Uses	Igbogene		Akenfa		Edepie		Swali		Biogbolo		Amarata		Total	
	F	%	F	%	F	%	F	%	F	%	F	%	F	%
Domestic	13	65.0	16	80.0	14	70.0	19	95.0	15	75.0	17	85.0	94	78.3
Agricultural	5	25.0	4	20.0	6	30.0	1	5.0	3	15.0	3	15.0	22	18.3
Industrial	2	10.0	0	0	0	0	0	0	2	10.0	0	0	4	3.4
Total	20	100	20	100	20	100	20	100	20	100	20	100	120	100

F- Frequency; % - Percent.

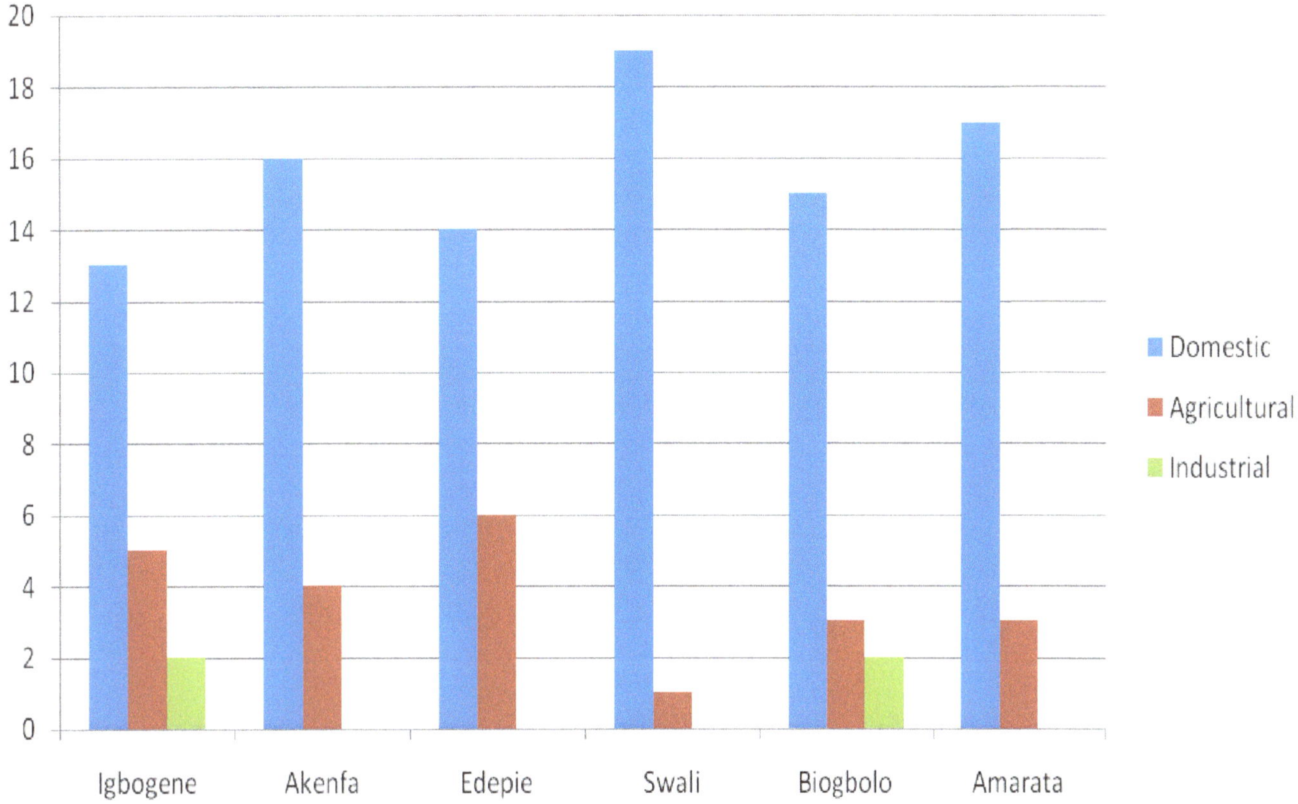

Figure 3. Water use among the selected communities.

14.2 and 4.1% believed that boiling, filtering and addition of chemicals are used to treat water respectively while 70% of total respondents do not treat water. These show that majority does not subscribe to water treatment in Yenagoa and this may pose serious impacts on their health status.

Impacts of water pollution on residents

The result of the analysis displayed in Table 5 presents the effects of water pollution on the health status of the residents in Yenagoa. In Igbogene, 35 and 20% believed that water pollution causes typhoid and diarrhea respectively while 45% believed that it causes skin irritation. However, in Akenfa, 25, 20 and 5% of total respondents perceived that pollutants in water can cause typhoid, diarrhea and cholera respectively while 50% perceived that it causes skin irritation. In Edepie, 25, 30, 15 and 30% of respondents perceived that water pollution leads to typhoid, diarrhea, cholera and skin irritation respectively. In Swali, 40% believed that water pollution results to typhoid while 25, 5 and 20% believed that water pollution cause diarrhea, cholera and skin irritation. 30% of respondents in Biogbolo believe that water pollution causes typhoid while 20% believed that it causes skin irritation.

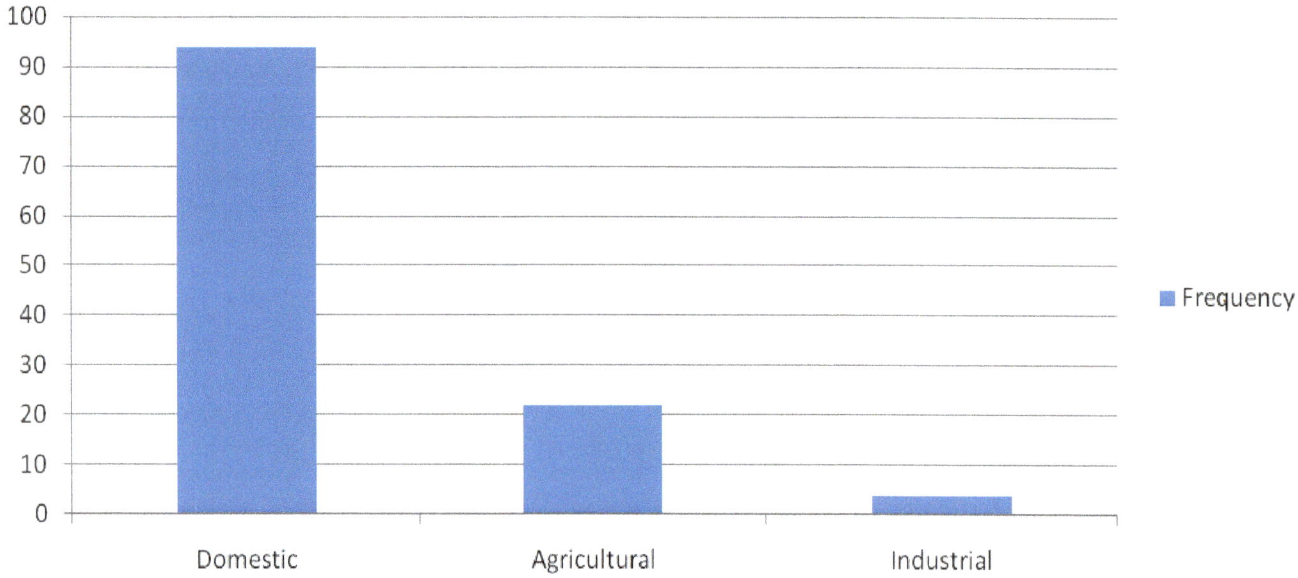

Figure 4. Water use in the entire study area.

Table 4. Water treatment methods in the study area.

Treatment	Igbogene		Akenfa		Edepie		Swali		Biogbolo		Amarata		Total	
	F	%	F	%	F	%	F	%	F	%	F	%	F	%
Boiling	2	10.0	1	5.0	2	10.0	2	10.0	3	15.0	4	20.0	14	11.7
Filtering	3	15.0	1	5.0	2	10.0	2	10.0	3	15.0	6	30.0	17	14.2
Addition of Chemicals	1	5.0	0	0	0	0	0	0	2	10.0	2	10.0	5	4.1
None	14	70.0	18	90.0	16	80.0	16	80.0	12	60.0	8	40.0	84	70.0
Total	20	100	20	100	20	100	20	100	20	100	20	100	120	100

N.B.: F- Frequency; % - Percent.

Table 5. Effects of water pollution on human health.

Health effects	Igbogene		Akenfa		Edepie		Swali		Biogbolo		Amarata		Total	
	F	%	F	%	F	%	F	%	F	%	F	%	F	%
Typhoid	7	35.0	5	25.0	5	25.0	8	40.0	6	30.0	2	10.0	33	27.5
Diarrhea	4	20.0	4	20.0	6	30.0	5	25.0	4	20.0	1	5.0	24	20.0
Cholera	0	0	1	5.0	3	15.0	1	5.0	2	10.0	0	0	7	5.8
Skin irritation	9	45.0	10	50.0	6	30.0	4	20.0	4	20.0	3	15.0	36	30.0
None	0	0	0	0	0	0	2	10.0	4	20.0	14	70.0	20	16.7
Total	20	100	20	100	20	100	20	100	20	100	20	100	120	100

N.B: F- Frequency; % - Percent.

In Amarata, 10% agreed that water pollution causes typhoid and 15% agreed that it causes skin irritation. As shown in Table 5 and presented in Figure 5, 27.5% of total respondents believed that water pollution causes typhoid while 20, 5.8 and 30.0% agreed that water pollution causes diarrhea, cholera and skin irritation respectively. It shows that residents in the study area are really suffering from typhoid, diarrhea and skin irritation diseases while the case of cholera is not frequent (Figure 6).

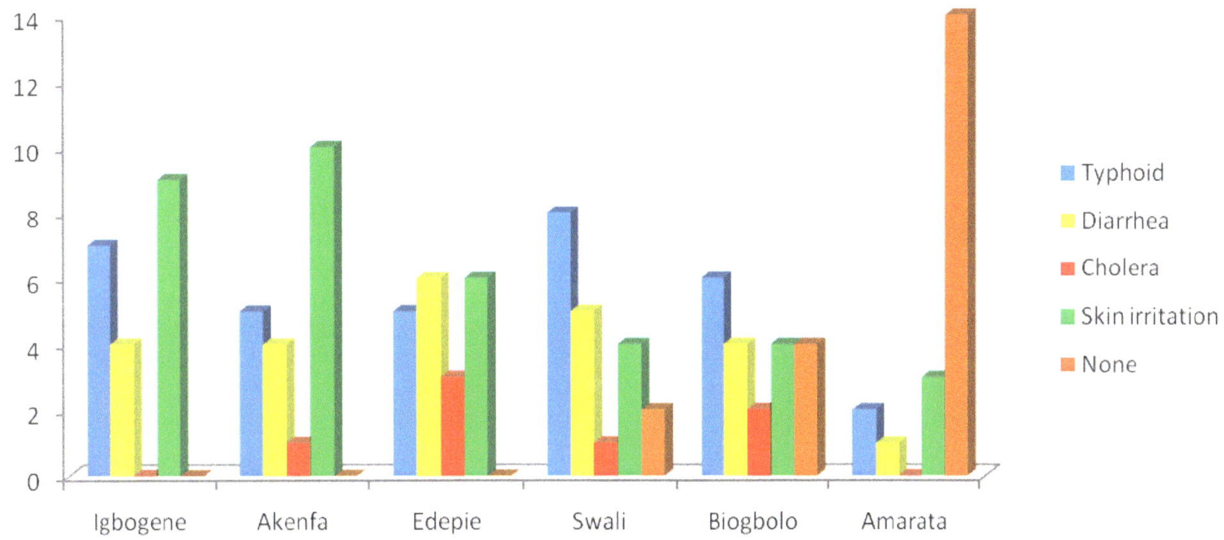

Figure 5. Levels of diseases caused by water pollution.

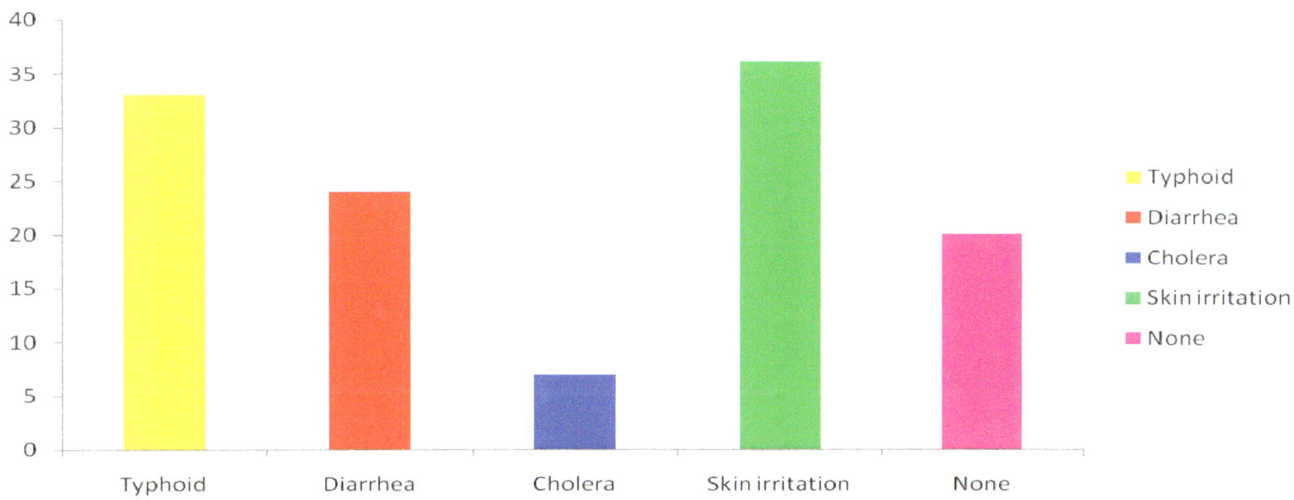

Figure 6. Total level of occurrence of diseases caused by water pollution.

Implications of results

The pH of the ground water was slightly acidic and slightly alkaline and the concentration fell within the WHO standard while the TDS in Yenagoa was low. These values are indications of the inland nature of the aquifers and their remoteness from the influence of any saline intrusion (Galley et al., 1975). The concentration of heavy metals (Lead, Iron, and Copper) in groundwater samples revealed that they were generally low. Lead ranged between 0.00 and 0.02 mg/L while Iron ranged between 0.001 and 0.073 mg/L and Copper which ranged between 0.0016 and 0.105 mg/L. The result corresponds to an Environmental Impact Assessment study by Shell Petroleum Development Company (SPDC) (2006) confirming that the concentration of these heavy metals in the ground water is still within acceptable levels.

The nitrate and sulphate concentrations were generally low in the ground and surface water but still higher in the latter. Their values were within the WHO limit for potable water. The low values may be due to the use of agricultural chemicals at a low rate in the study area. Chaturvedi (1976) confirmed that in many developing countries, agricultural chemical use has been low in comparison to levels in industrialized countries. The total coliforms, faecal coliforms and THC were high in the study area. This may be attributed to the direct discharge of untreated waste water in Yenagoa from the industries. Morris (1994) concluded that direct discharge of untreated wastewater has led to substantial increases in pollutants (NO_3^-, NH_4^+, Cl^-, faecal coliforms, and dissolved organic carbon) in the shallow aquifers. The pollution especially the faecal coliform may contribute to the occurrence of diseases like diarrhea and cholera in Yenagoa. As a result of the high presence of coliform in the ground and surface water of Yenagoa, should be subjected to adequate treatment before human consumption. A report by Shell Petroleum Development Company (SPDC) (2006) stated that the WHO requires that drinking water should be without faecal coliform bacteria.

It was discovered that majority of the people prefer ground water to surface water in the study area. This may be attributed to the colourless nature of ground water. Surface water unlike ground water supplies are affected by short term drought (although long term drought affects both). Groundwater is generally free of sediments, color, pathogenic (diseases) organisms, in severely polluted areas. In cases where ground water is polluted, such conditions are considered irreversible (Christopherson, 2002). Phiilp (1994) also confirmed that ground water is an important source of water supply for municipalities, agriculture and industry.

The perception of individuals to the cause of diseases due to polluted water in the study area confirmed that water pollution can lead to various forms of diseases like typhoid, skin irritation, cholera and diarrhea. The reason may be due to the concentrations of heavy metals like lead, copper, and iron which might have been released to the environment through industrialization and transportation. Horsfall et al. (2002) and Abia et al. (2002) corroborated that many of the industrial operations release various toxic heavy metals in their effluents. Heavy metals, arsenic and arsenical compounds, barium, chromium, lead, cadmium, manganese, zinc and other harmful materials are known to cause serious health effects. Such health problems includes but not limited to acute and chronic respiratory problems, gastro-intestinal tract infection, cardiovascular system disorder, nervous system and blood forming organisms malformation cases of skin and lung cancer. Pedley and Howard (1997) concluded that in many urban and peri-urban areas, the use of pit latrines and soak pits discharge wastes below the soil and weathered zone layers, thus polluting ground water.

Conclusions

The laboratory analyses carried out on the parameter of water samples reveals that the pH level of the water in the study area was slightly acidic and slightly alkaline. All water parameters tested in the area were lower in concentration to WHO standard except Iron (Fe) under surface water and THC, total coliform and faecal coliform in both the groundwater and surface water. Comparing the concentrations of water parameters between groundwater and surface water in Yenagoa, all the physico-chemical variables were higher significantly in the surface water than the groundwater. Yet chloride, nitrate, sulphate and calcium were very low in the study area in both the surface and groundwater. The proper analysis and efficient utilization of groundwater cannot be over emphasized and its importance cannot be overestimated. Its uses are very important in all aspects of human life. Its importance is as any other source of water, but due to its availability and proximity, it makes it seem more important as compared to other sources of water. Although the communities in Yenagoa Local Government Area are semi-urban settlements or also known as a developing urban areas which are surrounded by water channels but the use of the groundwater is still paramount compared to other sources of water available in this environment. The other sources of water (rain and the river) prior to this time were also much utilized as aforementioned due to their availability. The study gave a proper insight of the preference of groundwater (borehole) to the surface water and there was significant difference in the choice between the ground water and surface water. This may be due to the notion that groundwater is safer and cleaner than any other source of water Yenagoa.

The study also reveals that boiling, filtering and adding chemicals were the common methods of treating water in the area but by the minority. Majority still lack the knowledge about water treatment, especially for domestic use. Inadequacy of treating water was directly correlated to the issue of occurrence of diseases in the area and the relationship was found to be a positively strong correlation and water treatment could explain 81% of the occurrence of the diseases in the area. The study revealed that lack of water treatment has resulted to the occurrence of some diseases like typhoid, diarrhea, cholera and skin irritation.

Following the results of this study, groundwater is of better quality than surface water. This study therefore, recommend that more groundwater and surface water assessment studies in the study area and the entire state be carried out because the production, creation and the availability of these data through proper analysis ensures effective management and utilization of water resources in the study area. The government should employ the services of an interagency working group on groundwater management that would allow the UN agencies concerned to focus efforts on this course and develop mutually reinforcing normative and operational programs. More boreholes are needed to be sunk in the study area by the Federal, State and Local Governments.

REFERENCES

Abia AA, Horsfall M Jnr, Didi O (2002). Studies on the use of agricultural by-products for the removal of trace metals from aqueous solution. J. Appl. Sci. Environ. Manage. 6:89-95.

Burke JJ (2002). Groundwater for irrigation: productivity gains and the need to manage hydro-environmental risk. In R. Llamas & E. Custodio, eds. Intensive use of groundwater challenges and opportunities. Abingdon, U.K., Balkema, pp. 478.

Chaturvedi MC (1976). Second India studies: Water. New Delhi, Macmillan Company of India Limited.

Christopherson RW (2002). Geosystem. Prentice Hall Incorporation, New Jersey.

Galley WH, Eqan JL, Monjam RT, West PW, Widmanls (1975). Environmental Pollutants - Selected Analytical Methods (Scope 6,) Butherworts London.

Horsfall M (jnr), Abia AA, Didi O (2002). Studies on the use of Africulture by-products for the removal diff trace-metals from aqueous solution. J. Appl. Sci. Environ. Manage. 6(2):95-98.

Mogheir Y, Singh VP (2002). Application of Information Theory to Groundwater Quality Monitoring Networks. Water Res. Manage. 16:37-49.

Morris BL (1994). The Impact of Urbanisation on Groundwater Quality (Project Summary Report). Keyworth, British Geological Survey.

Nwankwoala HO, Udom GJ, Ugwu SA (2011). Some heavy metal investigations in groundwater sources in Yenagoa, Bayelsa State, Nigeria. J. Appl. Technol. Environ. Sanit. 1(2):163-170.

Pedley S, Howard G (1997). The public health implications of microbiological contamination of groundwater. Quart. J. Eng. Geol. 30(2):179-188.

Phillips FM (1994). Environmental tracers for water movement in desert soils of the American southwest. Soil Sci. Soc. Am. J. 58:15-24.

Shell Petroleum Development Corporation (SPDC) (2006). Environmental Impact Assessment of Rumuekpe (OML 22) and Etelebou (OML 28) 3D Seismic Survey, p. 393.

World Health Organisation (WHO) (2006). Environmental Health Criteria Cd. General, p. 134.

Evaluation of mountain-front recharge estimation techniques for Southern New Mexico basins

B. V. N. P. Kambhammettu[1*], James P. King[2] and Praveena Allena[3]

[1]New Mexico Water Resources Research Institute, MSC 3167, P. O. Box. 30001, Las Cruces, NM–88003, U.S.A.
[2]Civil Engineering Department, New Mexico State University, Las Cruces, NM–88003, U.S.A.
[3]Office of the State Engineer, 301 S Tin St., Deming, NM–88030, U.S.A.

Technically sound water management practices in arid to semi-arid regions require effective quantification of model input parameters. Mountain-front recharge is one of the most uncertain parameters in groundwater flow models due to the complexity of estimation. This paper evaluated three mountain-front recharge estimation techniques using a numerical model developed for the Southern Jornada Del Muerto Basin, New Mexico. Digital elevation model (DEM) and distributed precipitation maps were used to delineate recharge zones in the study area on geographic information system (GIS) platform. Statistical and residual parameters were used to assess the reliability of the calibrated recharge. An estimated annual recharge of about 3,360 enters the basin through ephemeral streams. Simulated heads from the three estimation techniques had approximately equal mean, with different variance. Analysis of variance on the simulated piezometric heads revealed that all the three empirical methods were consistent with the observed data. Waltemeyer equation proved to be more efficient in estimating mountain-front recharge as it gave the closest match between the observed and simulated piezometric heads and other basin-specific climatic characteristics.

Key words: Mountain-front recharge, numerical model, Southern Jornada Del Muerto Basin, GIS.

INTRODUCTION

Reliable and accurate estimation of groundwater recharge in southwest United States is a challenging task due to complexities in measuring the extremely low fluxes which highly vary in time and space. Variations in topography, geology, geomorphology, and magnitude and form of precipitation also make recharge flux one of the most uncertain parameters of groundwater flow models in the region (Scanlon, 2004). Mountain-front recharge (MFR) is the process of aquifer recharge through infiltration of surface flow from streams with headwaters in mountains (Anderholm, 2001). Common water balance approaches used in estimating the recharge flux may not be so practicable in arid regions as most of the input and output parameters fall within the measurement errors (Flint et al., 2004). The Chloride balance method produces low recharge estimates in this region probably due to the presence of additional chlorides in the unsaturated and groundwater zones, and lengthy travel times resulting from the large sized drainage basins (Anderholm, 2001). Hence groundwater flow models developed for southern New Mexico basins rely on empirical techniques to estimate recharge fluxes that are originally derived for similar catchments. However, uncertainties in the estimated recharge fluxes have not been quantified in most of the modeling works due to: i) a very low magnitude of recharge compared to other fluxes and ii) the absence of calibration targets near the recharge zones. Hence selection of proper method for estimating recharge flux is vital for the effective simulation of groundwater availability in the region.

In the present study, MFR and other fluxes in the Southern Jornada Del Muerto (SJDM) Basin were assessed using a numerical model. Geographic information system

*Corresponding author. E-mail: kbvnphanindra@gmail.com.

(GIS) tools were used to process and analyze the geologic, hydraulic and storage characteristics of the basin. The model simulation was done on a seasonal basis from 1968 through 2007 using distributed recharge fluxes in proportion to precipitation. The model was calibrated against static monitoring well data from which basis, the hydraulic and storage characteristics of the basin were updated. The calibrated model was then tested with three empirical estimates of MFR so as to: i) select a method that best reflects the basin characteristics based on residual statistical analysis and ii) estimate sub-zone-based recharge distribution in the SJDM Basin.

STUDY AREA

Jornada Del Muerto Basin is in the northern portion of the Chihuahuan Desert, in the highlands of the basin and range province of Mexico (Hawley, 1975). The parallel north-south mountain ranges are separated by broad valleys, filled with alluvial deposits. The local climate is characterized by high solar radiation, large variations in diurnal temperature, low relative humidity, extremely variable precipitation, and high potential evapotranspiration. The study area is limited to southern part of the Jornada Del Muerto Basin existing within Doña Ana County, New Mexico. The study area is limited on the north by the point of rocks (POR) in that region. The hydrogeology to the north of POR is poorly understood (Hawley and Kennedy, 2004). There is also relatively less human activity and water used in this region. The aerial extent of study area and of the model domain is 576 mi^2 or 1,492 km^2 (Figure 1). In 2007, the SJDM Basin was reported an annual yield of 8,000 ac-ft (9.878×10^6 m^3). It is therefore a potentially important groundwater source for Las Cruces city (LCC).

The SJDM Basin is characterized by the Santa Fe group deposit. This deposit is generally divided into three stratigraphic units on the basis of lithology, depositional environment and age (Hawley and Kennedy, 2004). The lower Santa Fe (LSF) unit consists primarily of fine-grained and partly consolidated sediments. The middle Santa Fe (MSF) unit consists of clean fluvial eolian sand, interbedded with silty clay on the basin floor. The upper Santa Fe (USF) consists of sand and gravel deposits from the ancestral Rio Grande River. This deposit forms the most productive aquifer system in the Santa Fe sequence.

MATERIALS AND METHODS

Model design

A three layer conceptual groundwater flow model that was developed for the study area is used in the present study (Kambhammettu et al., 2010). A digital elevation model (DEM) was used to delineate study area and the hydrogeologic boundaries on GIS platform. Using spatial and geostatistical analyst tools of GIS, continuous surfaces of hydrologic and storage properties of the study area were generated from the measured values. To account for the geologic faults, the elevation profile of each layer was manually mapped from the scanned contours. The initial water table profile was based on 1968 water table elevation data derived from 51 monitoring wells (King et al., 1971; Wilson et al., 1981). Pumping from domestic, stock and multi-utility wells, mountain-front recharge, and shallow groundwater evapotranspiration were modeled as head dependant boundary condition. All GIS files were projected into New Mexico State Plane Central FIPS coordinate system on North American Datum (NAD, 1983). The model was grided into 180 rows and 105 columns at a uniform cell size of 402.3 m. The model was run for 1968 through 2007 in 79 stress periods. The model grid was oriented at 24° in counter clockwise direction to optimize the number of active cells and to align the grids with the general flow direction in the Rincon-Mesilla Basin. The model was calibrated by comparing the simulated groundwater head with piezometric data from 72 monitoring wells.

Sensitivity analysis showed that the model was sensitive to hydraulic conductivity in the upper unconfined aquifer and to storativity in the central, eastern and north-eastern regions of the confined aquifer.

Empirical methods

The three empirical methods used in the present study to estimate the mountain-front recharge fluxes based on spatially distributed precipitation fluxes are: Maxey-Eakin (1949), Hearne-Dewey (1988), and Waltemeyer (1993). Average monthly precipitation was obtained from the Oregon State University's Parameter-elevation Regressions on Independent Slopes Model (PRISM) data. ArcHydro tool in GIS was used to delineate the sub-basins from a 30-m DEM of the region. The area outside the model boundary and up to the groundwater divide was divided into several sub-basins with designated areas. These sub-basins were intersected with precipitation and slope maps, resulting in multiple sub-polygons within each recharge zone with designated recharge parameters.

Maxey-Eakin method

Maxey-Eakin (1949) developed a method for estimating precipitation recharge based on water balance studies in many river basins in Nevada. After adjusting for surface runoff and evapotranspiration losses, recharge along the model boundary was estimated as a function of precipitation:

$$P_e = P \times C_{me} \tag{1}$$

Where P_e is average annual recharge depth; P is average annual precipitation depth; and C_{me} is Maxey-Eakin coefficient, which is exclusively a function of annual precipitation.

Hearne-Dewey method

Hearne-Dewey (1988) developed a regression equation based on data from 16 basins in northern New Mexico. This method assumes that aquifer recharge occurs through ephemeral streams in response to snowmelt runoff and local storms:

$$Q = (1.074 \times 10^{-5}) \times A^{1.216} \times P^{2.749} \times S^{0.536} \tag{2}$$

Where Q is mean annual recharge (cfs); A is drainage basin are

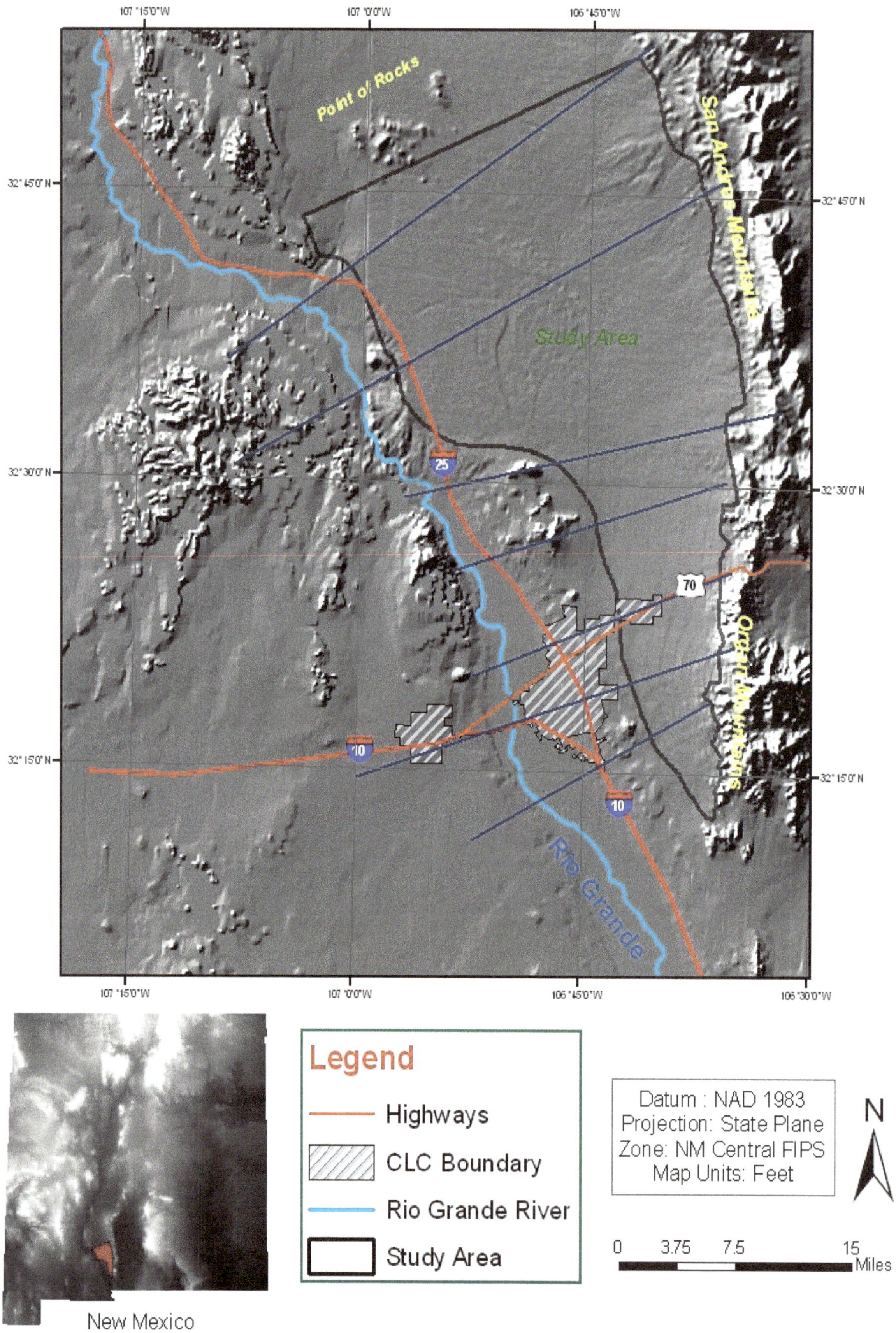

Figure 1. Location of study area showing hydrogeologic cross-sections.

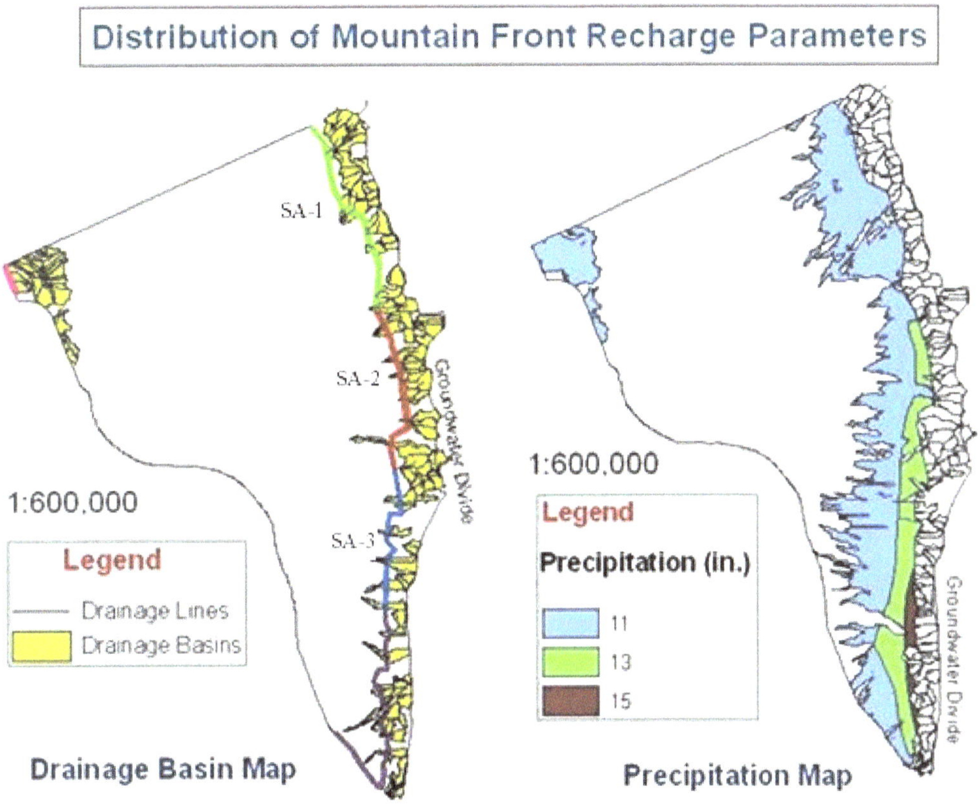

Figure 2. Maps of distributed precipitation and recharge parameters in the study area.

(mi^2); *P* is mean winter (October to April) precipitation (in); and *S* is basin average slope (ft/mi).

Waltemeyer method

Waltemeyer (2001) developed a regression equation based on basin climatic characteristics using data from 13 streamflow gage stations in Tularosa Basin in central New Mexico:

$$Q = (1.7 \times 10^{-4}) \times A^{1.35} \times P^{1.65} \quad (3)$$

Where *Q* is mean annual recharge (cfs); *A* is drainage basin area (mi^2); and *P* is mean annual precipitation (in).

RESULTS AND DISCUSSION

Recharge distribution within the basin along the mountain region was divided into three zones (San Andres, Organ, and Caballo) based on the delineated drainage sub-basins. The San Andres zone was further divided into three sub-zones (San Andres-1, 2 and 3) due to high variability in mean annual precipitation (Figure 2). A total of 216 finite difference cells (53 for San Andres-1, 46 forSan Andres-2, 39 for San Andres-3, 69 for Organ, and 9 for Caballo) were built along the basin boundaries for inputting recharge flux. Recharge rate for each stress period was estimated as the ratio of the fraction of mean annual recharge in a zonal area, which was equally distributed over the cells in a zone. Zone-wise distributed MFR, estimated by the three empirical precipitation recharge methods is given in Table 1. Due to the high coefficients, Maxey-Eakin estimated MFR was higher than that of the Waltemeyer method at high precipitations (San Andres-3, Organ), irrespective of the drainage area. The Hearne-Dewey method underestimated recharge flux. This was probably due to the low coefficient of proportionality (Equation 2) and use of mean winter precipitation instead of mean annual precipitation. Estimates of recharge by all the three methods showed that San Andres Basin is the major contributor of recharge in the region. It accounted for over 70% of the estimated basin recharge. This is primarily due to the higher precipitation index and the large drainage area.

To determine the best method for estimating MFR in the SJDM Basin, the model was re-calibrated with varying recharge fluxes against the data from 72 monitoring wells that were uniformly distributed across the study area. The residual statistics were evaluated in terms of mean error (ME), sum of squared errors (SSE) and root mean square error (RMSE). Zonal variations in the residual statistical parameters were estimated using the three empirical methods (Table 2). The Waltemeyer estimated recharge was noted to be in close agreement

Table 1. Estimated mountain-front recharge by three empirical methods.

Method	Zone wise Recharge distribution in ac.ft/yr (cfs)					Basin recharge in ac.ft/yr (cfs)
	San Andres-1	San Andres-2	San Andres-3	Organ mountains	Rincon - Caballo	
Maxey-Eakin	588.19 (0.81)	1381.44 (1.91)	1143.93 (1.58)	1128.44 (1.56)	240.77 (0.33)	4482.77 (6.19)
Hearne-Dewey	270.62 (0.37)	466.35 (0.64)	338.10 (0.47)	133.84 (0.18)	91.71 (0.13)	1300.62 (1.79)
Waltemeyer	734.90 (1.01)	1198.58 (1.65)	604.07 (0.83)	606.37 (0.84)	220.06 (0.30)	3363.98 (4.63)

Table 2. Residual statistics on recharge estimates using three empirical methods.

Recharge zone	Target wells	Method used	ME (ft)	SSE (ft^2)	RMS (ft)
San Andres – 1	4	Maxey-Eakin	-8.30	592.83	12.17
		Hearne-Dewey	-8.24	582.92	12.07
		Waltemeyer	8.33	596.38	12.21
San Andres – 2	2	Maxey-Eakin	-1.31	97.53	6.99
		Hearne-Dewey	-1.21	95.27	6.92
		Waltemeyer	-1.31	97.42	6.99
San Andres – 3	34	Maxey-Eakin	-3.45	1981.83	7.64
		Hearne-Dewey	0.89	1323.34	6.23
		Waltemeyer	-1.12	1429.48	6.50
Organ mountains	15	Maxey-Eakin	3.25	1712.49	10.70
		Hearne-Dewey	6.82	2150.30	11.98
		Waltemeyer	5.22	1328.72	9.42
Caballo mountains	0	---	---	---	---
Study area	72	Maxey-Eakin	-1.71	5256.42	8.53
		Hearne-Dewey	1.35	5404.87	8.66
		Waltemeyer	-0.07	5075.36	8.40

with the measured groundwater levels in the 72 monitoring wells. This is particularly true for the San Andres-3 and the Organ zones, where most of the target wells are in the vicinities of the recharge zones. Pumping in the southern part of the study area near San Andres-3 and Organ had a greater influence on budget flows and simulated heads. Hence an accurate estimation of recharge in this zone is essential for effective management of groundwater resources of the region. The calibrated models showed that Maxey-Eakin and Waltemeyer equations better estimated recharge in the SJDM Basin. This was primarily due to the increased amounts of recharge flux in the central and eastern parts of study area. Groundwater gradient in the region is primarily towards the west/northwest, but is apparently flattening out with time. The best agreement between the observed and simulated heads was produced by the Waltemeyer equation (Figure 3).

A higher discrepancy in groundwater elevations along the adjacent wells in the south central part of the study area resulted in poor residual statistics for the other models. To test the consistence of the estimated recharge by the three methods with the basin characteristics, one-way analysis of variance (ANOVA) was performed on the simulated heads. A graph of cluster averages of the simulated heads for the 49 monitoring wells near San Andres-3 and Organ zones is plotted in Figure 4. The null hypothesis (H_0) used in the ANOVA test assumed that the simulated heads resulting from the three empirical methods were consistent with the

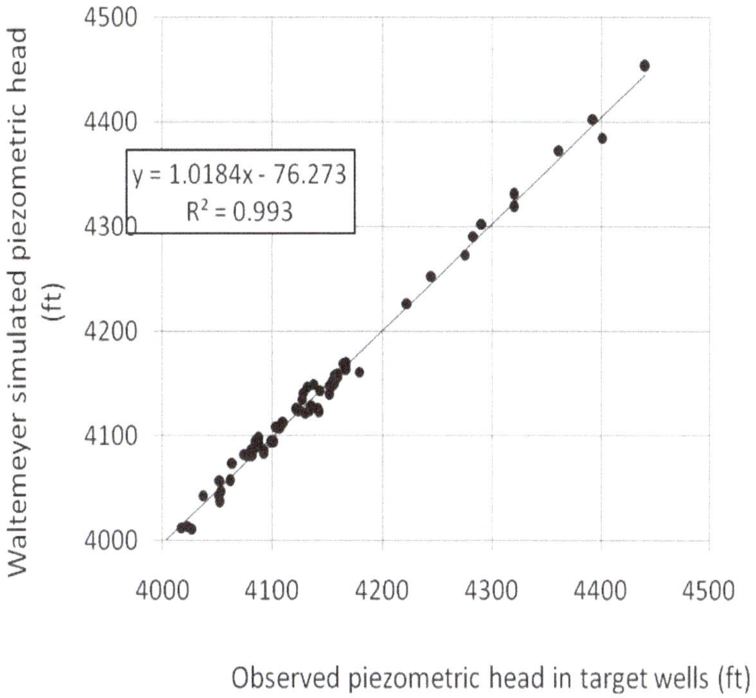

Figure 3. Comparison of observed and simulated heads by the Waltemeyer method.

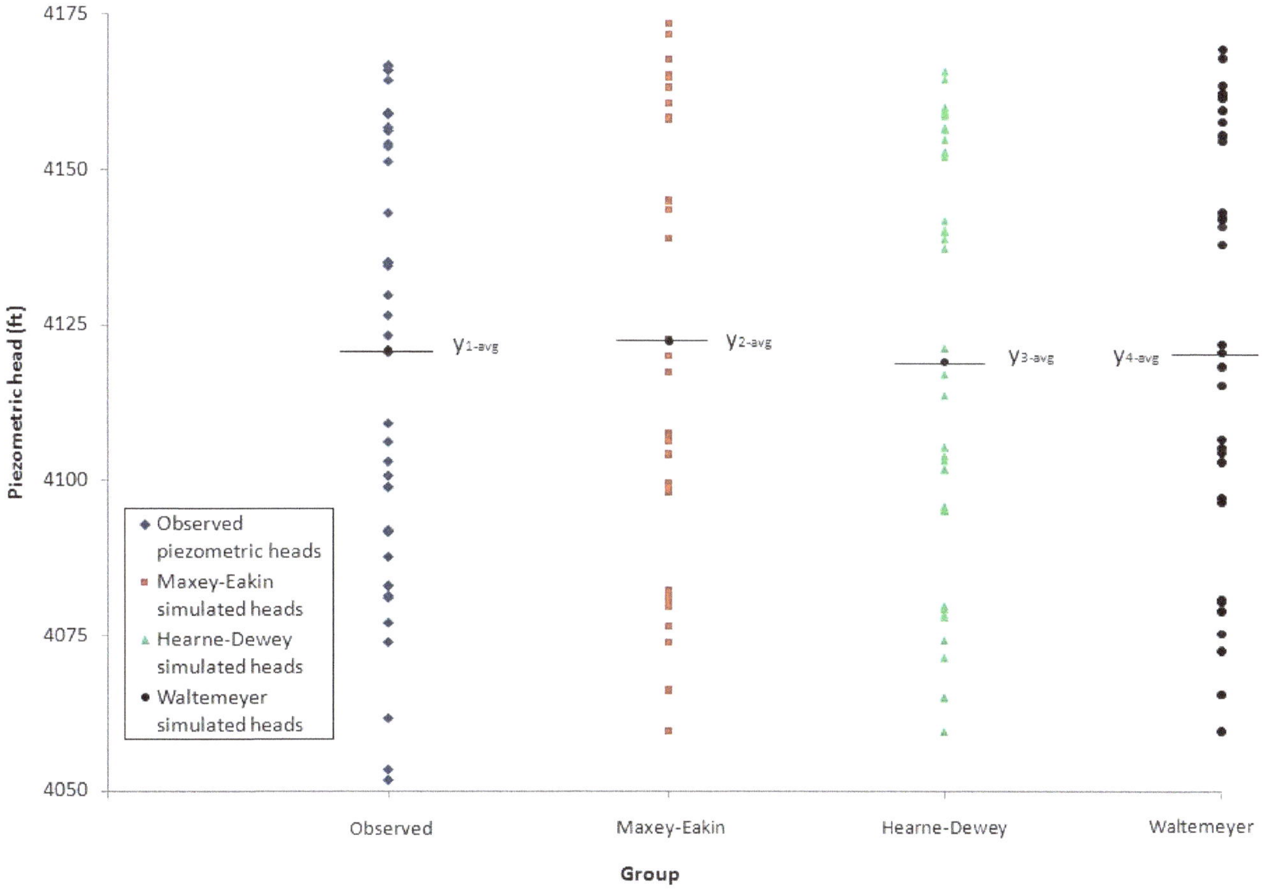

Figure 4. Deviation of piezometric heads among clusters (San Andres-3 and Organ zone data).

observed values, with probably based on a unique population. Results of the ANOVA test suggested that the null hypothesis was acceptable at all confidence levels. However, the Waltemeyer equation, which was originally developed for the Tularosa Basin in central New Mexico, seemed to be more appropriate for the SJDM Basin. This was mainly due to the strong similarities between the two basins in terms of climate and hydrogeologic characteristics.

Conclusions

A three-layer numerical groundwater recharge for the Southern Jornada Del Muerto Basin in New Mexico was used to assess current MFR distribution techniques in the region. The model was grided into 180 rows and 105 columns, and run for 1968 through 2007 on seasonal time scale. The hydrogeologic parameters of the upper aquifer were most sensitive in the simulation (Kambhammettu et al., 2010). Three empirical estimates of mountain-front recharge were evaluated via inverse modeling and residual statistics analysis. The Maxey-Eakin and Waltemeyer equations yielded approximately same amount of recharge because both use annual precipitation instead of average winter precipitation. The Maxey-Eakin method slightly overestimated recharge due to the high coefficients resulting from high precipitation in the southern recharge zones. The Hearne-Dewey equation appeared to underestimate recharge in the basin. Most of the target wells considered in the model calibration were in the vicinity of the San Andres-3 and Organ zones. The estimated recharge via the three methods differed significantly for these zones.

The Waltemeyer equation, which was originally developed for a basin with similar hydrogeologic characteristics to the study area, produced the least residual statistics. Analysis of variance on the simulated heads suggested that all the three empirical methods were consistent and reliable in terms of the estimated recharge. However, the Waltemeyer simulated heads were in the strongest agreement with measured groundwater elevations in the region. This close agreement was attributed to the strong similarities in the hydro-climatic characteristics of the basin. Based on the results, the Waltemeyer method was recommended for use in estimating recharge in the SJDM Basin.

REFERENCES

Anderholm SK (2001). Mountain-front recharge along the eastern side of the Middle Rio Grande Basin, central New Mexico. U.S. Geological Survey Water-Resources Investigations Report. 00-4010: 36 p.

Flint AL, Flint LE, Hevesi JA, Blainey JB (2004). Fundamental Concepts of Recharge in the Desert Southwest: A Regional Modeling Perspective. In: Hogan JF, Phillips FM, Scanlon BR (eds.). Groundwater Recharge in a Deseret Environment: The Southwestern United States. American Geophysical Union, Washington, DC: 159-184.

Hawley JW (1975). Quaternary history of Doña Ana county regions south-central New Mexico. New Mexico Geological Society, 26th Annual Field Conference Guidebook, pp. 138-150.

Hawley JW, Kennedy JF (2004). Creation of a digital hydrogeologic framework model of the Mesilla Basin and southern Jornada Del Muerto Basin. New Mexico Water Resources Research Institute Technical Report, 332: 105.

Hearne GA, Dewey JD (1988). Hydrologic Analysis of the Rio Grande Basin north of Embudo, New Mexico, Colorado and New Mexico. U.S. Geological Survey Water-Resources Investigations Report, pp. 86-4113.

Kambhammettu BP, Allena P, King JP (2010). Simulation of groundwater flow in the southern Jornada del Muerto Basin - New Mexico. New Mexico Water Resources Research Institute Technical Completion Report, 352: 74p.

King WE, Hawley JW, Taylor AM, Wilson RP (1971). Geology and Ground-water Resources of Central and Western Doña Ana County. New Mexico Bureau of Mines and Mineral Resources Hydrologic Report. 1: 64p.

Maxey GB, Eakin TE (1949). Ground water in White River Valley, White Pine, Nye, and Lincoln Counties, Nevada. Water Resources Bulletin No. 8. Nevada State Engineer, 59 p.

Scanlon BR (2004). Evaluation of Methods of Estimating Recharge in Semiarid and Arid Regions in Southwestern U.S. In: Hogan JF, Phillips FM, Scanlon BR (eds.). Groundwater Recharge in a Deseret Environment: The Southwestern United States. American Geophysical Union, Washington, DC., pp. 235-254.

Waltemeyer SD (2001). Estimates of mountain-front streamflow available for potential recharge to the Tularosa Basin, New Mexico. U.S. Geological Survey Water-Resources Investigations Report. 01-4013: 8 p.

Wilson CA, White RR, Orr BR, Roybal RG (1981). Water Resources of the Rincon and Mesilla Valleys and Adjacent Areas. New Mexico State Engineer Technical Report, 43: 514.

15

Study of double breakpoints during chlorination of river Yamuna water, Delhi, India

Lokesh Kumar

120 MGD Water Treatment Plant, Water Works Wazirabad, Delhi Jal Board, Municipal Water Supply Department, Government of NCT Delhi, Delhi 110054, India.

A study on chlorination of raw Yamuna River water is reported in this study. Samples were chlorinated with increasing doses of standard chlorine water and residual chlorine (Cl_2) was measured by Starch-Iodide method. For each sample, the chlorination curve (chlorine residual versus chlorine dose) was obtained. Curves showed the typical irregularity attributed to the formation and destruction of chloramines and transformation of toxic cyanobacteria (blue-green algae) by chlorine. It was observed that, after reactions with strong reductants and chloramines forming compounds, the remaining organic matter exerted a certain demand of chlorine. The evolutions of chlorination curves were studied. Till date, study on breakpoint chlorination revealed only single breakpoint but, in present study single breakpoints during chlorination of raw waters were not established in many cases. Double breakpoints have been reported in this study. The evolutions of different breakpoint curves might be attributed to formation and destruction of numerous chemical disinfection by-products (DBPs) and biological bacteria/alga, due to variant pollution conditions of raw water.

Key words: Ammonia, breakpoint chlorination, cyanobacteria, chlorine, pollution, nitrite.

INTRODUCTION

There has been limited research done on the breakpoint chlorination phenomenon. Initial research occurred in the last 60 years, when breakpoint was a hot topic and drew the attention of researchers. The latter half of this century has seen little done in the way of research into breakpoint chlorination. With the exception of a few papers (March and Gual, 2007), mostly plant specific studies have been performed. No study on double breakpoint during chlorination of water has been observed/reported in any research paper. There has however, been a great deal of research done in the field of chlorine with cyanobacterial blooms and chlorine-ammonia chemistry, as well as advances in the fields of chlorine and ammonia analysis. While this study certainly challenges all obtainable research on breakpoint chlorination, it also focuses on research done in the field of chlorine-ammonia chemistry, as it pertains to the breakpoint reaction. Establishment of double breakpoints during chlorination of water has been reported in this study. Chlorine and other disinfectants react with natural organic matter (NOM) and/or inorganic substances occurring in water to form various disinfection by-products (DBPs) such as trihalomethanes (THMs), haloacetic acids (HAAs) and other compounds (Rook, 1974; Singer, 1994). Cyanobacterial blooms are also a great challenge for drinking water production, for their occurrence in drinking water resource often causes several process disturbances in treatment plant, such as faster filter clogging and reactant (chlorine demand) consumption increase.

REVIEW OF LITERATURE

Initial research efforts into the mechanism of the breakpoint reaction are attributed to Calvert (1940), and later studies by Griffin and Chamberlin (1941), and Rossum (1943). Ensuing research by others (Palin, 1950; Wei and Morris, 1974; Weil and Morris, 1949) has led to an understanding of the stoichiometry and kinetics associated with the breakpoint process. A comprehensive study of the kinetics of breakpoint chlorination was performed by Saunier and Selleck (1976). The goal of their work was to develop a mathematical model, derived from laboratory observations, which would provide "a rational basis for the design and operation of the breakpoint process in order to achieve predictable ammonia removal" (Saunier and Selleck, 1976). Unfortunately, past research yielded little or no insight into the problem of successfully controlling the breakpoint process in a full-scale water treatment plant. Although Saunier and Selleck (1976) performed comprehensive pilot study work, the results were never incorporated into a full-scale plant application. Pressley et al. (1973) performed extensive pilot study research in order to provide design criteria for a full scale breakpoint operation at the Blue Plains wastewater treatment plant in Washington, D.C. Atkins et al. (1973) performed an extensive pre-design pilot study to provide information for full-scale breakpoint operations at the Owosso wastewater treatment plant in Michigan. The engineering firm of Camp, (Dresser and McKee, 1983) also performed bench scale. Figure 5 shows what happens when chlorine (either chlorine gas or a hypochlorite) is added to water. First (between points 1 and 2), the water reacts with reducing compounds in the water, such as hydrogen sulfide. These compounds use up the chlorine, producing no chlorine residual. Next (between points 2 and 3), the chlorine reacts with organics and ammonia naturally found in the water. Some combined chlorine residual is formed (chloramines). Note that if chloramines were to be used as the disinfecting agent, more ammonia would be added to the water to react with the chlorine. The process would be stopped at point 3. Using chloramines as the disinfecting agent results in little THMs production but causes taste and odor problems since chloramines typically give a "swimming pool" odor to water. In contrast, if hypochlorous acid is to be used as the chlorine residual, then chlorine will be added past point 3. Between points 3 and 4, the chlorine will break down most of the chloramines in the water, actually lowering the chlorine residual.

Finally, the water reaches the breakpoint, shown at point 4. The breakpoint is the point at which the chlorine demand has been totally satisfied; the chlorine has reacted with all reducing agents, organics, and ammonia in the water. When more chlorine is added past the breakpoint, the chlorine reacts with water and forms hypochlorous acid in direct proportion to the amount of chlorine added. This process, known as breakpoint chlorination, is the most common form of chlorination, in which enough chlorine is added to the water to bring it past the breakpoint and to create some free chlorine residual (White, 1992). Sodium hypochlorite or chlorine is used to oxidize the ammonia in raw, secondary and lime clarified municipal wastewaters. In systems with only ammonia as the chlorine demand, the breakpoint exhibits a minimum chlorine dose at approximately an 8:1 weight ratio of Cl:NH_3 --N in the pH range of 6 to 7. End products at the breakpoint are N_2, NCl_3, and NO_3^-.

Kinetic analysis of the oxidation of toxins in natural water revealed significant differences in their susceptibility to chlorine, saxitoxins being the easiest to oxidize, followed by cylindrospermopsin and microcystin-LR. Furthermore, concentrations of THMs and HAAs (<40 μg/L) and N-nitrosodimethylamine (<10 ng/L) as chlorination by-products were lower than the guideline values even at the highest chlorine exposure (CT) value (220 mg min/L). However, the DBP concentrations in environmental bloom conditions with very high cell numbers were over the guideline values (Arash et al., 2012).

Thus, there are numbers of intermediate chlorination by-products which are still unidentified and contribute to double breakpoint during chlorine demand of water.

Oxidations of chemical pollutants

More than 500 DBPs in drinking water have been detected and divided into several groups of compounds (Richardson, 1998; Richardson et al., 2000): halogenated organic by-products (e.g., THMs, HAAs), inorganic by-products (e.g., chlorate, chlorite, bromated and iodate ions, ammonia) and organic oxidation by-products (e.g., aldehydes, carboxylic acids, assimilable organic carbon). Several DBPs result from chlorine disinfection (that is, CBPs) such as THMs, HAAs, haloacetonitriles (HANs) and other halonitriles (that is, cyanogen chloride), haloketones (HKs), halophenols (that is, 2-chlorophenol, 2,4-dichlorophenol, 2,4,6-trichlorophenol), halonitromethanes (that is, chloropicrin), haloaldehydes (that is, chloral hydrate) and MX (3-chloro-4-(dichloromethyl)-5-hydroxy-2(5H)-furanone) (US.EPA, 1999; Richardson et al., 2000). These compounds are formed under different conditions of pH and temperature, nature and amount of NOM and/or inorganic substances in raw and treated water, contact time between water and the disinfecting agent and biological degradation for some compounds (Singer, 1994; Chen and Weisel, 1998; Williams et al., 1998; Rodriguez and Serodes, 2001; Liang and Singer, 2003; Baribeau et al., 2005). This variability of conditions related to the formation of these CBPs in drinking water makes it difficult to study their occurrence in drinking water distribution systems (Rodriguez et al., 2004).

Oxidation of creatinine and creatine

Creatinine is a cyclic compound containing two reactive nitrogen atoms. It forms relatively stable chlorinated

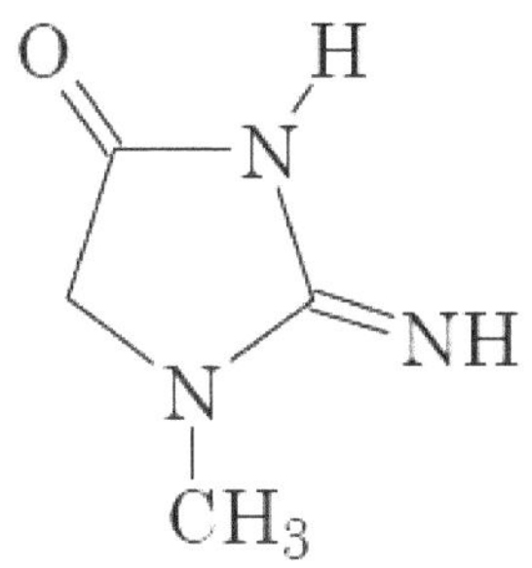

$$+ 10.5HOCl \rightarrow 4CO_2 + 1.5N_2 + 10.5HCl + 3.5H_2O$$

Reaction 1. Creatinine.

derivatives (Lomas, 1967) that are decomposed by excess available chlorine (Alouini and Seux, 1988). Under swimming pool conditions, chlorocreatinines decompose very slowly and can persist beyond the breakpoint in oxidation of ammonia and have been characterized as nuisance residuals. Creatinine and creatine (a linear compound $H_2NC(N{=}H)N(CH_3)CH_2COOH$), are closely related compounds. Dehydration of creatine yields creatinine (creatine → creatinine + H_2O). Oxidation of creatinine and creatine by excess available chlorine yields carbon dioxide, water, hydrochloric acid, and ammonia chloramines that are then oxidized to nitrogen. The overall reaction for creatinine is shown in Reaction 1.

In the absence of free chlorine, chlorocreatinines can decompose to intermediate products such as: creatinine, creatine, chlorocreatines, and 1-methylhydantoin.

Oxidation of uric acid

Uric acid can form mono-, di-, tri-, and terachloro derivatives. With excess chlorine, uric acid is oxidized to carbon dioxide, nitrogen, hydrochloric acid, and water. Meso-oxalic acid may form as an intermediate product that can be oxidized to carbon dioxide and water (Reaction 2).

Oxidation of biological pollutants

Cyanobacteria also known as blue-green bacteria, blue-green algae, and cyanophyta, are a phylum of bacteria that obtain their energy through photosynthesis. Some cyanobacteria produce toxins, called cyanotoxins. These include anatoxin-a, anatoxin-as, and some cyanobacteria produce toxins, called cyanotoxins. These include anatoxin-a, anatoxin-as, aplysiatoxin, cylindrospermopsin, domoic acid, microcystin LR, nodularin R (from Nodularia), or saxitoxin. Cyanobacteria reproduce explosively under certain conditions. These results in algal blooms, can become harmful to other species if the cyanobacteria involved produce toxins. Microcystins are cyclic nonribosomal peptides produced by cyanobacteria (e.g. *Microcystis aeruginosa* and Planktothrix). They are cyanotoxins and can be very toxic for plants and animals including humans. Their hepatotoxicity may cause serious damage to the liver. Sufficient chlorine doses are required to react with *Microcystins* and to satisfy chlorine demand of NOM which compete with toxins and potentially reduce chlorination efficiency (Kull et al., 2006) (Reaction 3).

Chlorination of microcystin

Cylindrospermopsin (abbreviated to CYN or CYL) is a cyanotoxin produced by a variety of freshwater cyanobacteria. CYN is a polycyclic uracil derivative containing guanidino and sulfate groups. It is also zwitterionic, making it highly water soluble. CYN is toxic to liver and kidney tissue and is thought to inhibit protein synthesis and to covalently modify DNA and/or RNA. CYN was shown to be vulnerable to chlorine (Senogles et al., 2000; Banker et al., 2001; Newcombe and Nicholson, 2004; Rodríguez et al., 2007a; Rodríguez et al., 2007b) (Reaction 4).

Chlorination of CYN

The bulk of the research on chlorine demand has been related to chlorination of potable waters. Feben and Taras (1950, 1951) found a definite relationship between chlorine demand and the complexity of the organic nitrogen compounds found in the city of Detroit water supply. Pilot study work by Pressley et al. (1973) correlated the level of pretreatment of the wastewater to the chlorine-to-ammonia ratio required to reach breakpoint. In their studies, water receiving high levels of treatment, e.g. waters with lower concentrations of organics present required lower $Cl:NH_3$-N dosages to reach the breakpoint, while all other factors were constant. The end products of the breakpoint reaction were found to be primarily nitrogen gas (N_2) and secondarily nitrate (NO_3^-) (Palin, 1950; Pressley et al., 1973; Saunier and Selleck, 1976). The introduction to above literature is cited to interpret formation of multi-chlorination by-products which are potentially responsible

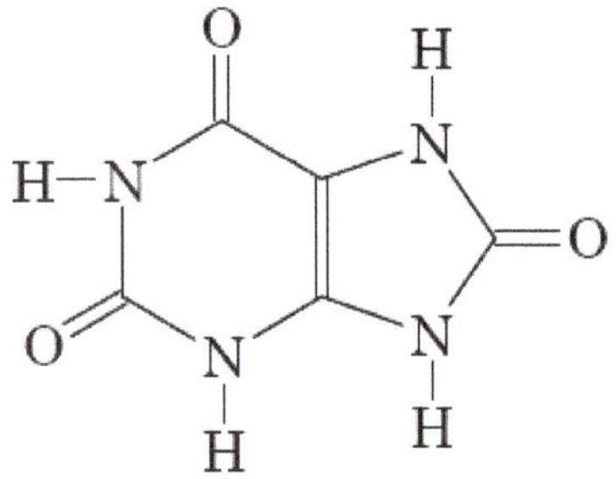

$$+ 9HOCl \rightarrow 5CO_2 + 2N_2 + 9HCl + 2H_2O$$

Reaction 2. Uric acid.

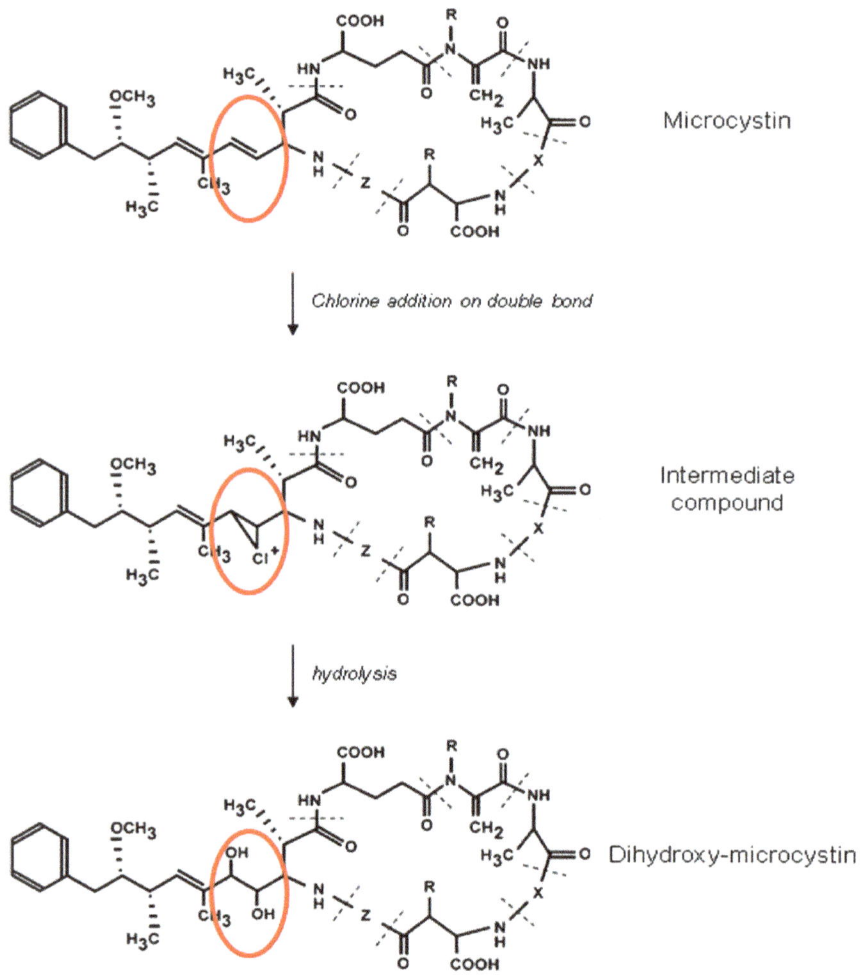

Reaction 3. Chlorination of microcystin.

for the cause of double breakpoint during chlorine demand of raw Yamuna River water.

MATERIALS AND METHODS

The physical-chemical process of ammonia oxidation with chlorine has been practiced in the water treatment field for over 60 years. As early as the 1920s, super chlorination was used as a successful means of controlling taste and odors in water treatment plants. In the 1930s, an unexplained phenomenon was being observed at water treatment plants using higher than normal chlorine dosages. These events prompted research into the chlorination reactions occurring at water treatment plants. Among the first researchers to explain these chlorine reactions, Griffin (1939) used the term breakpoint to describe the point where chlorine and ammonia concentrations were simultaneously minimized. The breakpoint reaction is defined as the chlorination of a water containing ammonia resulting in an initial increase in combined chlorine

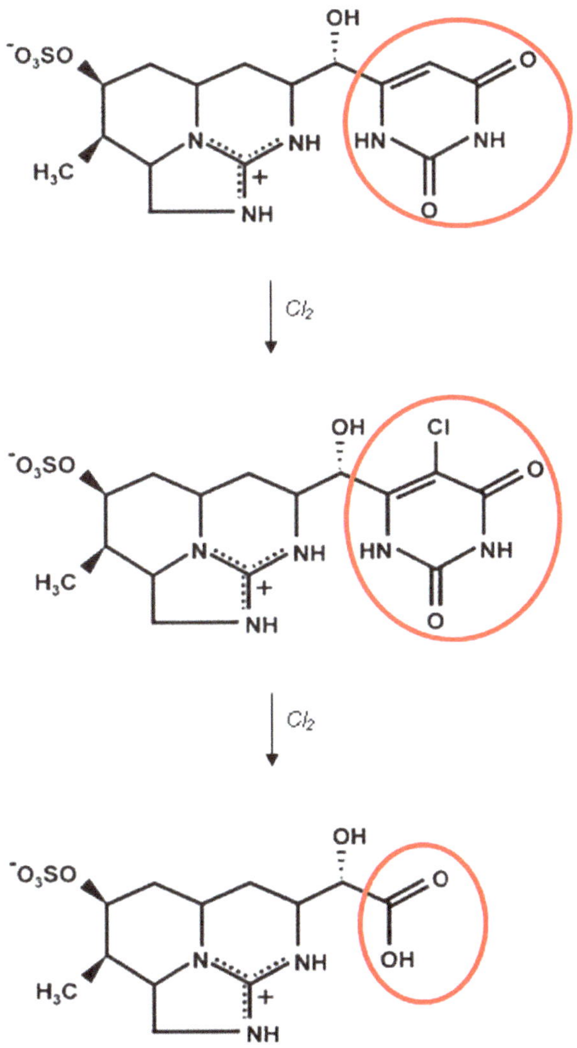

Reaction 4. Chlorination of CYN

residual, followed by a decrease in the combined chlorine residual along with ammonia concentrations, followed by an increase in free chlorine residual and near complete removal of ammonia as nitrogen gas. Figure 5 shows a hypothetical breakpoint curve for water with a dose requirement of 9:1 $Cl:NH_3$ (Jensen and Johnson, 1989).

Sample

The Wazirabad Water Works located in Delhi, India is one of the ancient water treatment plants running since 1966, which claims River Yamuna water for use in a potable water supply. The Yamuna River water is a reliable source of good quality water for a system serving over 50, 00000 (5 million) people in central and allied South Delhi of India National Capital territory. The Wazirabad Water Treatment plant is designed for operational reliability and redundancy in order to maintain high quality water under all circumstances. To this end, breakpoint chlorination process has been provided for nitrogen management. Nitrification involves the conversion of ammonia to nitrate. Breakpoint chlorination involves the use of chlorine (in the form of gas or sodium hypochlorite solution) to chemically oxidize ammonia and convert it to nitrogen gas. The water quality of River Yamuna kept deteriorating frequently exceeding its ammonia contents sometime 1.0 mg/L from 0.00 mg/L and nitrite contents to 0.10 mg/L from 0.006 mg/L. The results of this study are not likely to be repeatable with raw water that has not been polluted/deteriorated to the same degree as Yamuna River receiving at Wazirabad Barrage.

Data collection

Samples of River Yamuna water were collected and analyzed as per the guidelines laid in the book "Standard Methods" (1992) For Examination of Water and Wastewater prepared and published jointly by APHA, AWWA and WEF Washington, DC 20005. Tests were conducted during the year 2011 at 120 MGD Water Treatment Plant, Water Works Wazirabad, Delhi (India). The samples were tested in plant laboratory of water works as per "Standard Methods" book by Electrometric Method (4500-H^+ B) for pH, Argentometric Method (4500-Cl^- B) for chloride, Nesslerization Method (4500-NH_3 C) for Ammonia, Colorimetric Method (4500-NO_2^- B) for Nitrite, Chlorine Demand/Requirement Method (2350 B) for determining breakpoint, and Iodometric Method I (4500-Cl B) for residual chlorine. The raw water pH fluctuation of all the samples was in the range of 7.6 to 8.8 and chloride fluctuated in the range of 6.0 to 500 mg/L. Parametric values of pH, chloride, ammonia and nitrite were determined before performing detection of breakpoint in chlorine demand test. All tests were performed at normal room temperatures, fluctuated from 10 to 35°C in accordance with climate and season change from January, 2011 to December, 2011.

Breakpoint determination in chlorine demand test by Starch-Iodide method

Chlorine demand test were made daily on river water samples entering to Water Works Wazirabad, Delhi, India from January, 2011 to December, 2011. Standard chlorine water solution of 200 mg/L strength was added in general to a series of portions of the samples using increment of 1 to 2 ppm (part per million) chlorine. After 30 min contact, the residual chlorine was determined by the starch-iodide method according to "Standard Methods". Potassium iodide and starch were added to the sample and the liberated iodine was titrated with 0.0056338 sodium thiosulphate. The pH of the raw water samples was usually between 7.6 and 8.8. A stock solution of chlorine was prepared by mixing liquefied chlorine gas into 1 L distilled water. Chlorine dosage in increments of 0.1 mg/L for determining low chlorine demand/requirement and up to 1.0 mg/L or more for higher demands was used keeping in view ammonia and nitrite contaminations. Sample portions were dosed according to a staggered schedule that would permit determining the residual after predetermined contact time of half an hour. At the end of contact period of half an hour, residual chlorine of each of the bottle measured and noted. For determination of residual chlorine, 2 ml acetic acid was added followed by 2 ml KI solution (5% solution prepared freshly each time) to each of the bottle. Instantly each of the bottle was titrated with standard solution of sodium thiosulphate of 0.0056338 normality strength using starch solution as an indicator.

RESULTS AND DISCUSSION

Special investigations were performed to determine the specific chlorine demand of River Yamuna water (raw water) receiving at Wazirabad Barrage in Delhi, India. Two hundred (200) raw water samples were tested for studying breakpoint curves. Theoretical Chlorine Demand curve shows a sharp breakpoint as shown in Figure 5.

The plotting of applied chlorine dose versus residual chlorine gives a curve that can be depicted under four distinct areas or product formation as shown in Figure 5. The first area represents the oxidation of more reactive compounds than NH_3 by chlorine where residual chlorine found nil and is an inorganic demand phase. The second area represents an increase in combined chlorine residual (chloramines). The residual begins to drop because of destruction of the combined chloramines. At this point the nitrogen is given off as a gas and is lost to the atmosphere. Phase three continues until breakpoint. After breakpoint, all NH_3 products have been fully oxidized and free chlorine residual begins to form. But, in this present study, it was found that the theoretical chlorine demand curve as shown in Figure 5 did not always match but, differs at many times depending on the pollution condition of raw waters. It is widely held that N_2 and NO_3^- are the two main products of breakpoint and combined residue chlorination. Leao (1981) measured the amount of ammonia oxidized per chlorine reduced in the combined chlorine dose region after several days over considerable ranges of initial pH and Cl/N conditions and found no correlation to either of these parameters. Indeed, others have shown that at least one as yet unidentified product (Valentine et al., 1986; Hand et al., 1983; Leung, 1989; Valentine et al., 1990) and nitrate can be formed in the combined region. The unidentified product is relatively stable and contains chlorine, which is released as chloride ion upon reaction with sodium sulfite, indicating that it can act as an oxidant (Leung, 1989). During chlorine demand test of water, double breakpoints were found in the process of breakpoint chlorination as shown in Figures 1 and 2.

Here, water was heavily polluted with ammonia containing pollutants ($NH_3 - N > 0.40$ mg/L), while nitrite containing pollutants are in smaller magnitude (NO^{2-} N < 0.1 mg/L). The hypothetical reason for double breakpoints might be attributed to the formation of multi unidentified chlorine reaction products with ammoniacal contaminations and cyanobacterial bloom. The occurrence of cyanobacterial blooms is drastically increasing in temperate countries and drinking water resources are threatened. As a result, cyanotoxins should be considered in water treatment to protect human health. Chlorination efficiency on cyanotoxins alteration depends on pH, chlorine dose and oxidant nature. Microcystins and cylindrospermopsin are efficiently transformed by chlorine, with respectively 6 and 2 by-products identified. In addition, chlorination of microcystins and cylindrospermopsin is associated with a loss hump and dip of acute toxicity. Even though they have been less investigated, saxitoxins and nodularins are also altered by chlorine. For these toxins, no by-products have been identified, but the chlorinated mixture does not show acute toxicity. On the contrary, the fact that anatoxin-a has a very slow reaction kinetics suggests that this toxin resists chlorination. The typical profile of breakpoint curves can be attributed to the formation and destruction of chloramines and transformation of toxic cyanobacteria (blue-green algae) by chlorine that may have breakpoint earlier than disruption of ammonia compounds. Formation of unidentified products was estimated to account for up to 10 to 20% of the total observed reduction in chloramines oxidizing capacity under some conditions (Leung, 1989; Valentine et al., 1990). Nitrate formation, equivalent to approximately 15% of the loss in chloramines oxidizing capacity has also been observed in conjunction with the formation of unidentified product (Chad and Richard, 1992).

It is found that when raw water contaminated with nitrite containing pollutant heavily (NO^{2-}-N > 0.10 mg/L) and

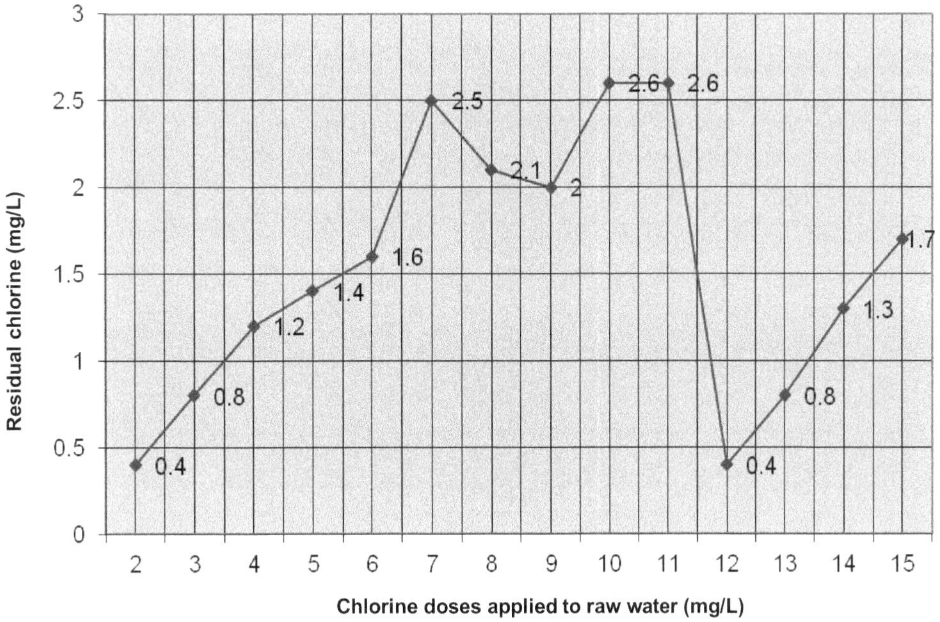

Figure 1. Chlorine residuals versus chlorine dose (Double breakpoints when ammonia is >0.4 mg/L and nitrate >0.1 mg/L).

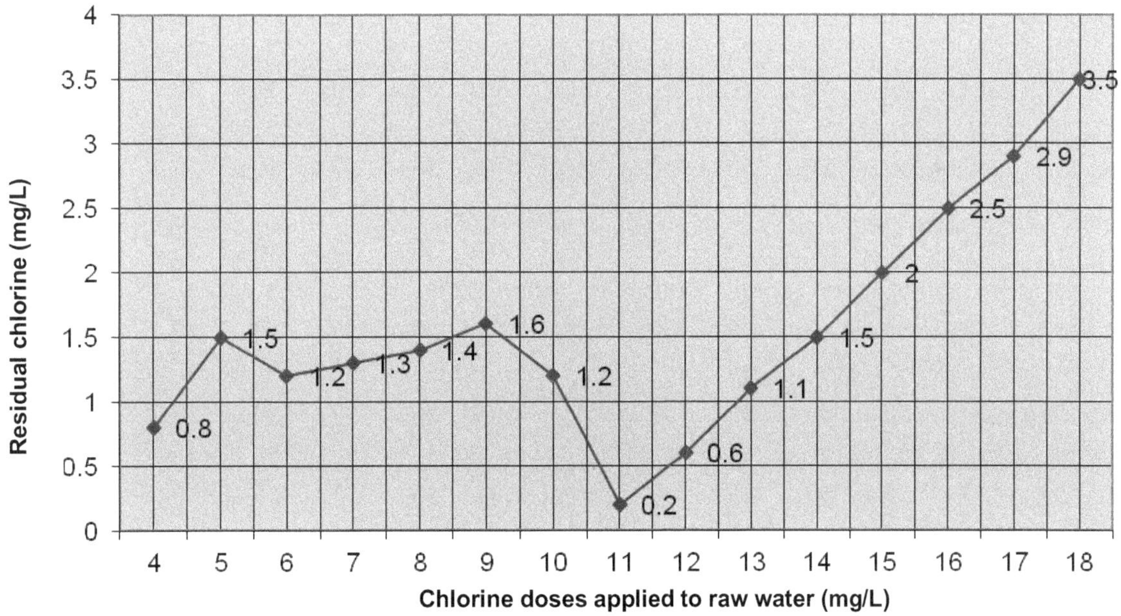

Figure 2. Chlorine residuals versus chlorine dose (Double breakpoints when ammonia is >0.4 mg/L and nitrate >0.1 mg/L).

ammonia containing pollutants are present to a lesser extent (NH_3 – N < 0.10 mg/L), a chlorine demand test curve does not have a breakpoint in it but, the graph is a straight line. The chlorine dose applied initially in Figure 3 is taken as 1.0 mg/L as minimum dose because applied chlorine below this amount does not show any residual and got consumed because of contaminants other than ammonia. The water samples under test might be contaminated by many others unidentified reducing compounds other than ammonia and nitrite.

This might be attributed to conversion of nitrite ion into nitrate, where no destruction of chloramines and other

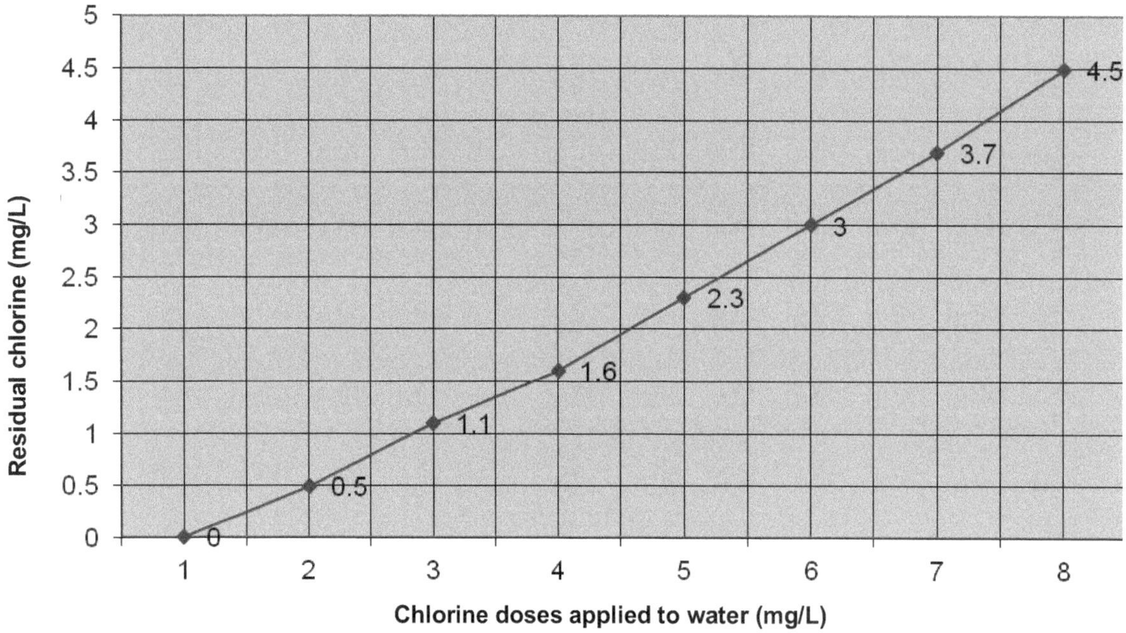

Figure 3. Chlorine residuals versus chlorine dose (No breakpoints when ammonia is <0.4 mg/L and nitrate >0.1 mg/L).

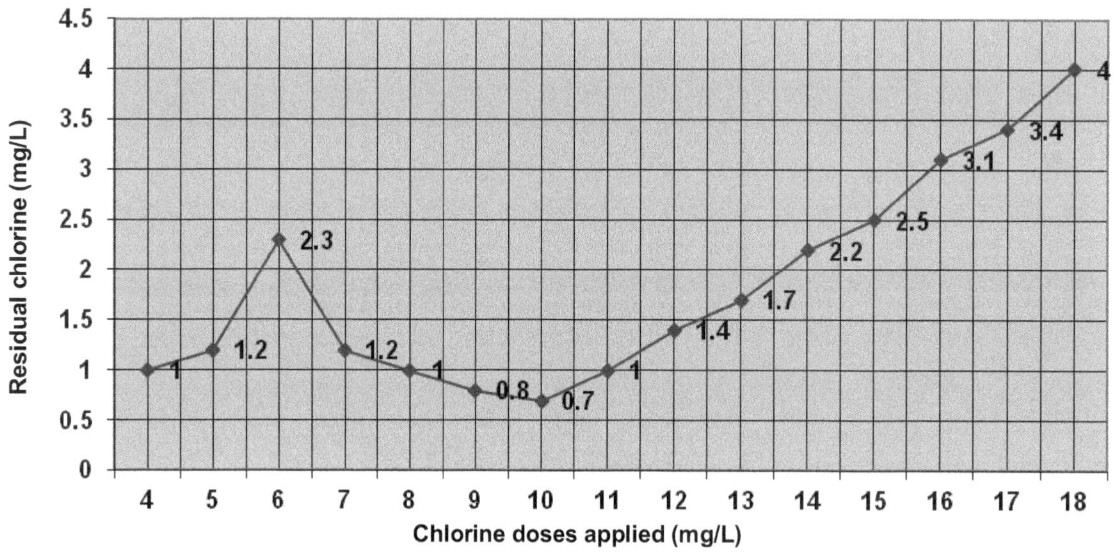

Figure 4. Chlorine residuals versus chlorine dose (Single breakpoint when ammonia is >0.4 mg/L and nitrate <0.1 mg/L).

organic- inorganic pollutants are required. No breakpoint has been observed with nitrite polluted water.

When raw water of River Yamuna was polluted with ammonia containing pollutants discharged from industries and non-functional/partially functional sewage treatment plants a graph similar to theoretical graph having single breakpoint was obtained.

Water samples collected and tested for ammonia and nitrite pollutants were found heavily polluted with ammonia (NH_3 – N > 0.40 mg/L) and nitrite (NO^{2-} -N > 0.10 mg/L). Here, a single breakpoint is observed with increasing doses of chlorine (Figure 4). Earlier results on breakpoint chlorination curves had shown only a single breakpoint during chlorination of water. However, any more specialized study especially on breakpoint chlorination curves has not been found in any journal.

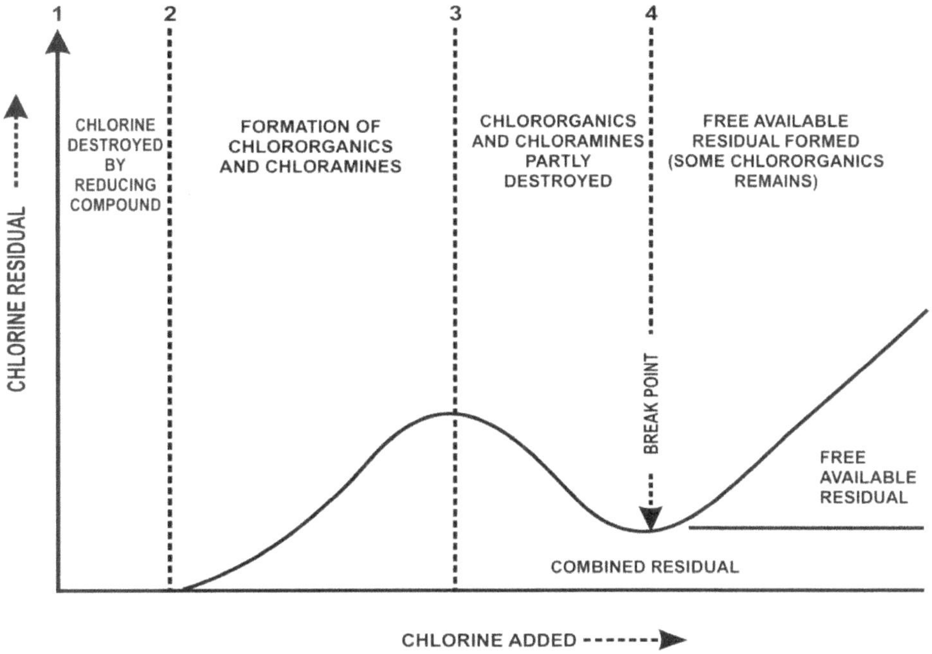

Figure 5. A breakpoint chlorination gragh.

Conclusion

It has been observed that "double breakpoints" also exists during chlorination of water which supersedes all studies indicating establishment of a single breakpoint during super chlorination of water. A single breakpoint was observed during chlorine demand test of water, when raw water was polluted/ contaminated with ammonia pollution. Double breakpoints were observed during chlorine demand test of water, when water was severely polluted with ammonia (>0.4 mg/L as N), nitrite (<0.01 mg/L as N), chloride and other suspected cyanobacterial blooms. A straight line/linear graph was observed during chlorine demand test between chlorine doses added versus residual chlorine for a contact period of half an hour when water was polluted severely with nitrite contents (>0.01 mg/L) and low ammonia containing pollutants (<0.1 mg/L). Ammonia and chlorine were both found co-existing in raw water. Despite what is known about reaction schemes between chlorine and ammoniacal pollutants in water, there is a considerable amount of critical information lacking, particularly information regarding reaction products. Double breakpoints in chlorination of River Yamuna water might be attributed to the oxidation of slower reacting organic chloramines, e.g., chlorinated derivatives of amino acids, creatinine, other organic nitrogen compounds and cyanobacterial blooms.

An extensive need of detailed study of kinetics of reaction between chlorine and different water pollutants are required to interprete causes of double breakpoints during chlorine demand of River Yamuna water. Additional research is necessary to identify all products over wide range of reaction conditions.

REFERENCES

Alouini Z, Seux R (1988). Kinetics and Mechanisms of Hypochlorite Oxidation of Creatinine. Water Res. 22:1519-1526.

Arash Z, Lionel H, Gayle N, Heriberto B, Michele P (2012). Fate of toxic cyanobacterial cells and disinfection by- products formation after chlorination. Water Res. 46:1524-535.

Atkins PF, Scherger DA, Barnes RA, Evans FL III (1973). Ammonia Removal by Physical-Chemical Treatment, J. WPCF 45(11):2372.

Banker R, Carmeli S, Werman M, Teltsch B, Porat R, Sukenik A (2001). Uracil moiety is required for toxicity of the cyanobacterial hepatotoxin cylindrospermopsin. J. Toxicol. Environ. Health: Part A 62(4):281-288.

Baribeau H, Krasner SW, Chinn R, Singer PC (2005). Impact of biomass on the stability of HAAs and THMs in a simulated distribution system. J. Am. Water Works Ass. 97(2):69-81.

Calvert CK (1940). Treatment with Copper Sulfate, Chlorine, and Ammonia, J. AWWA. 32(17):1155.

Camp D, McKee R (1983). Final Operability Report, Final Advanced Waste Treatment Plant, County of Fairfax, Virginia, Lower Potomac Pollution Control Plant.

Chad TJ, Richard LV (1992). Reaction Scheme for the Chlorination of Ammoniacal Water. Environ. Sci. Technol. 26(3):577-586.

Chen WJ, Weisel CP (1998). Halogenated DBP concentrations in a distribution system. J. Am Water Works Assoc. 90(4):151-63.

Feben D, Taras MJ (1950). Chlorine Demand in Detroit Water. J. AWWA 42:453.

Feben D, Taras MJ (1951). Chlorine Demand Constants. J. AWWA. 43:922.

Griffin AE (1939). Reaction of Heavy Doses of Chlorine in Various waters. J. AWWA 31(12):2121.

Griffin AE, Chamberlin NS (1941). Some Chemical Aspects of Breakpoint Chlorination. J. N. Engl. Water Works Assoc. 55:371.

Hand VC, Margerum DW (1983). Kinetics and mechanisms of the decomposition of dichloramine in aqueous solution. Inorg. Chem. 12:1449-1456.

Jensen JN, Johnson JD (1989). Specificity of the DPD and Amperometric Titration Methods for Free Available Chlorine: A Review, J. AWWA 81(12).

Leao SL (1981), University of California, Berkeley. Ph.D. Thesis

Leung SW (1989). The University of Iowa, Iowa City. Ph.D. Thesis

Liang L, Singer PC (2003). Factors influencing the formation and relative distribution of haloacetic acids and trihalomethanes in drinking water. Environ. Sci. Technol. 37(13):2920-2928.

Lomas PD (1967). Combined Chlorine residual of Swimming Pool Bath Water. J. Assoc. Public Anal. 5:27-36.

March JG, Gual M (2007). Breakpoint chlorination curves of grey water. Water Environ. Res. 79(8):828-832.

Newcombe G, Nicholson B (2004). Water treatment options for disolved cyanotoxins. J. Water Supply: Res. Technol.-AQUA 63(4):227-239.

Palin AT (1950). A Study of the Chloro Derivatives of Ammonia and Related Compounds, with Special Reference to Their Formation in the Chlorination of Natural and Polluted Waters, Water Wastewater Engr. 54:151-159, 189-100, 248-256.

Pressley TA, Bishop DF, Pinto AP, Cassel AF (1973). Ammonia-Nitrogen Removal by Breakpoint Chlorination. EPA report - 670/2-73-058.

Richardson SD (1998). Drinking water disinfection by-products. Ency. Environ. Anal. Remed. 3. John Wiley & Sons I; pp. 1398-1421.

Richardson SD, Thruston AD, Caughran TV, Chen PH, Collette TW, Schenck KM, Lykins BW, Rav-Acha C, Glezer V (2000). Identification of new drinking water disinfection by- products from ozone, chlorine dioxide, chloramine, and chlorine. Water Air Soil Poll. 123(1–4):95-102.

Rodriguez MJ, Serodes JB (2001). Spatial and temporal evolution of trihalomethanes in three water distribution systems. Water Res. 35(6):1572-1586.

Rodriguez MJ, Serodes JB, Levallois P (2004). Behavior of trihalomethanes and haloacetic acids in a drinking water distribution system. Water Res. 38(20):4367-4382.

Rodríguez E, Sordo A, Metcalf JS, Acero JL (2007a). Kinetics of the oxidation of cylindrospermopsin and anatoxin-a with chlorine, monochloramine and permanganate. Water Res. 41(9):2048-2056.

Rodríguez E, Onstad GD, Kull TPJ, Metcalf JS, Acero JL, von Gunten U (2007b). Oxidative elimination of cyanotoxins: Comparison of ozone, chlorine, chlorine dioxide and permanganate. Water Res. 41(15):3381-3393.

Rook JJ (1974). Formation of haloforms during chlorination of natural waters. Water Treat Exam. 23(2):234-43.

Rossum JR (1943). A Proposed Mechanism for Break-point Chlorination. J. AWWA 35:1446.

Saunier BM, Selleck RE (1976). "Kinetics of Breakpoint Chlorination and of disinfection." SERL 98 Report No. 76-2, University of California, Berkeley, CA.

Senogles P, Shaw G, Smith M, Norris R, Chiswell R, Mueller J, Sadler R, Eaglesham G (2000). Degradation of the cyanobacterial toxin cylindrospermopsin, from Cylindrospermopsis raciborskii, by chlorination. Toxicon 38(9):1203-1213.

Singer PC (1994). Control of disinfection by-products in drinking-water. J. Environ. Eng. 120(4):727-744.

Standard Methods for the Examination of Water and Wastewater (1992). 18th Edition pp. 4-78 Method 4500-NH_3-C (Nesslerization Method), AWWA.

United States Environmental Protection Agency (US EPA) (1999). Disinfectant use in water treatment. EPA Guidance Manual, Alternative Disinfectants and Oxidants, pp. 2, 54.

Valentine R, Brandt KI, Jafvert CT (1986). A spectrophotometric study of the formation of an unidentified monochloramine decomposition product. Water Res. 20:1067-1074.

Valentine RL, Wilber GG (1990). In Water Chlorination: Chemistry, Environmental Impact and Health Effects; Jolley RL et al, Eds.; Lewis Publishers, Inc.: Chelsea, MI; 6:63.

Weil I, Morris JC (1949). Kinetic studies on Chloramines. J. Am. Chem. Soc. 71:1664.

Wei I, Morris JC (1974). Dynamics of Breakpoint Chlorination in Chemistry of Water Supply Treatment and Distribution. A.J. Rubin, ed. Ann Arbor, Mich.: Ann Arbor Science.

White GC (1992). Handbook of Chlorination. 3rd ed.: Van Nostrand Reinhold Co., New York.

Williams DT, Benoit FM, Lebel GL (1998). Trends in levels of disinfection by-products. Environmetrics 9(5):555-63.

Strategies and techniques of providing adequate and affordable potable water in rural areas of Nigeria

S. Olu Adesogan

Civil Engineering Department, Federal University, Oye-Ekiti, Ekiti State, Nigeria.

Water is next to air in importance. The World Bank declared water as an economic good while endorsing the international demand for water supply. Human health depends on having access to safe, adequate and reliable water supply. In Africa, and of course in Nigeria, one half of the entire continent's people (particularly in rural areas/communities), suffer from one or more of the six main diseases associated with poor or polluted water. Statistics show that Africa has the highest occurrence of cholera and typhoid epidemics as well as child diarrhea. Of the 46 countries in which schistosomiasis are endemic, 40 are in Africa, of the 19 countries reporting guinea worm, 16 are in Africa. In September 2000, 147 heads of state and governments, and 189 nations in total, committed themselves to the Millennium Development Goals (MDGs). One of the targets defined for achieving the MDGs is to "halve by 2015, the proportion of people without sustainable access to safe drinking water and basic sanitation". This paper constitutes a source of information for water and sanitation coverage estimates in southwest geo-political zone of Nigeria. It provides information of the current status of water supply and sanitation in the zone. The paper attempts to look into the appropriateness of the use of the technology of integrated mini water scheme and infiltration gallery as a means of providing safe and adequate domestic water to rural community people to serve as the best preventive medicine against the prevalent water diseases.

Key words: Strategies, affordable, adequate, potable, water.

INTRODUCTION

Water is a ubiquitous chemical substance that is composed of hydrogen and oxygen and is vital for all known forms of life. In typical usage, water refers only to its liquid form or state, but the substance also has a solid state, ice, and a gaseous state, water vapor or steam. About 71% of the Earth's surface is covered by water. On Earth, it is found mostly in oceans and other large water bodies, with 1.6% of water below ground in aquifers and 0.001% in the air as vapor, clouds (formed of solid and liquid water particles suspended in air), and precipitation. Oceans hold 97% of surface water, glaciers and polar ice caps 2.4%, and other land surface water such as rivers, lakes and ponds 0.6%. A very small amount of the Earth's water is contained within biological bodies and manufactured products.

The impact of water use on our ecosystems should be an intricate issue of special concern in every area of the world as water is the one resource we cannot live without. Water is our most precious resource. Yet it is currently under attack by our waste, pollution, privatization, and the exacerbation of climate change. We must be aware of this and work to preserve and conserve water for future generations. This will be our legacy to our children (Harlander and Labuza, 2006).

It is good to finally see life cycle assessments being done for water use. More water is wasted (and polluted) in industry, yet they are not accountable for the water they use. And even though these assessments are not

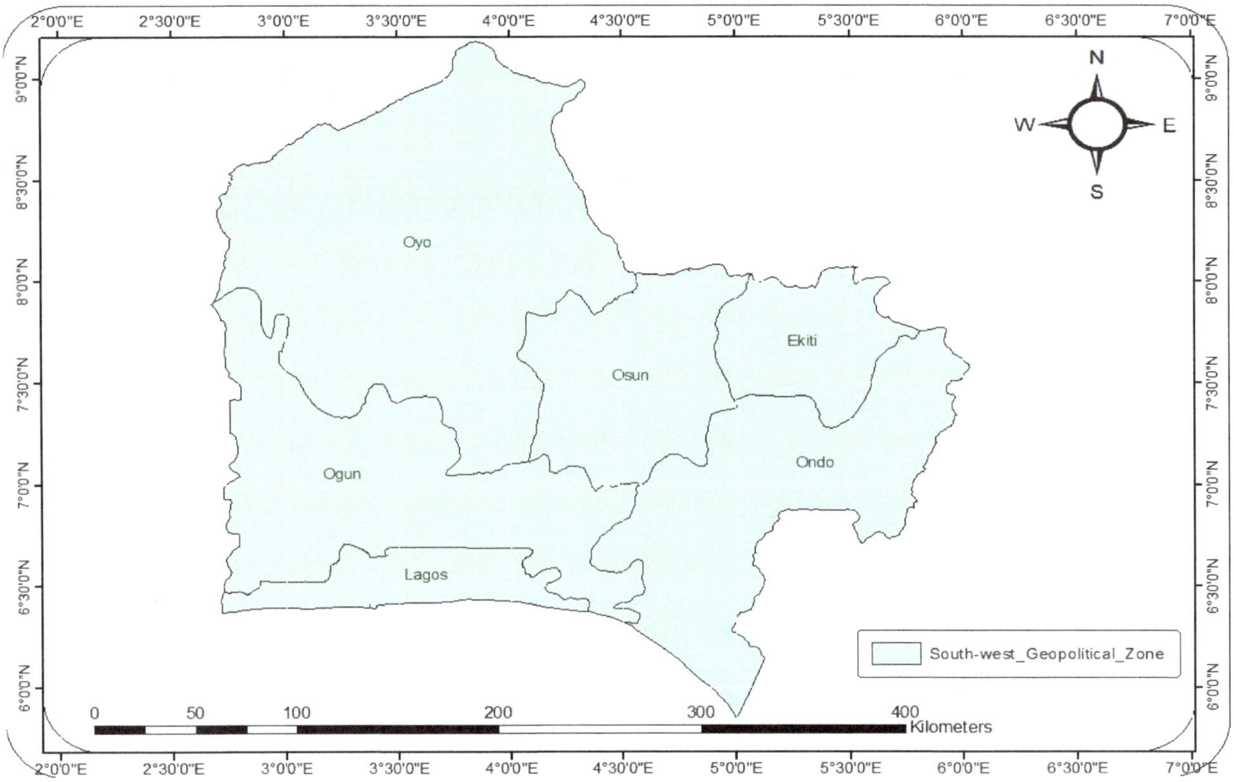

Figure 1. South-Western Nigeria.

based on changing factors over time, they at least give a good idea of what is being used, wasted, and how best to conserve water in different regions of the world experiencing different effects regarding that usage due to population, population growth, deforestation, agriculture, and now chiefly, climate change which is precipitating drought and melting glaciers more rapidly which absolutely effects the life cycle of water and all that depend on it (Osho and Dashell, 1997).

Africa is endowed with abundant water resources although its distribution and availability for use varies widely, with quite a number of countries facing water shortage and water stress. Regional and national water figures often conceal the dramatic effects of local water scarcity, limited or polluted supplies and inadequate distribution systems. Access to fresh water has been identified repeatedly as a key condition for development. National water policies and conservation efforts often tend to focus on the supply-side for domestic and agricultural use, and less commonly on industrial needs. Under these circumstances the uncontrolled use of a limited resource by water intensive industries takes on a special significance (UNEP, 2006). Potable water supply in our rural is achievable when we take all the necessary steps to achieving this, Adesogan (2008) stated that our world can be a world of plenty if we wish to make it so. In the past, many attempts have been made to solve the problem of water supply. These include the use of stream waters; construction of reservoirs and digging of shallow wells. These methods are not full proof as there are still records of potable water shortages even with these in place. The construction of reservoirs makes water to be susceptible to odour because the water in the reservoir may not be adequately aerated. Shallow wells, if too close to the septic tank can be polluted. These problems arising from poor water supply are crucial in the development of any community and efforts must be made to minimize or where possible eliminate them. An understanding of the strategies and techniques of providing potable water in our rural areas is necessary to be able to achieve this and therefore, there is need for this research. There is therefore the need to look into the appropriateness of the use of the technology of integrated mini water scheme and infiltration gallery as a means of providing safe and adequate domestic water to rural community people to serve as the best preventive medicine against the prevalent water diseases.

Study area

This study was carried out in Southwestern Nigeria comprising Oyo, Ogun, Lagos, Ekiti, Ondo and Osun states. The region has a bi-modal wind pattern with peaks occurring in April and August associated with rainstorm causing pollutions to water sources (Figure 1).

Table 1. Summary of water supply situation in Southwest Nigeria.

State	% Access to improved water supply			% Water supply coverage		
	Urban	Small towns	Rural	Urban	Small towns	Rural
Ekiti	56	68	57	29	33	27
Lagos	87	83	73	24	21	21
Ogun	84	66	55	49	36	27
Ondo	67	60	60	18	16	16
Osun	71	62	44	23	25	17
Oyo	72	59	54	27	22	21

Source: Survey (2009).

METHODOLOGY

Survey of water supply in South-Western Nigeria

The zone has six States out of which Ekiti State was selected as the pilot state for the research. The study approach involves the use of survey forms (questionnaire) designed the MIS consultant appointed by the Client. The forms are: Form 01 (Water Supply Facility Survey): To capture the location, attributes, and operational status of water supply facilities; Form 02 (Water Supply Agency Operational Survey): To collect data on the profile of water agencies in the state, in terms of production assets, capacity utilization, manpower; and financial sustainability; Form 03 (Sanitation Facility Survey): To capture the location, types and conditions of sanitation facilities; Form 04 (Water Related Diseases Survey): To collect data on reported cases of water related diseases from health institutions; Form 05 (Household Survey): To capture data on the proportion of households that have access to safe drinking water and sanitation facilities and prevalence of water related diseases in each community; Form 06: (Summary form for the Local Government Area).

Primary information was obtained through household survey carried out by recruited and well-trained enumerators. The survey was carried out in every ward in each of the six States of the geo-political zone. Within each ward, at least 24 houses were selected. Where the ward is a multi-community ward, the selected houses were shared among two randomly selected communities in the ward. However, for multi-ward communities, that is, communities with more than one ward within them, the selected houses were systematically spread among two randomly selected streets or quarters in the community. Facilities survey and determination of their coordinates were carried out by members of the project team with the aid of GPS. Secondary information on water supply and sanitation in the zone were obtained from the LGAs, State Water Boards/Corporations, State Ministries/Agencies of Environment, Ministries of Rural Development, Benin-Owena and Ogun-Oshun River Basin Development Authorities, water supply schemes, other sanitation offices and health institutions across the geo-political zone. Preliminary results of the study were discussed with stakeholders in each State de-briefing workshop where genuine concerns were addressed. State draft reports were circulated to respective State officials and a Zonal workshop organized to review the draft reports.

RESULTS AND DISCUSSION

Results of the water supply situation in the different settlement classes of the zone are summarized in the Table 1. With values mostly between 16 and 36%, water supply coverage is poor in the zone. The 49% value for urban Ogun State notwithstanding! However, access to improved water supply is much better all across the geo-political zone. Variations of these parameters within each Local Government Area are presented in individual State reports and in the zonal report. The results form the basis of calculating MDG targets as regards water supply by the year 2015. To meet this target, there is the need to substantially increase the water supply coverage from its present state all across the zone. In terms of sanitation, Table 2 presents a summary of percent access to sanitation facilities in the geo-political zone. Apart from Lagos, and to a lesser extent Ogun State, access to sanitation facilities is very poor in the zone. This result is further reinforced by the very little sanitation facilities mapped out during the survey as compared to water supply facilities. To accomplish the MDG target, there is a need to increase the number of sanitation facilities and pay more attention to people's access to the facilities. There are basically five categories of water supply promoters in the zone, with the State Governments being responsible for 100% of the surface water development. Even those developed by the Federal government are eventually handed over to their respective State governments. There are also the same five providers of sanitation facilities in the zone.

Disease survey

In the survey of more than 1,400 health institutions within the zone, malaria was found to be the most dominant water-related disease in each State of the zone. Typhoid, dysentery and diarrhea were found to be widely reported throughout the zone.

Strategies and techniques

Water supply schemes

Single-large and infective water schemes: In urban centres of developing countries with particular reference to Nigeria, Community water supply systems are

Table 2. Summary of access to sanitation facilities in Southwest Nigeria.

State	Total population	% Access to improved sanitation facilities
Ekiti	2,384,212	46.1
Lagos	9,013,534	84.8
Ogun	3,728,098	65.2
Ondo	3,441,024	36.7
Osun	3,423,535	36.5
Oyo	2,297,874	32.5

Source: Survey (2009).

fashioned after those that obtain in the industrialized countries with little or no adaptation. For example, Ado with a population of over 3.0 million is served from only a waterwork from Egbe with 180 million litres/day capacity. The water works combines Convectional treatment plant with mechanical pressure Filters in an arrangement that is very confusing resulting in frequent operational breakdowns. A disadvantage of this is that breakdown of the scheme will lay off all towns and villages supplied simultaneously. The distribution network is very scanty covering only about one-fifth of the entire city with complete neglect of the newly developed and economically viable areas of Omisanjana, Dallimore, Ajilosun, Opopogbooro etc. Only four reticulation reservoirs served the whole city with only three booster stations. Obviously, this cannot be an efficient system.

This same problem is likened to the old adage that "the strength of the chain is in the weakest link" in appraising the nature of the Ado water supply schemes. Ikere, Ise, Iju-Itaogbolu and many other villages are supplied from Ikere waterworks. Aramoko has only one waterworks, Irele, Itapaji, Ire, Oye, Agbado and Omuo were expected to be supplied from Ikole waterworks about 12 km from Irele. It is very clear and logical from the descriptions of some of the water scheme that acute water shortage has to be a citizen of our country. How do we "deport" this unwanted citizen?

Mini-water schemes

A study of 50 km radius of the city of Ado will reveal sources that could be harnessed separately for min-water schemes and which will meet the city's ultimate demand if integrated together. Adesogan (2000) in projecting the potentials of mini-water schemes identified over 60 streams in Ado land most of which take their sources from springs. Adeniran (1997) has proposed the use of 6 mini-water schemes as supplements to the two gigantic water schemes in Ibadan. These include Olodo, Omi-Adio, Odo Ona Elewe, Ogunpa behind St. Patrick's Grammar School at Idi-Ape Basorun, Tabielede at Ajibode and Sango/Oba at University of Ibadan. It is interesting to note that this report which was submitted at the end of Training course in the United States was never considered by the Water Corporation. It is doubtful if the report was ever brought to the notice of Government. It is however, gladdening that the same report was used by the University of Ibadan in developing the 6.0 millions litres/day mini-water scheme for the University. This scheme which serves a population of only 25,000 has made the University an oasis in the desert of Ibadan city. The U.I. scheme is maintained and operated by a total staff thus keeping overhead cost to the barest minimum. The scheme is also designed in modules such that any of the three modules can be operated without reference to others. This parallel arrangement makes water to be available at all times.

The importance of mini-water scheme can further be appreciated if an integrated water scheme is considered as is done in New York City. The entire city of New York with 5 Districts was discovered with 72 waterworks as at 1987. The largest of this waterworks only produced 20 Million litres/day. The water schemes are integrated in what was referred to as the "Pentagon Support Integration".

Each of the mini-water schemes is completely independent serving only a population of between 10,000 and 30,000. Each water scheme is under instruction to produce about 25% more than the water demand of the area that it serves. This is to take care of emergency fire hazard and to render immediate "support" to a sister water scheme that might be having a "breakdown" problem. This way all areas are covered effectively on a 24 h basis either from "internal supply" or from "support supply" obtained from all nearby pentagons. The staff kept by each water scheme is scanty. Of course, the revenue collection is not the business of the water scheme; this is carried out by private entrepreneurs. These business organizations only receive a the bill from water supply management authority on the basis of the quantity of water supplied to their area. These organizations pay immediately and arrange to collect the bills from consumers in their area. Water is thus a business in its own case and not a social service. Adequate records are kept of industrial developments, population changes etc. Any new development or expansion is expected to be self financing.

Private sector partnership

At present, Government, through the State water agencies seems to be singularly saddled with the responsibility of providing public water supply to the people. Water is often made a political issue. Such central institutions alone are not generally able to support large numbers of water supply schemes with the technical, human and financial resources required for successful long-term operation and maintenance.

Funds

Data collection, designing and construction of water supply schemes are very capital intensive, so also its operation and maintenance costs are very high. The high cost makes many governments to shy away from embarking on provision of water supply. The National Policy on Water Supply and Sanitation stated that 404.522 billion would be needed to address the absolute demand gap and increase by about 25 billion per year to increase the existing capacity of waterworks and construct new ones.

Capital cost

Due to high capital costs and lacking of securing of loans, there are many completely designed schemes gathering dusts in our offices and waiting to be constructed. For example, Ijan, \Aisegba, Ode and other towns and villages' water supply scheme was completely designed in 1991 but could be constructed due to lack of fund. As a result, of time-lag, the design is now obsolete and had to be re-awarded to another consultant who promised to secure loan for the construction of the scheme. This is yet to be actualized.

Others are mini-water supply schemes for eight local government headquarters in Ekiti State. The National Water Supply and Sanitation policy suggested that the capital cost of project be shared amongst the Federal, State and Local Governments and the community in the percentages shown in Table 3.

The National Policy also advocated cost distribution for operation and maintenance as shown in Table 4.

Also "business as usual" for provision of water supply must stop and "business unusual" must be adopted if all citizens of this Nation must have access to potable water by involving private sector participation in water supply. There are various options of Private Sector Participation as service, management, lease, Build-Operate and Transfer (BOT) and concession contracts which are available and being used by many countries including developing countries in the world. However, there must be a very strong political leadership before some of the options could be used.

Involvement of entreprenuers

The general discussion in the Nigerian water industry today is commercialization. Lending agents continue to emphasize that the water agencies become economically sustainable. This is a laudable and welcome development. However, from experience elsewhere in the world it is established that this thrust towards commercialization along will not be sufficient to change the culture and improve the efficiency without the autonomy of the water agencies. Waller (1998) has suggested that without the commercial freedom to employ staff, but chemicals, maintain plant and operate as a business outfits the SWA's will remain hindered however efficient they manage to be.

The Government may not want to release SWA's from their control except it is to the private entrepreneurs. However, the World Bank experience indicates that management contracts suffer from a similar fate to commercial operation without autonomy. The output of an organization cannot be improved without the ability of the management to make its own decisions. Hence, the involvement of private sector is critical step in achieving independence and hence a proper commercial capability.

It is obvious, from performance indices, that most of the Water Agencies in Nigeria and the developing world are poorly established in terms of physical and human resources that development by Government and technical assistance are the only option left for them. However, if strong management is introduced, many of the agencies can ensure that operating expenditure is at least brought into balance with revenue collection. This will often necessitate the increasing of tariffs in accordance with a well determined unit cost of production. The options for involving the private sector in the Nigerian water supply industries can be addressed by focusing on the development of assets information, encouragement of the private sector, regulation and reduction of running costs.

Health engineering with minimal resources

One of the themes running through this paper will be the intension of a growing understanding that aid or the direct giving of help is not development, and tends to be detrimental in the long term. In countries of Africa, this includes giving by Central Government as well as giving by external agencies. But there is also the realization of the need for pump-priming by outsiders, the urgent need for a catalyst to release the inherent abilities of the people. This can be seen most clearly in the requirement which gradually became clear of the need to ``invest'' on occasion in expensive structures which apparently served no particular purpose in improving health. For example, permanent materials Primary Health Care Clinic might appear an unnecessary luxury when compared to an improved mud, wattle and thatch building with a 20 to 1

Table 3. Water supply scheme.

Agency	Rural (%)	Small town (%)	Urban town (%)
Federal Government	50	50	30
State Government	25	30	50
Local Government	20	25	10
Community	5	5	10

Source: Federal Ministry of Water Resources.

Table 4. Water supply scheme.

Agency	Rural (%)	Small town (%)	Urban town (%)
Federal Government	-	-	-
State Government	10	-	100
Local Government	20	-	-
Community	70	100	-

Source: Federal Ministry of Water Resources (2008).

cost difference. However, it could serve a vital purpose in proving to the people that development was coming and would benefit hem. The other major lesson learnt was that although it is so easy to look down on people in such a situation, considering them underdeveloped with little initiative and little willingness to help, they were survivors. They were the ones who had managed to live through a long time in a hostile environment. They were therefore canny and wise and manipulative and in a very reasonable way were out to get the best for themselves, their people and their District. Not surprisingly this meant they wanted to see the largest amount of money invested in their District in the shortest possible time. They wanted structures and hardware and goods and medicine and were not enthusiastic to hear about community participation and the need to use their own resources.

Community involvement in water supply

Now it became possible to discuss alternative forms of improvement to existing water sources, particularly in areas where the people were widely scattered. The Community Health Workers responded enthusiastically to the opportunity to carry out a survey of water resources in their area and, with the people, to recommend ways of improvement. In the areas where unlined hand-dug wells were used already, concrete culvert rings being used in feeder road upgrading could be given to seal off the well from surface water intrusions. In other areas where there is stream a couple of bags of cement could allow a small masonry dam to be built on a stream to increase the storage of water in the sand during the dry season. Such sand reservoirs gave natural filtration, limited evaporation, prevented mosquito breeding on stagnant pools and by reducing the frogs' habitat reduced the number of snakes.

In an area which had proved fruitless for drilling because of the un-weathered fresh rock, part of an outcrop could serve as a natural reservoir which could be cleaned out and enlarged by a small masonry wall. This rock catchment would enable them to use the results of the rains for their water uses. Community involvement does not always have instant result of course. News of failure in any development project appeared to spread very quickly through a community. News of success was less obvious but people were watching and gauging what was going on and when an idea worked it picked up momentum and schemes appeared to multiply exponentially.

Artificial recharge and recovery method

There are natural ground water replenishment or recharge methods. It was noticed that this increases the volume of water in the soil, and raised the ground water level. The phenomenon was particularly noticed in the arid and shore or coastal areas. The subsequent rise in: (1) population; (2) surface water pollution, and (3) water consumption.

Definition of artifical recharge

Artificial ground water recharge could be defined as planned activities of man whereby surface (streams, rivers, lakes) waters are made to percolate and infiltrate the ground at rates and in quantities many times in excess of natural recharge. By judicious design of recharge and recovery cycle, corresponding increase in the amount of ground water abstracted is possible with impeccable bacteriological quality.

Methods of artifical recharge (AR)

Ground water reserves in the aquifer can be increased in two major ways.

Natural / artificial method: Which is generally referred to as indirect method? In this case, increased replenishment is obtained by locating the means of ground water abstraction (infiltration gallery, large or small diameter well), close to any surface water (stream, river or lake with porous or pervious banks within a well accompanying abstraction (by pumping) will result in increased in flow of water from both the surface water and the aquifer sources through the pervious soil. This method is otherwise called induced recharge.

Direct method: This is the method by which water from a distant (or near) surface source (stream, river, lake, pre-treated sewage water, storm drainage) is transported or conveyed to areas or point where it is spread over the surface of previous soil basins, ponds, ditches or furrows and by pits, shafts, channels or by mere flooding in less pervious soil.

Induced or indirect artificial recharge: Criteria of design

As described previously, induced AR is the situation whereby water is drafted into the soil by locating as infiltration gallery or a (small or large diameter) well at small but carefully designed distance parallel to a surface water source (stream, river or lake).

At beginning of the water recovery to the rural area, the water pumped out $q_o = q_n$. This is the case when the ground water table is higher than the water level in the surface source nearby (where q_n is the natural underground water level in the surface rainfall P). As the water demand increases, pumping rate increase, the ground water level will fall below the surface water level. At this time recovered water q_o could be expressed as follows:

$$q_o = \ + q_n$$

Where q_s is the water induced by pumping from the surface source through the soil into the gallery or well at distance, L form the source. This is the artificial groundwater supplied by downward percolation of the river water which infiltrates the pervious soil into the gallery or well or a battery of them.

Use of well

Induced ground water recharge with gallery is not only possible when the ground water is not at a great distance single well bear a surface may suffice but when the demand is large a line of wells parallel to the streamed is used.

Single well for small demand

When a single well is used, the only design factor is the distance L between the well and the shore line. The drawdown S_o is small and it is localized. Thus, it des not form a design limiting factor. Also, the entry rate V_e of the surface water into the aquifer is not a boundary condition as the aquifer is deep down. Thus detention time T for the flow of surface water to well is the only design requirement, the distance L must satisfy. And minimum required distance L of the well to the shore line. Therefore with the notion, the elevation h of the ground water is expressed as (using Darcy and continuity equations as applied to unconfined/and semi confined aquifer).

Battery of parallel wells for large capacity

When is needed to supply large amount of a cluster of villages or rural communities, urban segmented communities, or about 100,000 to 200,000 people, whose water demand is in the order of 0.23 m^3 /s (about 20,000 m^3/day) and above, induced recharge and recovery by battery of wells is the right solution. All the wells must be located at a distance "L" parallel to the shore line or streambed.

Design assumptions

The wells are each at a distance L from the shoreline/streambed. The distance of the wells from each other is a small distance, "b" compared with their distance L from the streambed. In the above two conditions, the flow pattern is assumed equal to that of a gallery with capacity. These three conditions simplify the design of this artificial induced recharge scheme to be exactly the same as for a gallery. A constant value of coefficient of transmissibility KH is assumed and that the river fully penetrates the aquifer thickness.

Conclusion

It is therefore not out of reason for water resources and environmentalengineerstore-access theabovesubmission (a mere reminder of what exists), then enlist the artificial recharge and recovery (AR & R) method in the list of water supply options to rural or small communities.

Mini water supply option based on small communities in all the local government area should replace the ineffective and inadequate gigantic option that is the vogue now. Decentralization of water supply scheme with monitoring, administrative, financial and technical authority shared between the community users (stakeholders), local and state governments. This will give and ensure efficient method for effective production of adequate hygienically reliable water that can ensure and give preventive medicine for the local/rural dwellers against water-borne disease.

REFERENCES

Adeniran AE (1997). "Mini Water Supply Scheme". Unpublished M.Sc.Project in Mechanical Engineering Department, University of Ibadan, Ibadan, Oyo State Nigeria

Adesogan SO (2008). "Soil and Water Conservation: A Challenge to Young Engineers". Unpublished paper contest at Obafemi Awolowo University, Ile Ife, Osun State Nigeria.

Harlander SK, Labuza I (2006). The Brewing Process. In: Biotechnology in Food Processing, Chapman-Hall, Ltd. University Press, London.

Osho M, Dashiell K (1997). Expanding Soya bean production, processing and utilization in Africa. Post harvest Systems. 1997; Online at http://www.iita.org.

United Nations Environment Programme (Industry and Environment) ABREW: (2006). Framework analysis country report for ABREW in Ghana, Uganda, Ethiopia and Morocco 2006.

17

The Role of Mara River Basin Wetland in Reduction of Nitrogen Load to Lake Victoria

Aloyce W. Mayo, Marwa Muraza and Joel Norbert

Department of Water Resources Engineering, University of Dar es Salaam, Tanzania.

The role of Mara River basin wetland in reduction of nitrogen load to Lake Victoria was investigated. Reconnaissance survey was carried out to identify the suitable sampling points in the wetland. Transects were developed in the inlet and outlet zones of the wetland through which three sampling points were established at each transect. Data for physical-chemical parameters such as pH, temperature and dissolved oxygen were determined *in situ* and nitrite-nitrogen, nitrate-nitrogen, ammonia-nitrogen and total Kjedahl nitrogen were determined in the laboratory at Mwanza. Sediment samples were collected from the field by using soil auger and samples were taken to the laboratory for examination of nitrogen content in the sediments. The determination of nitrogen mass in plants involved determination of plant dominance, plant density and plant biomass before samples were taken to determine nitrogen in biomass. The results indicate that Mara River wetland receives about 0.70 ~ 1.56 mg/L of nitrogen largely in the form of organic nitrogen (63.6%) and nitrates (29.1%). The wetland removes about 28.8% of this nitrogen largely through net-loss to sediments and uptake by wetland plants. It is estimated that about 75 tons of nitrogen is removed annually, which is equivalent to 3.67 kg/ha/year. About 0.38 gN/kg of dry sediments were trapped in the benthic layer and 67.9 gN/m^2 of wetland is trapped in plant biomass. It was concluded that Mara River wetland was effectively protecting Lake Victoria by reducing nitrogen load entering the lake.

Key words: Lake Victoria, Mara River wetland, nitrogen removal, nitrogen load.

INTRODUCTION

Lake Victoria, the largest fresh-water Lake in Africa and the second largest in the world, is a resource of great social-economic potential in East Africa (UNEP, 2006). Its basin, which is estimated to accommodate over 40 million people, has numerous economic opportunities including fishing, tourism, water, energy, agriculture, trade and industry. More than 80% of the population living in the basin engages in small scale agriculture and animal husbandry (Makalle et al., 2008). Unfortunately, basin community agricultural practices, livestock overgrazing has led to serious land degradation and deforestation (Raburu and Okeyo-Owuor, 2005; Makalle et al., 2008; Twesigye et al., 2011; Thenya et al., 2005; Bancy et al., 2005). Similar problems were reported in other lakes in the region including Lake Kivu (Bagalwa, 2005), Lake Naivasha (Gichuki et al., 2005; Mironga, 2005; Amondi et al., 2005) and Lake Nakuru (Gichuki et al., 2005; Raini, 2005). Deforestation and land degradation causes soil erosion, which is the major cause of deterioration of water quality of natural water bodies such as rivers and lakes (Amondi et al., 2005). Soil erosion is largely caused by run-off, which pick up and transport sediments and pollutants resulting from agricultural, industrial, mining or any other economic activity. Agriculture is known to cause non-point source pollution, which is diffuse in nature and has complicated spatial and temporal dimension (Andrew, 1990; Henry and Semili, 2005; Amondi et al., 2005).

As precipitation falls, run-off is created, which pick appreciable quantity of dissolved and adsorbed material that are carried away through hydraulic force of its flow (Kirkby and Morgan, 1980). Some of the adsorbed and dissolved substances found in run-off include nutrients from commercial fertilizers and animal manure (Rode and Lindenschmidt, 2000). Nutrients, sediments along with other pollutants such as heavy metals have contributed to ecological degradation of coastal marines and lakes including Lake Victoria (Henry and Semili, 2005; Omondi and Kusewa, 2005; Gichuki et al., 2005; Ongore et al., 2013). Globally, wetlands are known for their filtration capacity (Terer et al., 2005; Kansiime, 2004), maintenance of biodiversity (Hammer and Bastian, 1989; Muraza, 2013), retention of heavy metals (Marwa, 2013; Henry and Semili, 2005) and play a vital role as breeding ground for fish (Balirwa, 1995). Other benefits of wetlands include conserving wildlife, provision of products (such as fish, reeds, timber, firewood and medicines), microclimate stabilization, flood control and ground water recharge (CEC, 1995; Dugan, 1990; Maltby, 1990; Hogan et al., 1992). Wetlands are often cited as being effective at reducing nutrients loadings, acting as "the kidneys of the catchment" (Mitchell, 1994; Mitch and Gosselink, 1986) thereby reducing eutrophication in downstream water bodies. This has led to wetlands being managed or constructed as buffers or for treatment of domestic or industrial wastes (Allinson et al., 2000; Mayo and Bigambo, 2005; Marwa, 2013). The wetland vast buffering capacities and functions have attracted researchers' interest around Lake Victoria in Uganda (Kansiime et al., 1994; Kansiime and van Bruggen, 2001), Kenya (Terer et al., 2005) and Tanzania (Henry and Semili, 2005; Muraza, 2013).

The Mara River Basin faces serious environmental and water resources problems, primarily from the intensive settlement and cultivation in the Mara River Basin leading to loss of vegetation cover, widespread soil erosion, decreased water infiltration capacity, decreased soil fertility and increased sedimentation as well as water pollution in the river (WWF, 2006; Nile Basin Initiative, 2007; Bitala et al., 2009). The Mara River basin wetlands have been receiving pollutants from large and small scale gold mines, agricultural activities and animal husbandry (Henry and Semili, 2005; Muraza, 2013; Marwa, 2013). Furthermore, the Mara River is a home to the wild animals of the Serengeti National Park in Tanzania and Maasai-Mara Game Reserve in Kenya. The Serengeti-Maasai-Mara is a World's Heritage site and a Biosphere Reserve of global conservation significance and of great economic importance to the local communities. These economic and social activities adversely pollute the River. As a result, information on water quality of the river is needed for the planning of its management and the control of eutrophication in the river and Lake Victoria in general (WWF, 2006; Nile Basin Initiative, 2007). Such information is necessary for the efficient planning of long-term sustainable use of the Mara River basin wetlands (Nile Basin Initiative, 2007; Bitala et al., 2009).

This research was therefore carried out in order to determine the variation of water quality parameters in the Mara River basin wetland in order to address some of the unanswered questions.

MATERIALS AND METHODS

Study area

Mara River is an international river shared between Kenya and Tanzania. Its basin is about 13,750 km^2 of which about 65% is located in Kenya and 35% in Tanzania. The river originates from the forested Mau escarpment along the western rim of the Eastern Great Rift Valley in Kenya (at an altitude of 2,900 m above mean sea level), meanders through large scale agricultural farms, enters the Maasai-Mara and Serengeti National Parks in Kenya and Tanzania, respectively and ends its 395 km journey in Lake Victoria through Mara wetland (Figure 1). Annual rainfall on the Mara headwater watersheds ranges from 1400 to 1800 mm, while the lower regions receive only about 500 to 800 mm. The main hydrological processes in Mara wetland are mainly run-offs from Mara River inflows from upstream catchment that brings water into the wetland. From the wetland, water flows into Lake Victoria contributing to 37.5 m^3/s, which is about 4.8% of total discharge into Lake Victoria (LVEMP, 2005).

Mara wetland covers a total area of 204.46 km^2; its length is 36.8 km with a maximum width of 12.9 km (GLOWS, 2007). The wetland is situated between longitudes 34°00' and 34°25' East and between latitudes 1°08' and 1°39' South. At downstream part of the wetland at Kirumi Bridge, the wetland is about 6 km upstream of Lake Victoria. Administratively, the wetland lies between Tarime and Musoma rural districts of Mara region (GLOWS, 2007).

Sampling design

To establish the sampling points, the study site was surveyed to identify the suitable sampling points in the wetland (Figure 1). Some vegetation was to be cleared at some areas so as to provide accessibility and support during the sampling exercise. Sampling preparations also involved *in situ* identification of vegetation species located in various places so as to come up with the vegetation zonations. After the reconnaissance survey was done, transects were developed in the inlet and outlet zones in the wetland. Transect PT-1, which was largely covered by floating *Papyrus* mats, was located at the upstream end near Bisarwi village. Transect PT-2 was located downstream near Kirumi bridge at about 100 m upstream of the bridge and is dominated by mixed floating *Papyrus* and the rooted *Typha domingensis*. Through these transects, three sampling points were established at each transect. For each sampling location, a Global Positioning System (GPS) was used to record the sites' coordinates; hence, helped in finding the site whenever sampling was done. Water, plants and sediments samples were taken since they are required for the modeling processes having some state variables of water, plants and sediments.

Collection of water samples and laboratory analysis

Forty-five (45) sets of water samples were collected during the whole study period between April and July 2012. Collection of water samples was done through inserting a 1 L plastic container up to

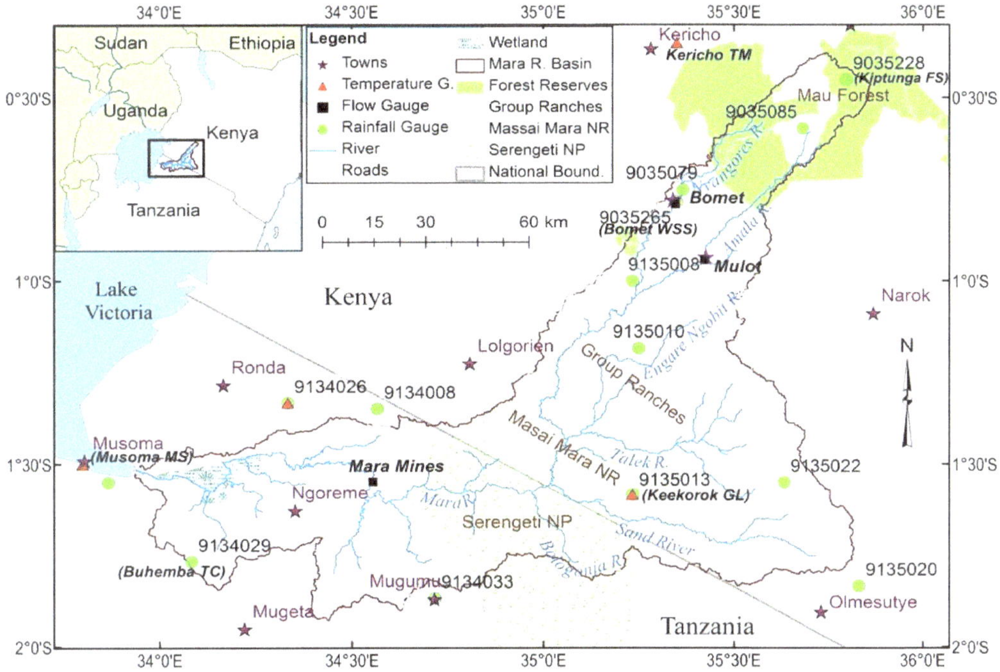

Figure 1. The location and relief of the Mara River Basin in Kenya and Tanzania.

the depth of about 30 cm deep and after the containers are filled with water, they were removed from the water column and kept in the cool box ready for transportation to the laboratory for analysis. Since prompt analysis of the samples was impossible, samples for analysis of ammonia nitrogen, total nitrogen and total Kjeldahl nitrogen were preserved with 0.8 ml conc. H_2SO_4 per litre of sample (to obtain a pH of 1.5 to 2.0), but those for analysis of nitrite nitrogen and nitrate nitrogen were not treated with an acid, they were stored at 4°C (together with those for analysis of ammonia nitrogen, total nitrogen and total Kjeldahl nitrogen after being treated with the acid) for one day in the Musoma Water laboratory before being transported to Mwanza water laboratory for analysis.

Water quality analysis

Parameters determined were physical parameters [pH, dissolved oxygen (DO) and temperature] and nutrients [nitrate nitrogen (NO_3^--N), nitrite nitrogen (NO_2^--N), ammonia nitrogen (NH_3-N), total nitrogen (TN) and total Kjedahl nitrogen (TKN)]. Temperature and pH were measured *in situ*. The pH was measured by a calibrated pH meter of Testo GmbH & Co. D-79849. Temperature was measured using temperature-meter Model HATCH HQ 30d. To obtain these parameters, probes were inserted into the water column (up to a depth of about 30 cm) and readings were taken after stabilization of the probes. Dissolved oxygen concentrations were analyzed in the Musoma Water Laboratory by Winkler titration method this is due to the fact that the DO-meter's probe did not stabilize even after a very long calibration; hence, we had to opt for the laboratory analysis. Water samples for dissolved oxygen determination were taken using special dissolved oxygen sampling bottles and preservation before analysis was done in accordance with Standard Methods (2012). Nutrients (NO_3^--N, NH_3-N, NO_2^--N, TN and TKN) analysis for water samples was carried out following procedures outlined in the standard methods for examination of water and wastewater (Standard Methods, 2012) at the Mwanza Water Laboratory. Both NO_3-N and NH_3-N were analyzed using Spectrophotometer with Cadmium Reduction and Turbidmetric methods respectively.

Nitra Ver.5 and Nessler reagents were used for NO_3^--N and NH_3-N analysis respectively. NO_2^--N was analyzed using calorimetric method with a spectrophotometry machine, UV-2001, TN was analyzed using the per sulphate digestion method and TKN was analyzed using the Semi-Micro Kjeldahl method.

Sediment sampling and analysis for nitrogen content determination

Sediments samples were collected eight times during the study period with the use of a soil auger. The obtained samples from the three points within each transect were packed in a cool box and transported to the laboratory in Mwanza. In the laboratory, samples were oven dried at 105°C. After drying, the samples were ground in a mortar and pestle, sieved through a 0.5 mm sieve size to obtain the dry powder. To analyze sediments for nitrogen contents, sub-samples with appropriate weights were taken and digested in a block using a concentrated sulphuric-salicylic mixture with selenium as a catalyst. Then, the analysis for nitrogen contents was done according to Standard Methods (2012).

Data and statistical analyses

Data analysis involved organizing the water quality data into the following data sets; the entire set of the two (inflow and outflow) systems, the inflow data separated into its three sampling points and the outflow separated into its three sampling points. For these data sets, the inflow and outflow data were kept separate to perform additional analysis for example to calculate the daily average values. Analysis of the data included statistical correlations and statistical differences between the data. The water quality,

Table 1. Characteristics of observed influent and effluent parameters.

Parameter	Influent value		Effluent value	
	Range	Mean	Range	Mean
pH	6.71- 7.27	6.95±0.16	6.34-7.17	6.85±0.29
Temperature (°C)	21.3-24.1	22.5±0.83	22.0-24.6	23.25±0.87
Dissolved Oxygen (mg/l)	1.3 - 4.8	3.16±0.99	2.6-4.9	3.85±0.74

Where number of samples (n) for all the parameters = 44.

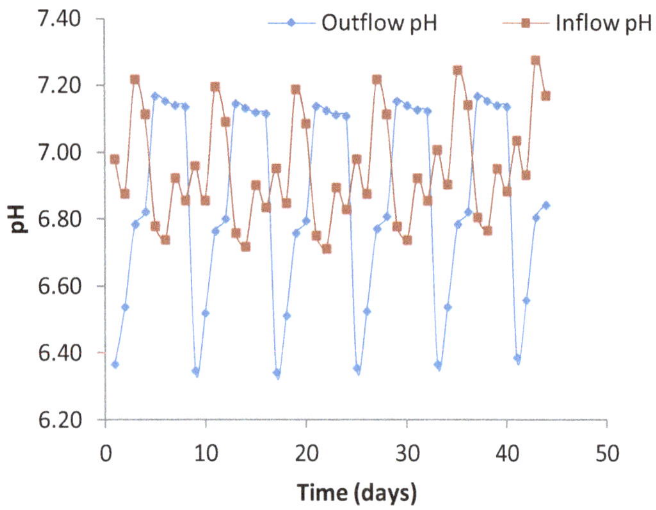

Figure 2. Variation of inflow and outflow pH with time.

sediments nutrients contents and plant nutrients contents data received from the laboratory were analyzed to check for their correctness. Constituents measured during the study were analyzed using Easy Fit 5.5 and Excel software systems, respectively. Spearman correlation tests were performed on various data sets to examine relationships between constituents. Correlation analyses measure the linear relationships between constituents and were performed using Spearman's rank-order linear relationships by looking at the coefficient, which ranges from -1.0 to +1.0. The stronger the relationship between constituents the higher the coefficient (±1 = a perfect correlation).

RESULTS

Variation of physical-chemical parameters

Table 1 shows that the pH of the water in the wetland ranged from 6.71 to 7.27 with a mean value of 6.95 ± 0.16 at the inlet zone while at the outlet zone it ranged from 6.34 to 7.17 with a mean value of 6.85 ± 0.29 at 95% confidence interval (α = 0.05). Figure 2 shows the variation of inflow and outflow pH in the wetland with time. The observed outflow pH value suggests that microbial activities in the wetland are favored since this pH is within their normal functioning pH range (Kadlec, 2009). The dissolved oxygen concentration ranged from 1.3 to 4.8 mg/L with a mean value of 3.16 ± 0.99 mg/L at the inlet zone whereas at the outlet zone it ranged from 2.6 to 4.9 mg/L with a mean value of 3.85 ± 0.74 mg/L at the 95% confidence interval (Table 1). Dissolved oxygen values suggest that the wetland is fairly aerated. Dissolved oxygen increased by 21.8% in the wetland suggesting that the water was significantly oxygenated by the biochemical activities taking place in the wetland, particularly the photosynthesis of the vegetations which gives out oxygen (Figure 3). Water temperature ranged from 21.3 to 24.1°C with a mean value of 22.5 ± 0.83°C at the inlet and 22.0 to 24.6°C with a mean value of 23.25 ± 0.87°C at the outlet zone (Table 1 and Figure 4). The pH had a very strong relationship with temperature (R = 0.937) as shown by Figure 5, suggesting that the variations of temperature in the wetland have a direct effect on the pH. This is because under limiting conditions, temperature which is influencing photosynthetic consumption of CO_2 causes bicarbonate to dissociate and release hydroxyl ions.

On the other hand, Kayombo et al. (2000, 2002) observed that dissolved oxygen, which is largely generated by photosynthesis, is a function of temperature and light intensity. The temperature and light intensity's influence on photosynthetic activities limit the maximum algal growth thus limits the maximum substrate utilization. Conversely, good correlation between dissolved oxygen with temperature (R = 0.773) indicates that temperature is one of the limiting factors for photosynthesis (Figure 5).

Ammonia nitrogen

Figure 6 shows the variation of inflow and outflow ammonia nitrogen concentrations with time. The concentration of ammonia nitrogen ranges from 0.005 to 0.632 mg/L with a mean value of 0.063 ± 0.017 mg/L and from 0.005 to 0.128 mg/L with mean value of 0.028 ± 0.023 mg/L for inflow and outflow zones, respectively (Table 2). The observed mean values gave an ammonia nitrogen removal efficiency of 55.6%, which was probably caused by nitrification of ammonia as a result of adequate dissolved oxygen favourable temperature and near optimum pH values. The evidence of strong correlation (R = 0.835) between the inflow and outflow concentration of ammonia is shown in Figure 7(a).

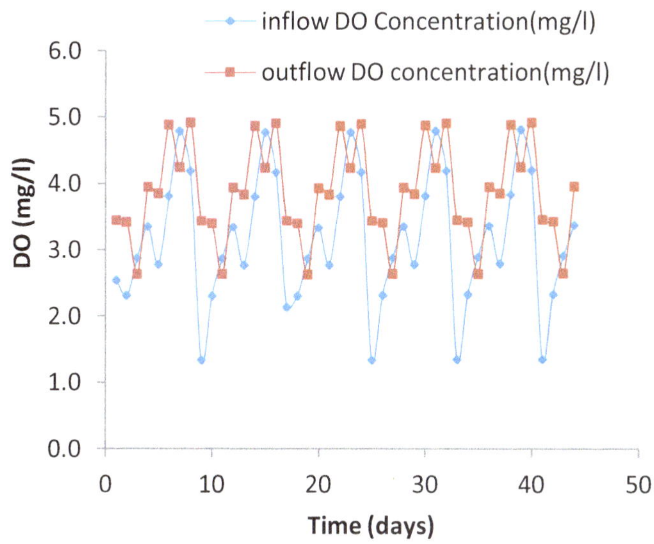

Figure 3. Variation of inflow and outflow DO concentrations with time

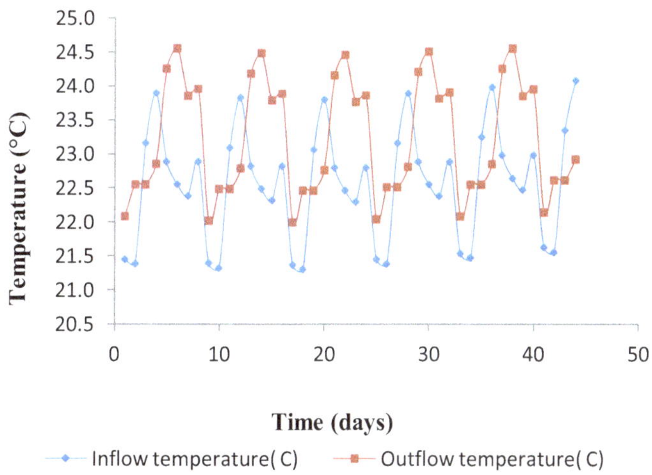

Figure 4. Variation of inflow and outflow temperature with time

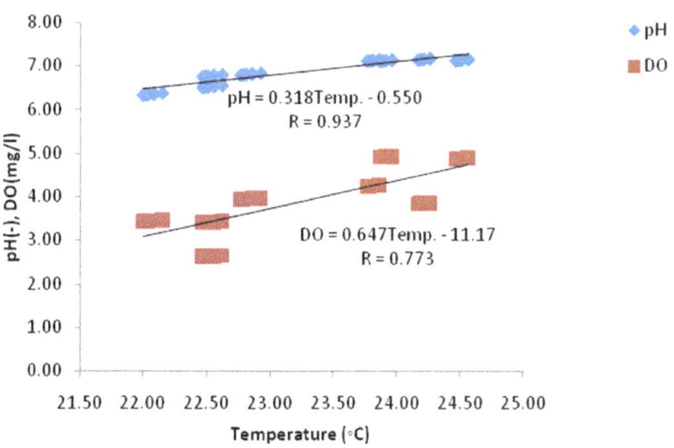

Figure 5. Variation of effluent pH and DO with temperature.

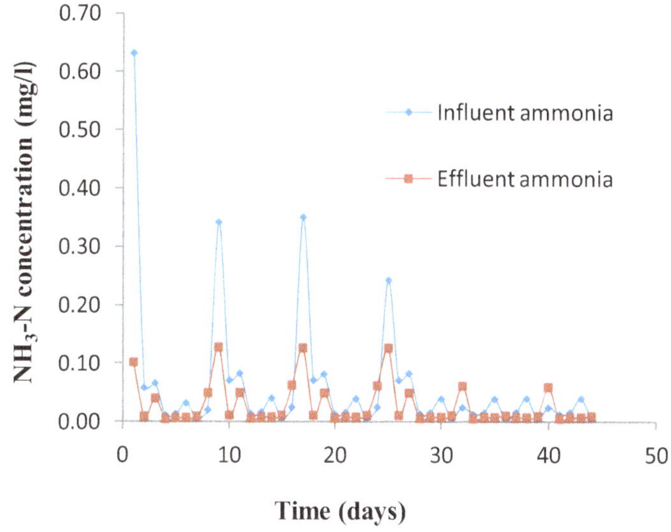

Figure 6. Variation of inflow and outflow NH_3-N concentrations with time.

Nitrite- and nitrate-nitrogen

Table 2 shows that concentration of nitrite nitrogen ranged from 0.002 to 0.062 mg/L with a mean value of 0.01 ± 0.007 mg/L and from 0.002 to 0.026 mg/L with a mean value of 0.009 ± 0.006 mg/L for inflow and outflow zones, respectively. Nitrite is usually obtained from biological oxidation of oxidation of ammonia, which is later followed by its oxidation to nitrates. The concentration of nitrites is generally low (Table 2) in oxygen-rich waters (Table 1) because degradation of ammonia to nitrite is usually the rate limiting step of nitrification. Figure 8 shows the variation of inflow and outflow nitrate nitrogen concentrations with time. The average concentration of nitrate-nitrogen ranged from 0.062 to 0.39 mg/L with a mean value of 0.284 ± 0.08 mg/L and from 0.132 to 0.507 mg/L with a mean value of 0.310 ± 0.133 mg/L for inflow and outflow zones, respectively (Table 2). As a result of nitrification, mean nitrate concentration increased by about 9.2% as water flows through the wetland. The Spearman's correlation tests between outflow nitrate nitrogen with dissolved oxygen, temperature and pH gave correlation coefficients of 0.469, 0.723 and 0.711, respectively. This is suggesting that the outflow nitrate nitrogen concentrations are strongly correlated with dissolved oxygen, temperature and pH. It was noted that temperature and pH depicted strong correlation with effluent nitrate concentration (Figure 9). However, correlation between oxygen and effluent concentration of nitrate was relatively poor, although it was generally observed that concentration of nitrate in the effluent was stimulated by increase in dissolve oxygen (relationship: NO_3-N = 0.076 DO with R = 0.469; DO and Nitrate in

Table 2. Variation of nitrogen in wetland system.

Parameter	Influent value		Effluent value	
	Range	Mean	Range	Mean
Nitrate nitrogen (mg/l)	0.062-0.39	0.28±40.08	0.132-0.507	0.310±0.133
Nitrite nitrogen (mg/l)	0.002-0.062	0.010±0.007	0.002- 0.026	0.009±0.006
Ammonia nitrogen (mg/l)	0.005-0.632	0.063±0.017	0.005-0.128	0.028±0.023
Organic nitrogen (mg/l)	0.181-2.29	0.621±0.535	0.08-0.746	0.349±0.152
TN (mg/l)	0.699-1.563	0.977±0.243	0.457-1.015	0.695±0.178
TKN (mg/l)	0.195-2.63	0.684±0.625	0.153-0.715	0.377±0.168

Where number of samples (n) for all the parameters = 44.

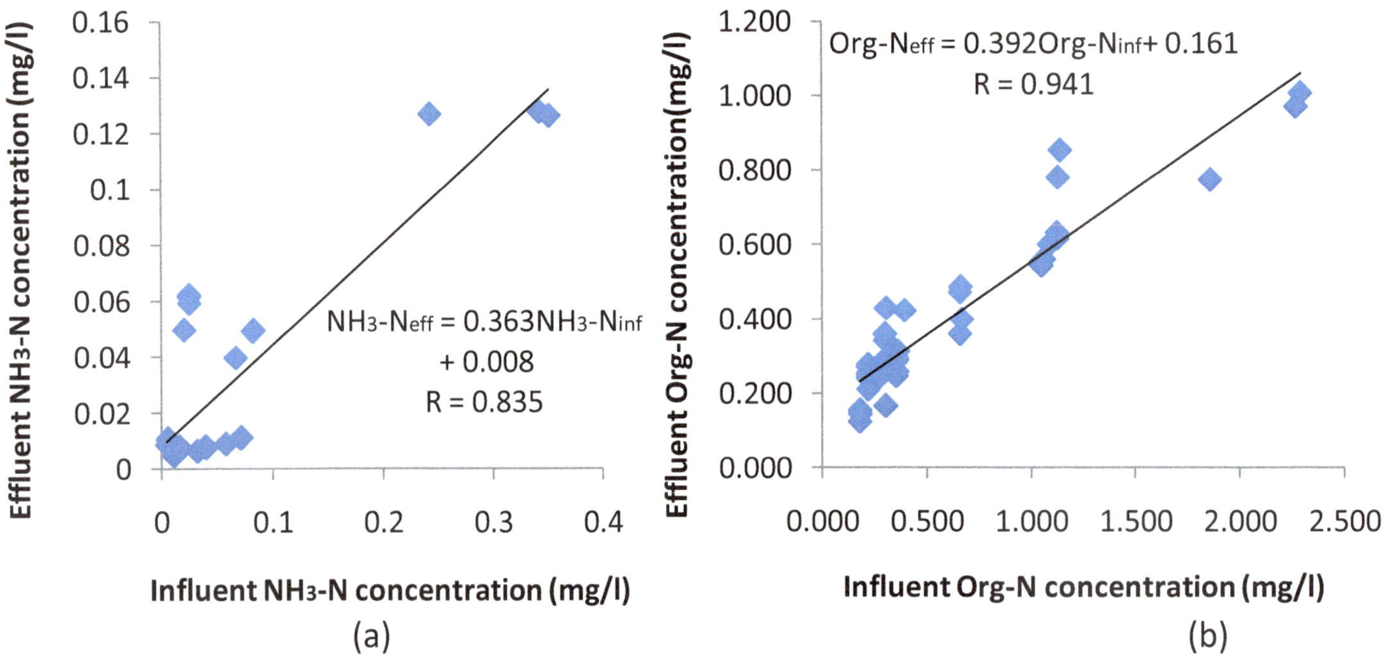

Figure 7. Variation of (a) Influent NH_3-N and Effluent NH_3-N and (b) Influent Organic Nitrogen and Effluent Organic Nitrogen.

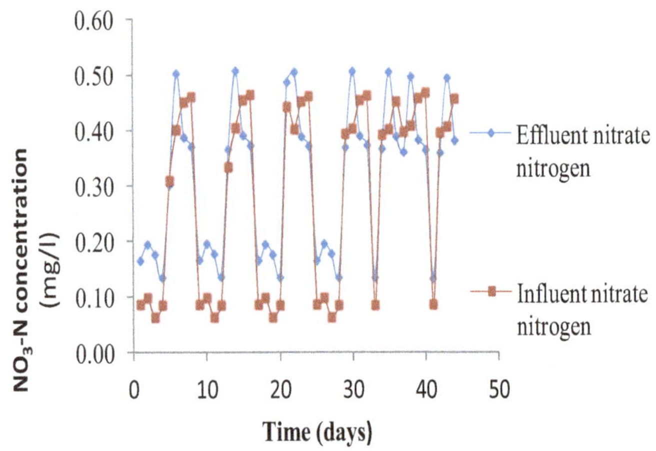

Figure 8. Variations of inflow and outflow NO_3-N concentrations with time.

mg/l).

Nitrification is affected by pH, temperature, dissolved oxygen and substrate concentrations. At dissolved oxygen values of less than 2 mg/L, nitrification is hindered while the optimum temperature for nitrification to take place is 20 to 30°C and pH optimum for this is 7.2 to 8.5, though nitrification under some conditions can proceed even below pH of 7.2 (Seitzinger, 1988). The observed values of pH, temperature and dissolved oxygen are at more or less optimal conditions for nitrification. On the other hand, denitrification in many wetland ecosystems is largely regulated by nitrification rates (Reddy et al., 1989; Vymazal, 2007). When the river water is not only nitrate-rich, but also oxygen-rich, denitrification will only occur if oxygen concentrations will drop to sufficiently low levels (Seitzinger, 1988). The aforementioned observations and discussions suggests

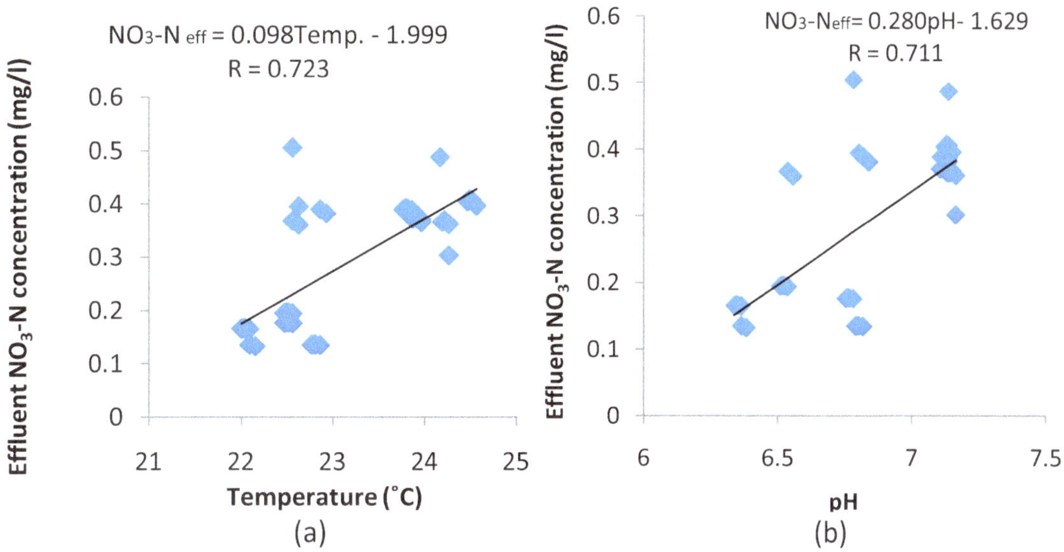

Figure 9. Graph of effluent NO_3-N concentration vs (a) Temperature and (b) pH.

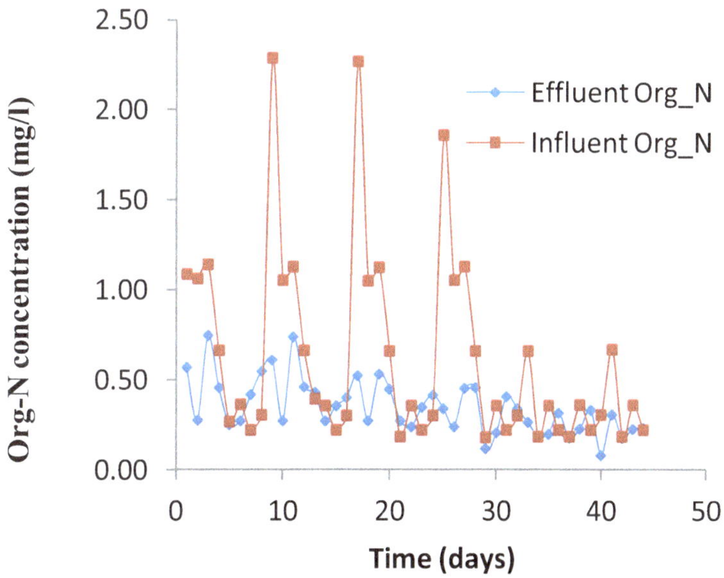

Figure 10. Variation of inflow and outflow Org-N concentrations with time.

that the nitrate nitrogen formation in the wetland is being favored considering the optimal levels of dissolved oxygen, pH and temperature, and the decreasing trend of ammonia nitrogen from which nitrate nitrogen emanates.

Organic nitrogen

Concentrations of organic nitrogen ranged from 0.181 to 2.29 mg/L with a mean value of 0.621 ± 0.535 mg/L and from 0.08 to 0.746 mg/L with a mean value of 0.349 ± 0.152 mg/L for inflow and outflow zones, respectively (Figure 10). The observed organic nitrogen removal efficiency is 43.8%, which is substantial and can be regarded as one of the major transformation processes observed in the wetland. It is worth noting that effluent concentration of organic nitrogen was strongly correlated (R = 0.941) to the influent concentration of organic nitrogen (Figure 7b).

Total nitrogen (TN)

Total nitrogen concentrations ranged from 0.699 to 1.563

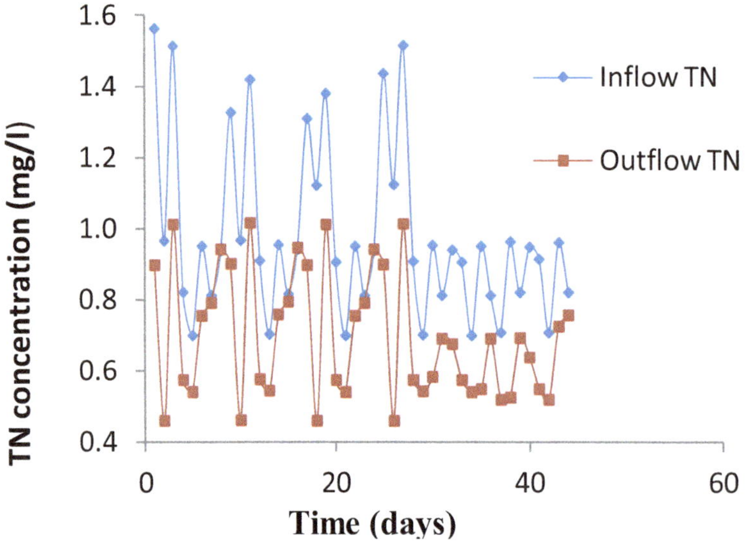

Figure 11. Variation of inflow and outflow TN with time.

mg/L with a mean value of 0.977 ± 0.243 mg/L and from 0.457 to 1.015 mg/L with a mean value of 0.695 ± 0.178 mg/L for inflow and outflow zones, respectively (Table 2), which is equivalent to 28.8% removal efficiency (Figure 11).

Estimating nitrogen content for the wetland plants

Plants nitrogen content determination was estimated by the application of the plants biomass (DWm^{-2}) and nitrogen content (%DW) data respectively, as detailed under the methodology part. After the plants biomass content (in $kgDWm^{-2}$) and the plants nitrogen content (in %DW) were determined, the final plants nitrogen content was determined (in gm^{-2}) as a product of biomass and nitrogen content. Mean plants nitrogen content found was 67.88 gNm^{-2}; this was used as a model input for the plants nitrogen. Table 3 shows the detailed nitrogen content determination process.

Nitrogen content in sediments

The results of the sediments' nitrogen content from the three transects shows that transect close to inlet of wetland has 0.22 ± 0.045 gN/kg, at the middle of wetland has 0.362 ± 0.063 gN/kg and near the exit has 0.577 ± 0.2 gN/kg of dry sediments. It is evident that more nitrogen accumulates in the sediments as water approaches the exit end of the wetland. This can be attributed by the increase in plants density whose roots and below ground formations in general helps in sediment trap; hence, nutrient retention in the sediments. The average nitrogen content of the wetland is about 0.38 ± 0.06 gN/kg of sediments, which translates to 201.26 ± 30.78 g/m^2 of wetland.

DISCUSSION

The concentration of total nitrogen in Mara River at Bisarwi village ranged from 0.70 to 1.56 mg/L, which was largely contributed by organic nitrogen (63.6%) and nitrates (29.1%). The concentration of organic nitrogen ranged from 0.18 to 2.29 mg/L, nitrates from 0.06 to 0.39 mg/L and ammonia from 0.005 to 0.63 mg/L. These values are lower than those reported by Kulekana (2004) who observed nitrate concentration of 0.91 mg/L near the water surface, which decreased to 0.35 mg/L at the bottom 0.5 m of River Mara. It is possible that dilution effect during April to July, which is just after rainy season in the area, reduced concentration to low levels in the river. Kulekana (2004) also reported dilution effect during rainy season reduced concentration of nutrients in Lake Malimbe, Burigi and Mara River. Mara River wetland removed total nitrogen by about 28.8% from an average of 0.98 mg/L at the inlet of wetland to 0.70 mg/L at the exit of wetland near Kirumi Bridge. This is substantial removal efficiency for a wetland that retained water for only 36 h. The removal was largely due to deposition of organic nitrogen in the wetland sediments. Organic nitrogen decreased from 0.62 mg/L at the inlet of the wetland to 0.35 mg/L at the exit. Some nitrification of ammonia was also observed as ammonia-nitrogen decreased from 0.063 to 0.028 mg/L in the wetland system, which resulted in a slight increase of nitrate from 0.28 to 0.31 mg/L.

The Spearman's correlation test between nitrate concentration at the exit of the wetland with temperature

Table 3. Plants nitrogen content.

Plant	Biomass (KgDWm-2)			Nitrogen content (%DW)			Nitrogen content (gm-2)		
	AG	BG	Total	AG	BG	Total	AG	BG	Total
Papyrus	2.52± 0.5	1.78± 0.39	4.3± 0..4	2.48± 0.2	0.87± 0.12	3.35± 0.22	62.49	15.49	77.98
Phragmites	2.3± 0.38	1.44± 0.19	3.74±0.23	2.72± 0.5	0.91± 0.2	3.63± 0.7	62.56	13.10	75.66
Typha	1.53± 0.25	1.17± 0.4	2.7± 0.55	2.61± 0.25	0.86± 0.1	3.47± 0.6	39.93	10.06	49.99
Average	**2.12± 0.52**	**1.46± 0.31**	**3.58±0.8**	**2.60± 0.12**	**0.88± 0.03**	**3.48± 0.14**	**55.00± 13**	**12.88± 2.71**	**67.88± 15.53**

AG= above ground organs; BG= below ground organs.

and dissolved oxygen gave correlation coefficients of 0.71 and 0.469, respectively. However, the concentrations of nitrate, ammonia and organic nitrogen in the influent were poorly correlated with temperature. This is suggesting that an increase in temperature, which is also strongly correlated with increase in dissolved oxygen (R = 0.773), resulted in increase of nitrate concentration. Having such a strong correlation and taking into account that temperature is one of the limiting factors for microbial activities in the nitrification/denitrification processes, it can be suggested that if other limiting factors like dissolved oxygen concentration are kept constant, temperature is at the optimal microbial functioning range as Vyamazal (2007) suggested that the activities of nitrifying and de-nitrifying bacteria in treatment wetlands proceed most efficiently at 20 to 25°C. Vandelannoote et al. (1996) reported higher nutrient levels in Ntahangwa River during the dry season compared to rainy season. Slight increase of nitrate levels may also be associated with the increased rate of deposition and mineralization of organic matter (Battle and Mihuc, 2000; Mayo and Mutamba, 2004) and subsequent increased nitrification as a result of high levels of dissolved oxygen in the wetland.

Muraza (2013) reported that flow rate in Mara River at Bisarwi village ranged from 5.8 to 14.2 m^3/s with a mean value of 8.5 ± 0.3 m^3/s, which is equivalent to 734,400 ± 229,279 m^3/d. Assuming the outflow rate of river near Kirumi Bridge is equal to the inflow rate, 75 tons of nitrogen is removed by the wetland annually, which is equivalent to 3.67 kg/ha/year. Nitrogen may be removed through nitrification, denitrification, uptake by wetland plant, settling of organic matter to sediments or ammonia volatilization (Bigambo and Mayo, 2005). The later mechanism is unlikely to occur in Mara River wetland because wetland water pH is near neutral (Table 1), whereas ammonia stripping is possible only when pH exceeds 9.4 (Reddy and Graetz, 1981). Samples from sediments indicated mean nitrogen levels of 0.38 ± 0.06 g/kg dry sediments, with wetland outlet zone sediments having 2.62 times more nitrogen than the inlet zone. This is in line with the observed organic nitrogen removal rate of 43.8%. It is worth mentioning that total nitrogen removal in the wetland was only 28.8%, which suggests that part of the organic matter deposited in the sediments was regenerated back to the water body and/or some organic matter was transformed to ammonia through mineralization. The analysis of plant nitrogen suggests that significant amount of nitrogen is contained in plant biomass (Table 4). Taking into consideration of plant biomass, the estimated nitrogen content is 678.8 kg/ha, which is assumed to be removed from the river water. Since the area of Mara River wetland is 204.46 km^2, the total nitrogen in wetland plants is about 13,879 tons. This is substantial amount of nitrogen that has been prevented from entering Lake Victoria waters.

From the findings of this study, it is suggested that for better functioning of the wetland, best management practices should be undertaken, since the wetland's functionality is a function of the biogeochemical relationships between the micro organisms, the geological features and plants complexes. For the wetlands to function as anticipated, the best management strategies should address the issues for making sure that this ecosystem's functioning is at its best level. To achieve this, strategies proposed for sustainable use while maintaining ecological and biodiversity quality include, but are not limited to, nutrient load reduction from the point and non point sources, creation and maintenance of buffer zones and wetland's participatory monitoring over its functionality (Muraza, 2013).

Conclusions

From the results of this research work, the following conclusions are made:

i) Mara River wetland receives about 0.70 ~ 1.56 mg/L of nitrogen, which is largely contributed by organic nitrogen (63.6%) and nitrates (29.1%). The wetland removes about 28.8% of the nitrogen it receives within 36 h of water retention in it.

ii) The wetland removes about 75 tons of nitrogen annually, which is equivalent to 3.67 kg/ha/year. It is estimated that about 0.38 g of nitrogen is trapped in each kg of dry sediments and about 67.9 g/m^2 is trapped in plant biomass. It can be concluded that Mara River wetland is effectively protecting Lake Victoria by reducing nitrogen load entering the lake.

Table 4. Nitrogen biomass in plants.

Plant	*Cyperus papyrus*	*Phragmites australis*	*Typha domingensis*
Rhizomes + roots (%DW)	0.87± 0.12	0.91± 0.08	0.86± 0.50
Culm (%DW)	1.09± 0.05	1.17± 0.10	1.13± 0.06
Umbel (%DW)	1.47± 0.03	1.55± 0.40	1.49± 0.18
Total (%DW)	3.35±0.22	3.63± 0.70	3.47± 0.45
Above ground (%DW)	2.37±0.22	2.42± 0.17	2.39± 0.20
Below ground (%DW)	1.03± 0.10	0.80± 0.40	0.90± 0.10

REFERENCES

Allinson G, Stagnitti F, Salzman S, Hill RJ, Cordell S, Smith L (2000). Strategies for the sustainable management of industrial wastewater: Determination of the chemical dynamics of a cascade series of five newly constructed ponds. Phys. Chem. Earth B 25:629-634.

Amondi OR, Onyango JO, Chemelil MC, Becht R (2005). Analysis of sediments, surface run-off and nutrient dynamics using agricultural non-point source pollution model (AGNPS) in Mirera sub-catchment of Lake Naivasha drainage basin, Kenya. Proceedings of the 11th World Lakes Conference, Nairobi, Kenya, 31st October to 4th November, Odada O, Olago DO, Ochola W, Ntiba M, Wandiga S, Gichuki N, Oyieke H (Ed.), pp. 606-614.

Andrew G (1990). The human impact on the natural environment. The MIT Press, Cambridge, Massachusetts.

Bagalwa M (2005). Environmental impacts of land use change on water quality of inflowing tributaries of Lake Kivu. Proceedings of the 11th World Lakes Conference, Nairobi, Kenya, 31st October to 4th November, Odada O, Olago DO, Ochola W, Ntiba M, Wandiga S, Gichuki N, Oyieke H (Ed.), pp. 379-383.

Balirwa JS (1995). The Lake Victoria environment: Its fisheries and wetlands: A review. Wetland Ecol. Manage. 3:209-224.

Bancy MM, Mutie S, Home P, Mtalo F, Gadain H (2005). Impacts of land use change on the hydrology of the transboundary Mara River. Proceedings of the 11th World Lakes Conference, Nairobi, Kenya, 31st October to 4th November, Odada O, Olago DO, Ochola W, Ntiba M, Wandiga S, Gichuki N, Oyieke H (Ed.), pp. 432-438.

Battle JM, Mihuc TB (2000). Decomposition dynamics of aquatioc macrophytes in the lower Atchafalaya, a large flood plain river, *Hydrobiologia,* 418:123-136.

Bigambo T, Mayo AW (2005). Nitrogen transformation in horizontal subsurface flow constructed Wetlands II: Effects of biofilms. J. Phys. Chem. Earth 30:668-672.

Bitala M, Kweyunga C, Manoko M (2009). Levels of heavy metals and cyanides in soil, sediments and water from the vicinity of north Mara gold mine in Tarime district, Tanzania, presented to CCT.

CEC (1995). Wise Use and Conservation of Wetlands, Communication from the commission to the council and European parliament, Commission of the European Communities. Com 95:189,154.

Dugan PJ (1990). Wetland Conservation: A review of current issues and required actions, IUNC, Gland, Switzerland.

Gichuki NN, Oyieke HA, Terer T (2005). Status and root causes of biodiversity loss in the Eastern Rift Valley Lakes, Kenya. Proceedings of the 11th World Lakes Conference, Nairobi, Kenya, 31st October to 4th November, Odada O, Olago DO, Ochola W, Ntiba M, Wandiga S, Gichuki N, Oyieke H (Ed.), pp. 511-517.

Global Water for Sustainability Program (GLOWS) (2007). *Maps,* http://globalwaters.net/glows/Publications/Maps/tabid/109/Default.aspx. Retrieved on November 4th, 2011.

Hammer DA, Bastian RK (1989). Wetland ecosystem: Natural water purifiers? In Hammer D.A. (Ed). Constructed wetlands for wastewater treatment: Municipal, industrial and agricultural. Lewis Publishers, Chelsea, pp. 5-19.

Henry L, Semili P (2005). Levels of heavy metal pollution in water and sediments in Simiyu wetland of Lake Victoria basin (Tanzania). Proceedings of the 11th World Lakes Conference, Nairobi, Kenya, 31st October to 4th November, Odada O, Olago DO, Ochola W, Ntiba M, Wandiga S, Gichuki N, Oyieke H (Ed.), 549-553.

Hogan DV, Maltby E, Lloyd JW, Baker CJ (1992). Water resources management and the protection of the wetland functioning, *NRA:* Research and Development Priorities, NRA Wessex Region.

Kadlec RH (2009). Wastewater treatment at Houghton Lake, Michigan: Hydrology and water quality. J. Ecol. Engr. 35:1287-1311.

Kansiime F (2004). Functioning and dynamics of wetland vegetation of Lake Victoria. In Ecotools. Tools for wetland ecosystem resource management in eastern Africa. EC RTD INCO-DEV Programme, pp. 5-11.

Kansiime F, Nalubega M, Bugenyi FWB, Tukahirwa EM (1994). The role of Nakivubo swamp in maintaining the water quality of Inner Murchison Bay- Lake Victoria. J. Trop. Hydrobiol. Fisher. 5:79-87.

Kansiime F, van Bruggen JJA (2001). Distribution and retention of faecal coliforms in Nakivubo wetland in Kampala Uganda. Water Sci. Technol. 44:11-12.

Kayombo S, Mbwette TSA, Mayo AW, Katima JHY, Jørgensen SE (2002). Diurnal cycles of variation of physical-chemical parameters in waste stabilization ponds. J. Ecol. Engr. 18:287-291.

Kayombo S, Mbwette TSA, Mayo AW, Katima JHY, Jørgensen SE (2000). Modelling diurnal variation of dissolved oxygen in waste stabilization ponds. J. Ecol. Modell. 127:21-31.

Kirkby MJ, Morgan RPC (1980). Soil Erosion, John Wiley and Sons, New York.

Kulekana JJ (2004). Levels of nitrate and phosphate in some satellite lakes within the Lake Victoria basin, Tanzania. Tanzania J. Sci. 30(1):1-9.

Lake Victoria Environmental Management Program (LVEMP) (2005). Lake Victoria Environmental Management Project Tanzania, Second Supplemental Credit, Musoma, Tanzania.

Makalle AMP, Obando J, Bamutaze Y (2008). Effects of land use practices on livelihoods in the transboundary sub-catchments of the Lake Victoria Basin. Afr. J. Environ. Sci. Technol. 2(10):309-317.

Maltby E (1990). Water logged Wealth: Why waste the World's wet places? Earth Science Press, London, U.K.

Marwa A (2013). Treatment of Acid Mine Drainage from North Mara Mine Using Constructed Wetlands. *MSc* (Eng) Thesis, Department of Chemical and Mining Engineering, University of Dar es Salaam, Tanzania.

Mayo AW, Bigambo T (2005). Nitrogen transformation in horizontal subsurface flow constructed wetlands I: Model development. J. Phys. Chem. Earth 30:658-667.

Mayo AW, Mutamba J (2004). Effect of HRT on nitrogen removal in a coupled HRP and unplanted subsurface flow gravel bed constructed wetland. J. Phys. Chem. Earth 29:1253-1257.

Mayo AW, Mutamba J (2005). Modelling nitrogen removal in a coupled HRP and unplanted horizontal flow subsurface gravel bed constructed wetland. J. Phys. Chem. Earth 30:673-679.

Mironga JM (2005). Environmental implications of water hyacinth infestation in Lake Naivasha. Proceedings of the 11th World Lakes Conference, Nairobi, Kenya, 31st October to 4th November, Odada O, Olago DO, Ochola W, Ntiba M, Wandiga S, Gichuki N, Oyieke H (Ed.), pp. 580-591.

Mitchell DS (1994). Flood plain wetlands of the Murray- Darling basin: Management, issues and challenges, NSW, Australia.

Mitsch WJ, Gosselink JC (1986). Wetlands, Van Nostrand Reinhold, New York, USA.

Muraza M (2013). Modelling nitrogen transformation and removal in Mara river basin wetlands. MSc (Water Resources Eng) dissertation, Department of Water Resources Engineering, University of Dar es Salaam, Tanzania.

Nile Basin Initiative (2007). Transboundary Environmental Action Project, National Nile Basin Water Quality Monitoring Baseline Report for Tanzania, http://www.nileteap.org/docs/publications/WQ/WQ_Baseline_report_Tanzania.pdf. Retrieved on 3rd November, 2011.

Omondi R, Kusewa M (2005). Macrophytes of Lake Victoria and succession after invasion of water hyacinth. Proceedings of the 11th World Lakes Conference, Nairobi, Kenya, 31st October to 4th November, Odada O, Olago DO, Ochola W, Ntiba M, Wandiga S, Gichuki N, Oyieke H (Ed.), pp. 600-602.

Ongore CO, Okuku EO, Mwangi SN, Kiteresi LI, Ohowa B, Wanjeri VO, Okumu S, Kilonzi J (2013). Characteristics of nutrients enrichment in the estuaries and related systems in Kenya coast. J. Environ. Sci. Water Res. 2(6):181-190.

Raburu PO, Okeyo-Owuor JB (2005). Impact of agro-industrial activities on the water quality of River Nyando, Lake Victoria Basin, Kenya. Proceedings of the 11th World Lakes Conference, Nairobi, Kenya, 31st October to 4th November, Odada O, Olago DO, Ochola W, Ntiba M, Wandiga S, Gichuki N, Oyieke H (Ed.), pp. 307-314.

Raini JA (2005). Long-term trends in water quality, water quantity and biodiversity at Lake Nakuru, Kenya. Proceedings of the 11th World Lakes Conference, Nairobi, Kenya, 31st October to 4th November, Odada O, Olago DO, Ochola W, Ntiba M, Wandiga S, Gichuki N, Oyieke H (Ed.), pp. 57-62.

Reddy KR, Graetz A (1981). Use of shallow reservoir and flooded organic soils for wastewater treatment: nitrogen and phosphorus

Reddy KR, Patrick Jr. WH, Lindau CW (1989). Nitrification-denitrification at the plant root sediment interface in wetlands. J. Limnol. Transform. J. Environ. Qual. 10:112-119.

Oceanograph. 34:1004-1013.

Rode M, Lindenschmidt KE (2000). Distributed sediment and phosphorus transport modeling on a medium sized catchment in Central Germany. Elsevier Sci. J. 26(7-8):635-640.

Seitzinger SP (1988). Denitrification in freshwater and coastal marine ecosystems: Ecological and geochemical significance. J. Limnol. Oceanogr. 33:702-724.

Standard Methods for Examination of Water and Wastewater (2012). American Water Works Association (AWWA), American Public Health Association, Water Pollution Control Federation, Water Environment Federation, 22nd edition, Washington, DC.

Terer T, Gichuki NN, Ndiritu GG (2005). Role of wetlands in maintaining stability of tropical lakes: A case study of River Nyando delta in Lake Victoria basin, Kenya. Proceedings of the 11th World Lakes Conference, Nairobi, Kenya, 31st October to 4th November, Odada O, Olago DO, Ochola W, Ntiba M, Wandiga S, Gichuki N, Oyieke H (Ed.), pp. 560-567.

Thenya T, Wassmann R, Verchot L, Mungai D (2005). Degradation of Riparian wetlands in the Lake Victoria basin- Yala swamp case study. Proceedings of the 11th World Lakes Conference, Nairobi, Kenya, 31st October to 4th November, Odada O, Olago DO, Ochola W, Ntiba M, Wandiga S, Gichuki N, Oyieke H (Ed.), pp 483-494.

Twesigye CK, Onywere SM, Getenga ZM, Mwakalila SS, Nakiranda JK (2011). The impact of land use activities on vegetation cover and water quality in the Lake Victoria watershed. Open Environ. Engr. J. 4:66-77.

UNEP (2006). Lake Victoria Basin Environmental Outlook: Environment and Development, UNEP, Nairobi, Kenya.

Vandelannoote AH, Robbertecht H, Deelastra F, Bitetera, Ollivier F (1996). The impact of the River Ntahangwa, the most polluted Burundian affluent of Lake Tanganyika, on the water quality of the lake. *Hydrobiologia* 328:161-171.

Vymazal J (2007). Removal of nutrients in various types of constructed wetlands. Sci. Total Environ. 380:48-65.

WWF-Tanzania Program Office (TPO) (2006). Mara River Basin Management Initiative, Musoma, Tanzania.

Assessment of water quality in Noyyal River through water quality index

Mariraj Mohan S.[1] and Vanalakshmi P.[2]

[1]Civil Engineering Department, Alagappa Chettiar College of Engineering and Technology, Karaikudi, Tamilnadu, India.
[2]Civil Engineering Department, Kalasalingam University, Srivilliputhur, Virudhunagar, India.

Assessment of water quality in Noyyal River from 20/12/2010 to 27/12/2011 was carried out in this study. Water samples were collected from the Noyyal River at a stretch of 3800 m. The sampling locations have been fixed at every 50 m. The parameters were estimated such as dissolved oxygen (DO), pH, Temperature, chemical oxygen demand (COD), total dissolved solids (TDS), Sulphate, Chloride, total hardness etc. This study had two phases. In the first phase, the estimation of water quality parameters was carried out and in the second phase water quality index have been determined based on the existing standards. The effects of municipal sewage on river water quality have also been investigated. The depletion of DO concentration due to the simultaneous effect of water pollution, thus leads to more uncertainty about the survival of DO dependent aquatic species. From the study, it revealed that TDS, sulphate, chloride, and hardness were estimated to be high concentration at sampling location 10 (S_{10}). Among the sample locations, in most of the places, high concentration of TDS, Hardness, sulphate and chloride and low level of DO were observed. Our findings highlighted the deterioration of water quality in the river and are due to human activities. This analysis reveals that the surface water needs some degree of treatment before consumption.

Key words: Noyyal River, dissolved oxygen (DO), water quality index, water quality parameters.

INTRODUCTION

"Pollution is the introduction of contaminants into a natural environment that causes instability, disorder, harm or discomfort to the ecosystem that is, physical systems or living organisms" (Merriam Webster online dictionary, 'http://www.merriam-webster.com/dictionary/pollution', 2010). The term pollutant in a broad term refers to a wide range of compounds, from a superabundance of nutrients giving rise to the enrichment of ecosystems to toxic compounds that may be carcinogenic, mutagenic, or teratogenic. Pollutants can be divided into two major groups, namely, those that affect the physical environment and those that are directly toxic to organisms, including human beings.

Rapid industrialization in the twentieth century had led to the generation of vast amounts of gas, liquid, and solid waste that were introduced into the environment without much thought by the manufacturers of that waste. This has affected the ecosystem and has caused health problems for the first habitants residing near the factories. Economical status of the country directly corresponded to the industrial growth. This is also reflected in the increase in rate of pollution (Rainwater harvesting, 'http://www.rainwater harvesting.org/crisis/river-noyyal.html', 2007).

Overview of textile industry

Textile industry in India is one of the oldest and largest organized sectors. There are over 700 large textile mills concentrated in Ahmedabad, Bombay, Coimbatore, Kanpur and Delhi. Textile processing industries nowadays are widespread sectors in developing countries. Among

*Corresponding author. E-mail: mari_sundar@haoo.com.

the various processes in the textile industry, dyeing process consumes large volume water for dyeing, fixing and washing. Thus, the wastewater generated from the textile processing industries contains suspended solids, high amount of dissolved solids, un-reacted dyestuffs (color) and other auxiliary chemicals that are used in the various stages of dyeing and other processing. The presence of even a small amount of dye in water (for example, 10 to 20 mg/L) is highly visible and affects the water transparency and the gas solubility of water bodies. Among several classes of textile dyestuffs, the reactive dyes contribute about 50% of the total market share and the most common group used as chromophore is the azo, followed by anthraquinone. Textile industries are found in most countries and their numbers have increased. These industries have shown a significant increase in the use of synthetic complex organic dyes as the coloring material. The annual world production of textiles is about 30 million tons requiring 700,000 tonnes of different dyes which causes considerable environmental pollution problems. During the last few decades, substantial global shifts have occurred in textile production and export.

Toxic effects of dyes

The effluent discharged with high temperature will increase the temperature of the receiving body, thereby reducing the solubility of oxygen in water. High alkalinity of the waste water causes increases in the pH value of the receiving stream. If the pH value exceeds 9 or falls below 5 on the pH scale, it will have an adverse effect on the aquatic biota. The soluble colors and dyes present in the wastewater will persist in the stream and interfere with the penetration of sunlight essential for photosynthesis. The colloidal organic matter in the wastewater will increase its turbidity along with dyes. Oily scum formed on the surface of the water will interfere with the mechanism of oxygen transfer at the air-water interface.

Keeping the above discussion in mind the objectives of this study has been formulated as follows:

(i) To analyses the water quality parameters.
(ii) To evaluate the variations of water quality parameters with respect to existing pollution.
(iii) To compute water quality index based on the existing standards.

METHODOLOGY

Study area

Coimbatore is the one of the major cities in Tamil Nadu and it is well known for industrial activities. The textile industries in and around Coimbatore, disposes the waste into the Noyyal River. The quantity of the wastewater discharged into the river is also getting increased. As a result, the quality of the river gets deteriorated. Hence the Noyyal River has been chosen as the study area. The Noyyal River rises from the Vellingiri hills in the Western Ghats in Tamil Nadu, southeastern India and drains into the Kaveri River. The river basin is 180 km long and 25 km wide and covers a total area of 3,500 km^2. Cultivated land in the basin amounts to 1,800 km^2 while the population density is 120 people per km^2 in the countryside, and 1000 people per km^2 in the cities (Lenin and Saseetharan, 2008). The area is known for its scanty rainfall and the development of the Noyyal River Tanks System holding any overflow from the rains plus the water in the Northeast and Southwest monsoon season been ecologically important. The 173 km long tributary of the Kaveri River filled 32 tanks. These interconnecting tanks held the water flowing from the Noyyal. One of the major and critical issues is the pollution of the rivers Noyyal and Nallaru originates and flowing in the Kongu region. It is a big menace troubling the people and the Kongunadu region as a whole, and also it is a disgrace to not just people all over India but all over the world constantly visiting Tirupur for industrial activity (Rainwater harvesting, 'http://www.rainwater harvesting.org/crisis/river-noyyal.html' 2007).

As like the many other hazardous changes due to the lifestyle changes and change in culture, the Noyyal is also down the way to death. The deterioration in various water quality characteristics clearly indicates the possibilities of pollution due to industrial activities such as coffee vegetables oils, leather tanning, textiles and foundries in and around Coimbatore city. The population of Coimbatore has also a strong impact on the Noyyal River with regard to pollution and due to this Noyyal River acts as a carrier for the pollution. During the non-flow period of the river, water can be stagnated and the pollution may enter into the ground water. So the ground water quality also gets depleted (Lenin and Saseetharan, 2008).

Sampling program, location and analysis

Schematic representation of sampling sites is as shown in Figure 1. In this study, it has been chosen that the confined stretch of the Noyyal River which passes through Perur, since the over exploitation of the river occurred over there. The samples were collected at different locations with a stretch of 3800 m. The distance between each sampling location is taken as 50 m, but the distance between 6th and 7th is 1 km, between 8th and 9th is 500 m and also between 9th and 10th sampling site is 2 km. The irregularities between the sampling locations are due to the bushes and sluggish in the river as shown in Figure 1. This condition is practically not possible to collect water samples.

The samples were collected during the period of December 2010 to February 2011. 11 sets of samples were collected in 10 locations designated as S_1 to S_{10}, which gives the total number of 110 samples. The time interval maintained for the collection of samples is one week.

The samples were collected by grab sampling (Islam and Alam, 2007). Each sample was analyzed for eight parameters such as dissolved oxygen (DO), total dissolved solids (TDS), chemical oxygen demand (COD), Sulphate, Temperature, pH, Chloride, and Hardness etc. All water quality parameters were estimated by standard methods (13). When compared to all the sites, at S_4, S_5 and S_{10} site activities like fishing, bathing, washing were observed. In the prescribed sampling location S_2, at the bank of the river, temple is located; the discharge of Pooja items has been observed at the river bank.

Water quality index

In this study, it was attempted to evaluate the quality of Noyyal River. It can be done using water quality index.

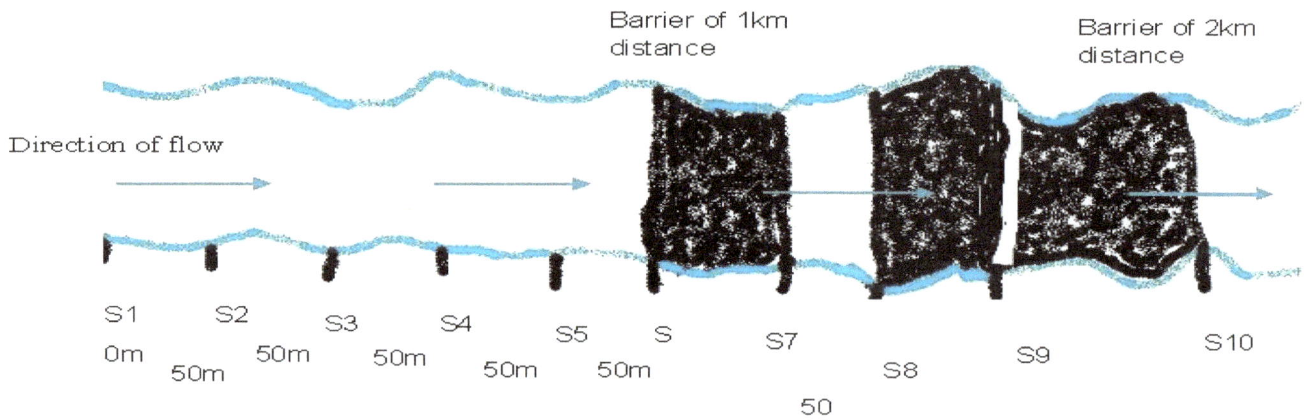

Figure 1. Overview of sampling location.

Water quality index (WQI) is defined as a rating reflecting the composite influence of different water quality parameters. WQI is calculated from the point of view of the suitability of ground water for human consumption Equation (1) (Ramakrishnaiah and Sadashivaiah, 2009).

$$Wi = \frac{wi}{\sum_{i=1}^{n} wi} \quad (1)$$

Where, W_i is the relative weight, w_i is the weight of each parameter and n is the number of water quality parameters.

$$qi = (Ci / Si) \times 100 \quad (2)$$

Where, q_i is the quality rating scale, C_i is the concentration of each parameter in each sample water in mg/L and S_i is the Indian drinking water standard for each chemical parameter in mg/L according to the guidelines of the IS 10500:1991.

$$SI_i = W_i \cdot q_i \quad (3)$$

$$WQI = \sum SI_i \quad (4)$$

SI_i is the subindex of ith parameter; qi is the rating based on concentration of ith parameter and n is the number of parameters. The computed WQI values are classified into five types, "excellent water, good water, poor water, very poor water and unsuitable for drinking".

Weighting has been given, based on the effects which will occur due to the variation of particular parameter. Iron is low in surface water. Hence it is given least weight age, TDS is the major importance in assessing the water quality so given top value

RESULT AND DISCUSSION

The samples were collected during the period of December 2010 to February 2011. Parameters such as DO, TDS, COD, Sulphate, temperature, pH, Chloride, Hardness, iron, fluoride, calcium, magnesium, phosphate, nitrate, nitrite, manganese, sodium have been determined. Out of seventeen parameters eight have been considered as important water quality parameters such as DO, TDS, COD, Sulphate, temperature, pH, Chloride, and Hardness for the classification of surface water. The values of water quality parameters are shown from the Tables 1 to 11.

Due to practical considerations water quality parameters such as iron, fluoride, calcium, magnesium, phosphate, nitrate, nitrite, manganese and sodium could not be measured from 17/01/2011 to 27/02/2011. When going through the results other than eight, all other parameters are within its permissible limits. So analysis were restricted to those eight parameters, the variation is quite more and all are greater than its permissible limits.

Variation of dissolved oxygen

The variation of DO over a period of time is shown in the Figure 2. The DO recorded from 20 /12/2010 to 10 /01/2011 is between 1.3 to 5 mg/L. The lowest DO level recorded was due to high organic load of sewage discharge into the river.

On 17/01/2011, S_1 to S_3 sampling locations recorded as increasing DO levels. In the sampling locations S_1 to S_3, at that period sampling there was no discharge of waste witnessed. In addition to this, the increase in turbulence of the river might have increase DO concentration from S_1 to S_3. It is decreased at sampling location to four (S_4) due to the exploitation such as washing, bathing etc. At S_5 DO concentration gets increased comparing to S_4. The DO was ranging between 6.4 to 10.8 mg/L. The lowest DO level recorded at sampling location S_4 due to high organic load of sewage discharge at that point (Basanta and Chitta, 2009; Sajidu and Henry, 2007).

On 23/01/2011, S_1 to S_3 sampling locations are observed to have a predominant level of DO. As discussed previously, there was no exploitation activity observed from S1 to S_3, which could be the reason for having a predominant level of DO concentration. On 23/01/2011 the DO varies from 7.2 to 10.4 mg/L.

Table 1. Water quality parameters on 20/12/2010.

Sampling locations	Water quality parameter															
	DO (mg/L)	COD (mg/L)	pH	T (°C)	TDS (mg/L)	Hardness (mg/L)	Cl^- (mg/L)	SO_4^{2-} (mg/L)	Fe (mg/L)	F (mg/L)	Ca^{2+} (mg/L)	Mg^{2+} (mg/L)	PO_4^{2-} (mg/L)	NO_3 (mg/L)	NO_2 (mg/L)	Mn (mg/L)	Na (mg/L)
S_1	4	240	7.55	17.5	352	130	80	40	0.24	0.4	34	11	3	22	0.44	0	50
S_2	4	20	7.49	18	348	126	80	38	0.24	0.4	34	10	2.5	20	0.2	0	48
S_3	4.4	40	7.48	18	346	126	78	40	0.24	0.4	34	10	2.5	22	0.2	0	48
S_4	5.8	1500	7.52	18.5	356	130	82	40	0.24	0.4	34	11	2.5	20	0.2	0	50
S_5	3.8	80	7.54	18.5	358	140	82	36	0.24	0.4	37	12	2.4	22	0	0	46
S_6	5	60	7.51	18.5	358	140	72	38	0.1	0.4	37	12	2.4	22	0.4	0	46
S_7	5	20	7.51	17	358	140	72	38	0.1	0.4	37	12	2.4	22	0.02	0	46
S_8	5.6	40	7.51	17.5	358	140	72	38	0.1	0.4	37	12	2.2	22	0.2	0	46
S_9	6.4	100	7.53	18.5	360	144	76	36	0.1	0.4	38	12	1.8	22	0.01	0	47
S_{10}	4.2	20	7.74	18.5	1575	540	290	246	0	0.6	144	43	1.5	46	0.2	0	250

Table 2. Water quality parameters on 27/12/2010.

Sampling locations	Water quality parameter															
	DO (mg/L)	COD (mg/L)	pH	T (°C)	TDS (mg/L)	Hardness (mg/L)	Cl^- (mg/L)	SO_4^{2-} (mg/L)	Fe (mg/L)	F (mg/L)	Ca^{2+} (mg/L)	Mg^{2+} (mg/L)	PO_4^{2-} (mg/L)	NO_3 (mg/L)	NO_2 (mg/L)	Mn (mg/L)	Na (mg/L)
S_1	2.8	1500	7.53	18	354	130	80	40	0.24	0.4	34	11	3	22	0.2	0	50
S_2	3	120	7.47	18.5	368	130	86	40	0.1	0.4	34	11	2.5	20	0.2	0	53
S_3	3	140	7.46	18.5	364	128	80	40	0.24	0.4	34	10	2.5	22	0.2	0	56
S_4	2.8	180	7.50	18.5	368	130	82	40	0.2	0.4	34	11	2.5	20	0.4	0	54
S_5	3.2	400	7.51	18	378	140	82	36	0.24	0.4	37	12	2.4	22	0	0	52
S_6	2.6	460	7.49	18.5	375	140	76	42	0.24	0.4	37	12	2.4	24	0.4	0	54
S_7	3.2	540	7.48	18.5	359	140	72	38	0.1	0.4	37	12	2.4	22	0.02	0	46
S_8	3	20	7.46	18	358	140	72	38	0.1	0.4	37	12	2.2	22	0.2	0	48
S_9	3	260	7.52	18	362	144	76	36	0.1	0.4	38	12	1.8	22	0.01	0	47
S_{10}	3	320	8.52	18.5	1214	490	230	190	0.34	0.6	131	39	4	40	0.6	0	160

On 01/02/2011, it was recorded that the lowest DO concentration occurred at S_7. On that particular day, dumping of used mats, pillows, garlands at the site was observed. In addition to that, mortal fish were floating on the river at sampling location S_7. This can be interpreted that lack of DO level at S_7 is due to dumping of inert materials. On that date of sampling the DO was varied from 1.2 to 8.8 mg/L.

On 07/02/2011, the DO concentration was recorded from 3.2 to 7.6 mg/L. The average DO concentration of this week was found to be 6.24

Table 3. Water quality parameters on 03/01/2011.

Sampling locations	Water quality parameter																
	DO (mg/L)	COD (mg/L)	pH	T (°C)	TDS (mg/L)	Hardness (mg/L)	Cl^- (mg/L)	SO_4^{2-} (mg/L)	Fe (mg/L)	F (mg/L)	Ca^{2+} (mg/L)	Mg^{2+} (mg/L)	PO_4^{2-} (mg/L)	NO_3 (mg/L)	NO_2 (mg/L)	Mn (mg/L)	Na (mg/L)
S_1	4	200	7.55	24	363	130	80	44	0.24	0.4	34	11	3	22	0.44	0	54
S_2	5	660	7.49	25	368	140	80	50	0.24	0.4	37	12	2.5	20	0.2	0	50
S_3	6	300	7.48	25	365	132	80	46	0.24	0.4	35	11	2.5	22	0.2	0	52
S_4	4.2	60	7.52	26	375	136	82	40	0.24	0.4	36	11	2.5	20	0.4	0	54
S_5	3.2	100	7.54	26	377	138	82	36	0.24	0.4	37	11	2.4	22	0	0	55
S_6	3.6	160	7.51	26	376	140	80	40	0.24	0.4	37	12	2.4	23	0.4	0	52
S_7	4.4	500	7.48	25	370	140	80	40	0.1	0.4	37	12	2.4	22	0.02	0	50
S_8	3.6	940	7.46	25	367	140	82	40	0.1	0.4	37	12	2.2	22	0.2	0	50
S_9	6	20	7.53	26	373	144	80	38	0.1	0.4	38	12	2	24	0.02	0	52
S_{10}	3.6	980	8.42	26	1463	540	280	240	0	0.6	142	42	1.5	44	0.2	0	220

Table 4. Water quality parameters on 10/01/2011.

Sampling locations	Water quality parameter																
	DO (mg/L)	COD (mg/L)	pH	T (°C)	TDS (mg/L)	Hardness (mg/L)	Cl^- (mg/L)	SO_4^{2-} (mg/L)	Fe (mg/L)	F (mg/L)	Ca^{2+} (mg/L)	Mg^{2+} (mg/L)	PO_4^{2-} (mg/L)	NO_3 (mg/L)	NO_2 (mg/L)	Mn (mg/L)	Na (mg/L)
S_1	3.8	40	7.21	21	367	210	210	14	0	0.3	56	17	4	7	0.2	0	20
S_2	4.2	160	7.32	21	361	200	200	12	0	0.3	54	16	3.5	7	0.2	0	24
S_3	5	80	7.22	21	358	198	198	14	0	0.3	53	16	3.2	7	0.2	0	22
S_4	3.2	180	7.41	22	370	210	210	13	0	0.3	56	17	3	7	0.2	0	20
S_5	3	700	7.22	21	377	56	56	24	0	0.3	56	17	3	7	0.2	0	24
S_6	2.6	900	7.35	21	376	210	210	14	0	0.3	56	17	2.8	7	0.2	0	24
S_7	2.4	180	7.38	21	382	212	212	15	0	0.3	56	17	2.5	7	0.2	0	24
S_8	2.6	40	7.32	21	378	210	210	14	0	0.3	56	17	2.3	7	0.2	0	24
S_9	2.6	20	7.25	22	382	212	212	15	0	0.3	56	17	2.2	7	0.2	0	24
S_{10}	3.4	40	6.95	22	1400	400	400	180	0	0.6	106	32	3	40	0.4	0	260

mg/L. This average DO is less as compared to other weeks of subsequent sampling. On that day, the strewing of funeral materials along the river was witnessed. This could have been due to the low average DO concentration as compared to the other sampling period.

On 13/02/2011, the DO concentration recorded was from 4.4 to 7.2 mg/L. The lower DO concentration of 4.4 mg/L was recorded due to washing activities at S_5. The higher DO concentration of 7.2 mg/L was recorded at S_4 and S_7 because temperature of water at this sampling

Table 5. Water quality parameters on 17/01/2011.

Sampling locations	Water quality parameter						
	DO (mg/L)	COD (mg/L)	pH	T (°C)	TDS (mg/L)	Hardness (mg/L)	SO_4^{2-} (mg/L)
S_1	8	140	7.34	25	600	186	2123.7
S_2	8.8	150	7.11	23	330	116	2007
S_3	10.8	160	7.06	24	412	118	2217.1
S_4	6.4	110	7.4	23	500	136	2240.4
S_5	7.6	20	7.7	25	300	128	1864
S_6	6.8	80	7.34	23	495	136	1890.4
S_7	8	50	7.3	24	600	124	1820.3
S_8	7.6	110	7.4	24	700	118	1703.7
S_9	8	100	7.2	25	412	130	1610.34
S_{10}	8.4	50	7.94	22	1319	294	2473.8

Table 6. Water quality parameters on 23/01/2011.

Sampling locations	Water quality parameter						
	DO (mg/L)	COD (mg/L)	pH	T (°C)	TDS (mg/L)	Hardness (mg/L)	SO_4^{2-} (mg/L)
S_1	7.6	30	7.4	24	303	142	1563.66
S_2	10.4	40	7.48	22	379	132	1376.96
S_3	9.6	40	7.63	24	600	155	1330.28
S_4	8	100	7.57	23	500	175	1587
S_5	7.6	150	7.42	25	300	165	1423.6
S_6	7.6	140	7.35	22	379	185	1400.3
S_7	8	140	7.36	25	400	165	1283.6
S_8	7.2	70	7.3	23	313	140	1400.3
S_9	8	70	7.38	24	303	160	1190.25
S_{10}	7.6	100	7.99	22	1364	210	1470.31

Table 7. Water quality parameters on 01/02/2011.

Sampling locations	Water quality parameter							
	DO (mg/L)	COD (mg/L)	pH	T (°C)	TDS (mg/L)	Hardness (mg/L)	SO_4^{2-} (mg/L)	Cl^- (mg/L)
S_1	5.6	1500	7.5	26	1000	150	1376.9	60
S_2	8.8	80	7.5	24	939	195	1376.9	60
S_3	7.6	50	7.5	25	700	100	1260.2	70
S_4	8.4	200	7.5	23	768	135	1213.5	90
S_5	7.2	90	7.5	26	900	140	1166.9	70
S_6	8.4	300	8.5	23	700	140	1376.9	90
S_7	1.2	120	7.5	23	1000	160	1283.6	70
S_8	6	50	7.5	26	800	160	1283.6	70
S_9	3.6	120	7.5	24	300	155	606.79	90
S_{10}	2	330	7.5	25	1700	230	1376.9	140

sites were recorded as 23°C which was less as compared to other sampling locations on this day.

On 20/02/2011, the DO varies between 8.8 to 11.6 mg/L, this DO concentration is comparatively higher than the previous weeks. On that day, no discharge of sewage was waste observed. Since the river is free flowing in nature, previously discharged pollution might have been diluted.

Table 8. Water quality parameters on 07/02/2011.

Sampling locations	Water quality parameter							
	DO (mg/L)	COD (mg/L)	pH	T (°C)	TDS (mg/L)	Hardness (mg/L)	SO_4^{2-} (mg/L)	Cl^- (mg/L)
S_1	6.8	120	8	26	407	130	1516.9	90
S_2	6.4	20	7.5	23	489	125	1516.9	80
S_3	6.8	40	7.5	24	600	145	1423.6	90
S_4	5.6	70	7.5	25	400	145	1400.3	250
S_5	7.2	120	7	23	600	210	1260.2	100
S_6	6.4	40	7	26	400	225	1446.9	270
S_7	3.2	100	8.5	25	700	130	1516.9	290
S_8	7.6	140	7	23	300	140	1143.5	60
S_9	6.4	80	7	24	500	120	1493.6	70
S_{10}	6	50	7.5	26	1385	260	1680.36	70

Table 9. Water quality parameters on 13/02/2011.

Sampling locations	Water quality parameter							
	DO (mg/L)	COD (mg/L)	pH	T (°C)	TDS (mg/L)	Hardness (mg/L)	SO_4^{2-} (mg/L)	Cl^- (mg/L)
S_1	5.6	80	7.5	25	1100	105	1750.37	80
S_2	6.4	150	7.5	24	800	150	1797	90
S_3	6.8	180	7.5	26	1300	95	1516.9	90
S_4	7.2	40	7.5	23	800	110	1633.6	80
S_5	4.4	20	7.5	24	700	125	1446.9	100
S_6	5.6	110	7.5	25	1200	110	1283.6	80
S_7	7.2	120	7.5	23	800	120	1563.6	110
S_8	6.4	90	7.5	24	954	155	1306.9	130
S_9	6.8	70	7.5	26	700	120	1260.2	250
S_{10}	6.4	160	8.5	25	2147	230	2007	340

Table 10. Water quality parameters on 20/02/2011.

Sampling locations	Water quality parameter							
	DO (mg/L)	COD (mg/L)	pH	T (°C)	TDS (mg/L)	Hardness (mg/L)	SO_4^{2-} (mg/L)	Cl^- (mg/L)
S_1	10	70	7.5	28	400	165	2380.5	90
S_2	10.8	50	7.5	26	400	185	1633.6	100
S_3	10.4	60	7.5	28	600	165	1493.6	80
S_4	11.2	60	7.5	27	500	150	1750.37	80
S_5	8.8	60	7.5	28	500	160	1563.6	90
S_6	11.6	60	7.5	27	300	185	1423.6	90
S_7	10	30	7.5	26	500	180	1750.37	120
S_8	11.6	30	7.5	28	700	185	1820.39	80
S_9	8.8	50	7.5	26	800	265	2147.129	300
S_{10}	8.4	50	7.5	28	1300	280	2147.129	310

On 27/02/2011, the DO has been recorded higher, between 6.8 to 10 mg/L. This high DO level is due to the recent rainfall before the day the sampling set has was collected. Based on the above discussion it can be said that the water is not fit for drinking purpose. The prime requirements for DO arise in connection with fish life and it is generally true that if water quality is suitable for fish life; the cardinal point about the solubility of oxygen in

Table 11. Water quality parameters on 27/02/2011.

Sampling locations	Water quality parameter							
	DO (mg/L)	COD (mg/L)	pH	T (℃)	TDS (mg/L)	Hardness (mg/L)	SO_4^{2-} (mg/L)	Cl^- (mg/L)
S_1	10	120	7.5	25	1500	200	1797	300
S_2	10	320	7.5	27	900	300	2077	600
S_3	9.2	360	7.5	25	900	200	1867	400
S_4	10	120	7.5	26	900	200	1727	600
S_5	7.6	400	7.5	27	1400	300	1633.6	700
S_6	7.2	240	7.5	25	700	300	1563.6	500
S_7	8.4	1560	7.5	24	800	200	1843.7	600
S_8	6.8	160	7.5	26	400	400	1703.7	500
S_9	7.2	50	7.5	25	1300	100	1493.6	240
S_{10}	8.4	110	8.5	25	1300	150	1913.7	280

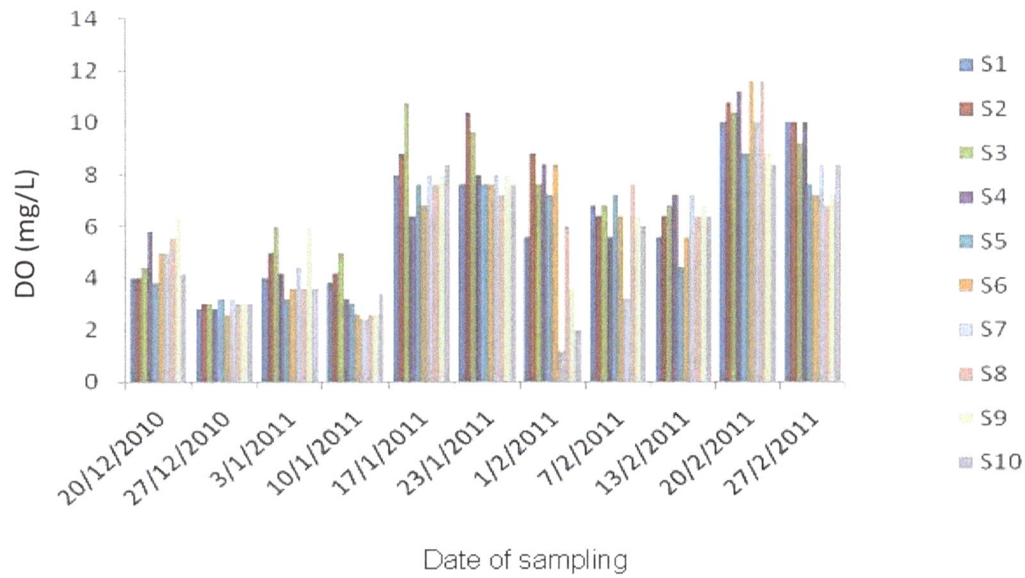

Figure 2. Variation of DO over a period of time.

water is that it has an inverse relationship with temperature.

Variation of temperature

The temperature was recorded between 17 to 19℃ on 20/12/2010 and 27/12/2010. This lower temperature of about 17℃ has been recorded due to light drizzling and weather change dramatically (Dhage et al., 2006). The temperature was recorded between 22 to 28℃ from 3/01/2011 to 27/02/2011. This normal temperature is recorded due to the normal atmospheric temperature and the high temperature will affect the DO concentration.

Variation of pH

The variation between the pH of samples collected from 20/12/2010 to 17/01/2011, S_1 to S_9 is 7.4 to 7.52. The pH for S_{10} is 7.7 to 8.6 over the same period of three weeks. This is comparatively higher than that of S_1 to S_9 for the period of same three weeks, because the S_{10} is highly utilized by people for domestic purposes such as washing, bathing etc. This increases the soap content in water at S_{10}. This might have increased pH at that sampling location. The pH was recorded in the range of 7 to 8 for all samples (S_1 to S_9) collected from 23/01/2011 to 27/02/2011. The pH of S_{10} is above 8. The high pH shows that the water is alkaline due to the addition of soap and sewage intrusion at S_{10}.

In generally accepted sense, Dissolved gases such as CO_2, H_2S, and NH3 also affect the pH of water. The industrial waste water will be strongly acidic or basic and their effect on the pH value of receiving water depends on buffering capacity of water. Higher value of pH has a bitter taste. Higher values of pH hasten the scale

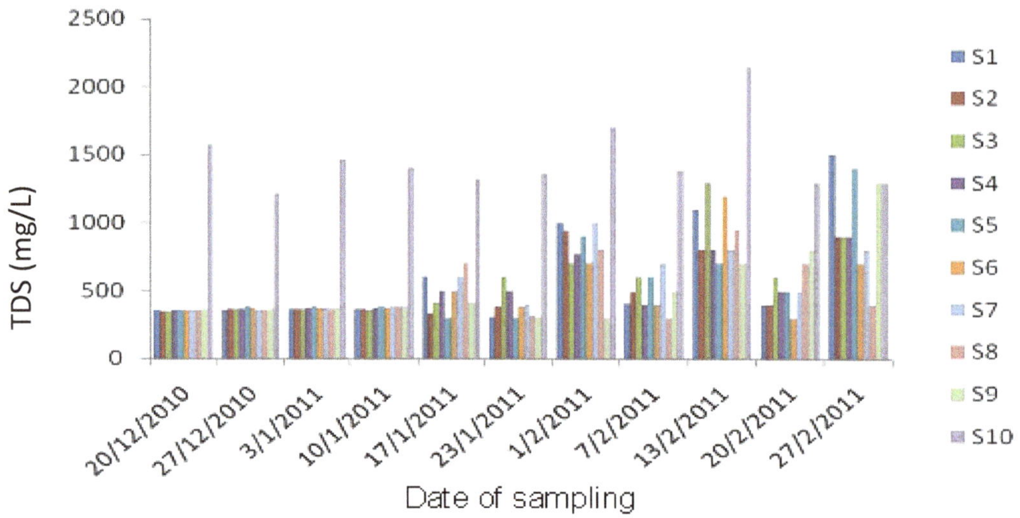

Figure 3. Variation of TDS over a period of time.

formation in water heating apparatus.

Variation of total dissolved solids

The variation of TDS over a period of time is shown in the Figure 3. The TDS concentration recorded for S_1 to S_9 from 20/12/2010 to 27/02/2011 is between 300 to 1000 mg/L. The TDS for S_{10} up to the same period of time is between 1214 to 2147 mg/L. This change is due to high exploitation of that location of the local people for washing, domestic purpose and intrusion of sewage.

The permissible limit of TDS is 500 mg/L for drinking water as per the IS 10500: 1991. More number of values are not within the permissible limit, only few values are within the permissible limit. Many dissolved substances are undesirable in water. Dissolved minerals, gases and organic constituents may produce aesthetically displeasing color, taste and odour. Some dissolved organic chemicals may deplete the DO in receiving waters and some may be inert to biological oxidation, yet others have been identified as carcinogens. Water with higher solids content often has a laxative and some time reverse effect upon people whose bodies are not adjusted to them.

Variation of hardness

The variation of hardness over a period of time is shown in the Figure 4. Hardness recorded between 98 to 540 mg/L for S_1 to S_{10} from 20/12/2010 to 27/01/2011. The high concentration of hardness was recorded due to the intrusion of sewage, pseudo hardness by utilization of soap for washing (Gupta and Sahara, 2009).

It denotes that the water is slightly and moderately hard. The high concentration of hardness above 350 mg/L indicates that the water is excessively hard. The hard water is useful to growth of children due to the presence of Ca^{2+} and Mg^{2+}. Moderately hard water is preferred to soft water for irrigation purpose. Cardiovascular diseases are reported in soft water areas (Jurgen, 1991). The permissible limit of hardness is 300 mg/L as per the IS 10500:1991.

Since Ca and Mg are hardness produced species, their variation over a period of time is discussed here. The Calcium was recorded between 34 to 144mg/L. At S_{10}, the Calcium concentration is high. When compared with other sampling location, the increased value of Calcium was found on 10/01/2011 sample. Its concentration was found to be 144 mg/L. Calcium is one of the most common constituents present in natural water. The high concentration of Calcium causes hardness in water and incrustation in boilers. It was within the permissible limit of 75 mg/L except S_{10}.

The magnesium was ranging between 10 to 43 mg/L and is the entire sampling range. All the Mg^{2+} values are within the permissible limit of 30 mg/L except S_{10} values. Magnesium is a common constituent in natural water. Mg^{2+} salts are important contributors to the hardness of water which break down heated, forming scale in boilers. The increase in Mg^{2+} concentration at S_{10} may be due to sewage disposed to that sampling location. Softening, reverse osmosis, electro dialysis, or ion exchange reduces the Mg^{2+} and associated hardness to acceptable limits.

Variation of sulphate

The variation of sulphate over a period of time is shown in the Figure 5. The sulphate was recorded between 12 to 2473.8 mg/L at S_1 toS_{10} during 20/12/2010 to 27/02/2011. The reason for this is sewage discharge by the local area people near river. Comparing to all sampling locations, value at S_{10} have been recorded as

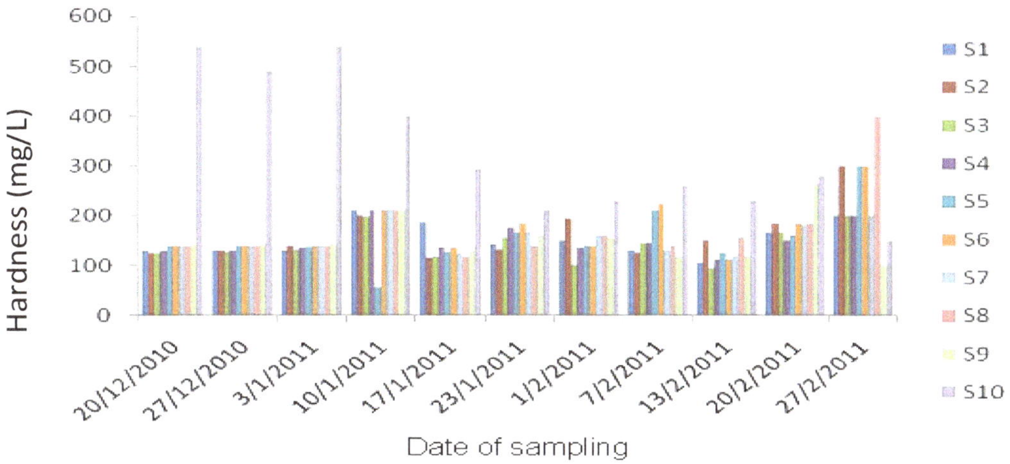

Figure 4. Variation of hardness over a period of time.

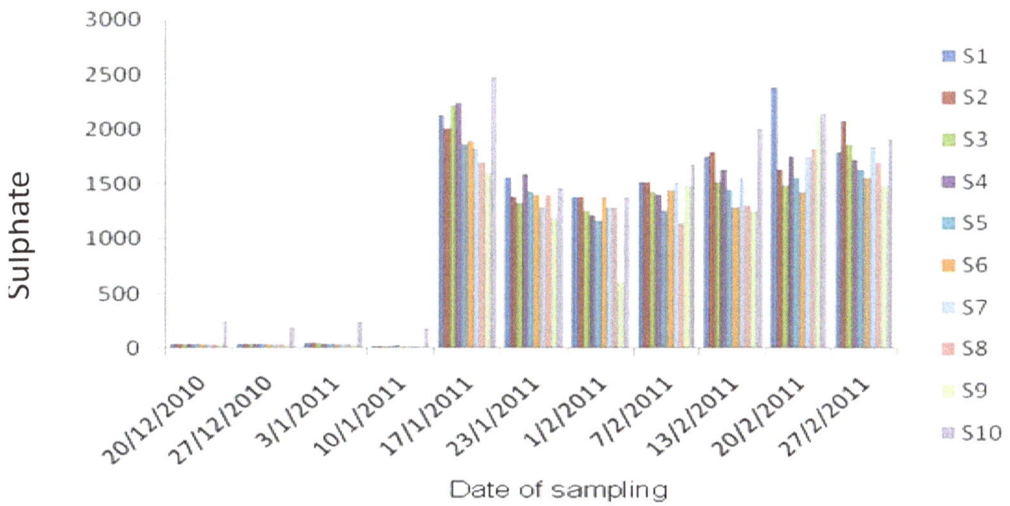

Figure 5. Variation of Sulphate over a period of time.

high value of sulphate. Most of the readings which were recorded are not within the permissible limit of 200 mg/L as per IS 10500:1991. The sulphate content in water is important in determining the suitability of water for public and industrial use. Sulphate may also contribute to the corrosion of pipelines in the distribution system. The major physiological effects resulting from the ingestion of large quantities of sulphate are catharsis, dehydration and gastrointestinal irritation. The Mg^{2+} and Na^{+} are present in most sources of their combination with sulphate and will have an enhanced laxative effect.

Variation of chloride

The variation of chloride over a period of time is shown in the Figure 6. The chloride concentration ranged from 36 to 700 mg/L was recorded during 20/12/2010 to 27/02/2011. The chloride concentration always exceeded the permissible limit at S_{10}. If the water with high chloride is used for construction purpose, this may corrode the concrete. The highest concentration of chloride at S_{10} is due to the intrusion of sewage water into the river. At S_{10}, the chloride concentration is not within the permissible limit of 250 mg/L. High chloride content may harm metallic pipes and structures as well as growing plants. It will not develop any adverse effect once the human system becomes adapted to the water. The sewage is a rich source of chloride, a high value at S_{10} in all sampling periods indicating pollution of water by a sewage effluent. On 27/02/2011 before the day of sampling, rain was over there. Hence runoff water into the river might be contributing to chloride concentration in the river.

Variation of chemical oxygen demand

The variation of COD over a period of time is shown in

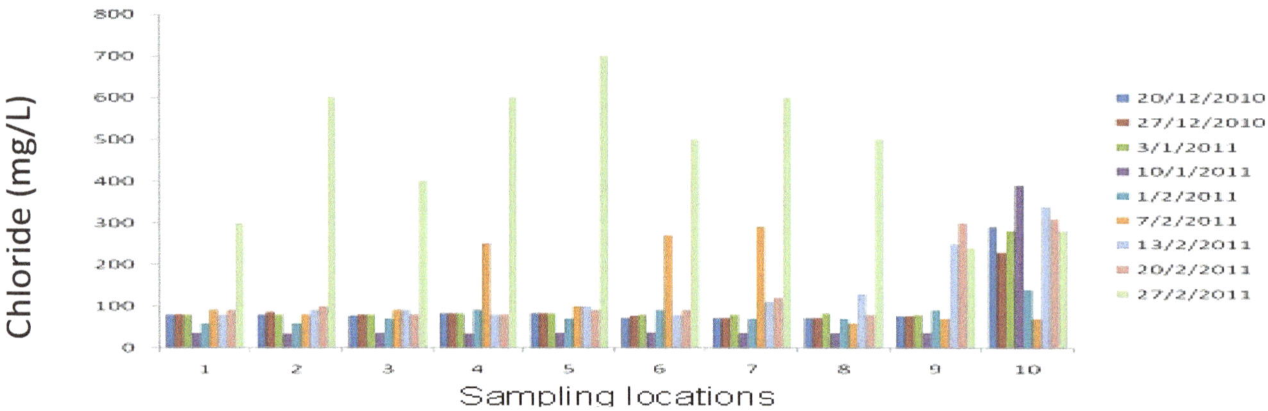

Figure 6. Variation of Chloride over a period of time.

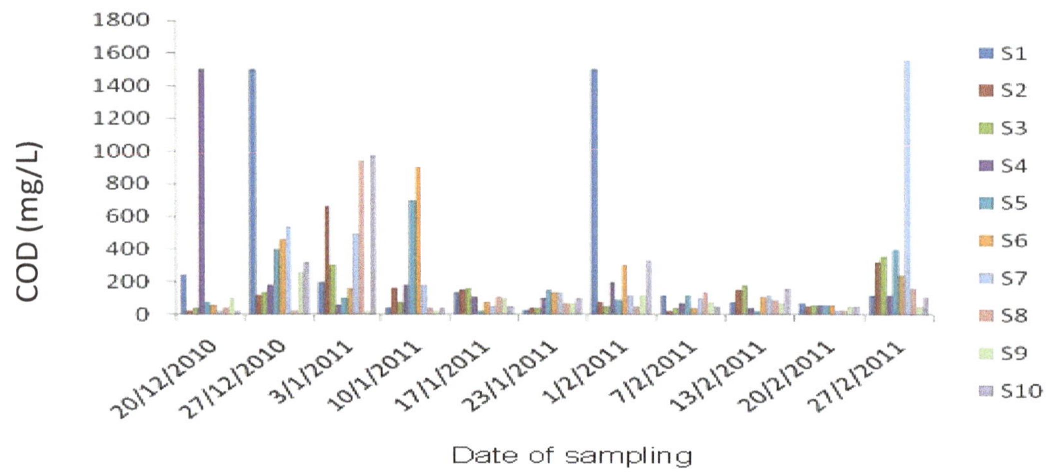

Figure 7. Variation of COD over a period of time.

the Figure 7. The COD concentration was recorded between 20 to 1500 mg/L. In the S_4 locations, it has a maximum value of COD which is due to the washing and bathing activities on the location. The lower concentration of COD was recorded as 20 mg/L. In the sampling location S_1, the COD concentration was found to be as high as 1500 mg/L so DO get decreased in that location of 2.8 mg/L. It is obvious that COD has an inverse relationship with DO. Whenever the sewage discharge is observed in S_6 the COD concentration is found to be high. This COD concentration was recorded as 900 mg/L.

Variation of iron

The values of iron were between 0.1 to 0.34 mg/L. All the values are within the permissible limit of 0.3 mg/L except S_{10} of 27/12/2010. On 27/12/2010 at S_{10} high iron value of 0.34 mg/L was observed. Iron is essential nutrition to the human health. Turbidity and color will develop in piped systems at levels above 0.05 to 0.1 mg/L. Long time consumption of drinking water with a high concentration of iron can lead to liver disease. Iron also promotes the growth of iron bacteria in water and it gives a rusty appearance to the water. The highest concentration of iron at S_{10} is due to the sewage intrusion. On 10/01/2011 for all the samples (S_1 to S_{10}) iron concentration is zero. The nil iron concentration observed was due to no discharge of sewage; since the river is free flowing, previously discharged pollution might have been diluted.

Variation of phosphate

The variation of phosphate over a period of time is shown in the Figure 8. The values of phosphate were between 1.5 to 4 mg/L. Phosphate occurs as trace in natural water. Stream receiving raw or treated sewage, agricultural drainage and certain industrial water normally

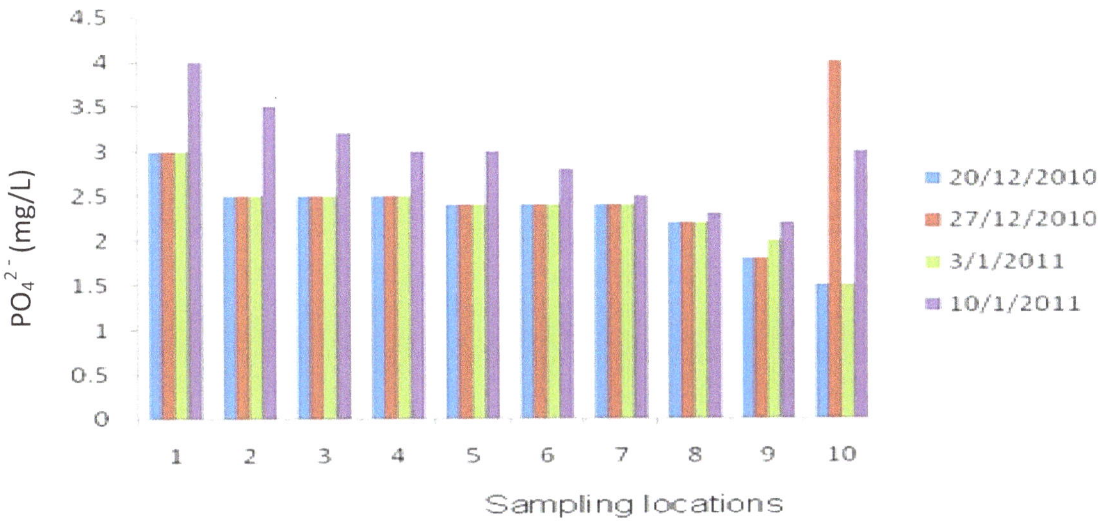

Figure 8. Variation of Phosphate over a period of time.

contain significant concentrations of phosphate. Trace of phosphate increase the tendency of troublesome algae growth. The highest concentration of phosphate was found to be 4 mg/L and is due to detergent leached into the streams through the waste water generated industrially, domestically or from cloth dyeing and garment industries.

Variation of manganese

The Mn recorded was zero for all samples from 20/12/2010 to 10/01/2011. The main source of Mn pollution is from steel industry but this river is polluted with textile effluent so the Mn pollution is not discharged into the river. As a result, Zero Mn concentration was obtained. The intake of manganese can be high as 20 mg/day without apparent ill effects. It should be noted that manganese may be objectionable to the consumer if it is deposited in water mains and cause discoloration. Concentrations below 0.1 mg/L are usually acceptable to consumers. This may vary with local circumstance.

Variation of nitrate and nitrite

In this study, the nitrate concentrations were between the 20 to 44 mg/L. The nitrite was recorded between the 0.01 to 0.6 mg/L. At S_5 during 20/12/2010, 27/12/2010, 3/01/2011 the nitrite concentration was zero. Nitrate generally occurs in trace quantities in surface waters but may attain high levels in some ground water. Nitrite in water is either due to oxidation of ammonium compounds or due to a reduction of nitrate. It can be toxic to certain aquatic organism even at concentrations of 1 mg/L. Excessive limits of it contributes to the illness known as methanoglobinemia in infants. All NO_3 concentrations are within the permissible limit of 50 mg/L as per the IS 10500:1991.

Variation of sodium

The sodium was recorded between 20 to 260 mg/L. This is due to the discharge of sewage and it took place at S_{10}. Sewage is the main source of sodium salts. The permissible limit for sodium in Drinking Water is 200 mg/L as per the IS 10500:1991.

Variation of fluoride

The fluoride was ranging between 0.3 to 0.6 mg/L. Trace of fluorides is present in many waters. All the fluoride concentrations are within the permissible limit of 1 mg/L. The fluoride is designated as two edge sword because if it is less than limit and also greater than limit it will cause harm to the human and aquatic organisms (Jurgen, 1991). Presence of large amounts of fluoride is associated with dental and skeletal fluorosis (1.5 mg/L) and inadequate amounts with dental caries (< 1 mg/L).

Water quality index

Water quality index (WQI) is defined as a rating reflecting the composite influence of different water quality parameters. WQI is calculated from the point of view of the suitability of ground water for human consumption. In computing WQI three steps are followed. In the first step, each of the ten parameters has been assigned a weight (w_i) according to its relative importance in the overall quality of water for drinking purposes. It is shown in the Table 12. The maximum weight of 5 has been assigned to the parameter TDS due to its major importance in water quality assessment. Iron which is given the weight of 1 due to the low value was observed in surface water. In the second step, the relative weight, (W_i) is computed

Table 12. Relative weight of water quality parameters.

Chemical parameter	IS10500: 1991	Weight (w_i)	Relative weight (W_i)
TDS	500 – 1000 mg/L	5	0.1724
pH	6.5 - 8.5	4	0.1379
Hardness	300 - 600 mg/L	3	0.1034
Chloride	250 - 1000 mg/L	3	0.1034
Sulphate	200 - 400 mg/L	3	0.1034
Nitrate	50 mg/L	3	0.1034
Fluoride	1 – 1.5 mg/L	3	0.1034
Magnesium	30 - 100 mg/L	2	0.0689
Calcium	75 - 200 mg/L	2	0.0689
Iron	0.3 – 1.0 mg/L	1	0.0344
Summation		$\Sigma w_i = 29$	$\Sigma W_i = 1.000$

Table 13. Water quality classification based on wqi value.

WQI value	Water quality	Percentage of water samples
< 50	Excellent	1.8
50 – 100	Good water	33.6
100 – 200	Poor water	62.72
200 – 300	Very poor water	1.8
>300	Water unsuitable for drinking	-

from the Equation 1. In the third step, a quality rating scale (q_i) for each parameter is assigned by dividing its concentration in each water sample by its respective standard according to the guidelines laid down in the IS 10500:1991. For computing the WQI, the SI is first determined for each water quality parameter, which is then used to determine the WQI as shown by the Equation 4. Table 13 shows the percentage of water samples that falls under different water quality.

In this study, the computed WQI values from 46.56 to 217.8 therefore, can be categorized into five types "Excellent Water" to "Water unsuitable for drinking". The high value of WQI at sampling locations has been found to be mainly from the higher values of TDS and Sulphate. Among the total sampling, 62.72% of collected sample fell in the range of 100 to 200 on water quality index scale. Since the most percentage fell in that category it can be concluded that the Noyyal River has poor water quality. On the entire sampling, 1.8% fell in the range of 200 to 300 this indicated the very poor water quality in Noyyal River.

Conclusion

The following conclusions were drawn on the basis of the physico-chemical analysis of surface water for Noyyal River in Coimbatore city:

1. Comparing all the sampling locations, at S_{10} the samples have the highest concentration of TDS, sulphate, hardness, pH, chloride due to the sewage disposal and utilization of soap for clothes washing.
2. The DO concentration was found to be low at sampling location S_7 and is due to the dumping of used mats, pillows, garlands at that location. It leads to mortified fishes and were floating on the river.
3. COD concentration was found to be high due to the domestic waste discharge into the river.
4. Water quality index (WQI) for 110 samples ranges from 46.56 to 217.8. Among the total samples, 62.72% of the samples exceed WQI of 100. The high values of WQI at these sampling locations have been found to be mainly from the higher values of sulphate and TDS. These 62.72% of water samples were found to be poor quality. It needs some degree of treatment for human consumption. So, it is concluded that the Noyyal River has poor water quality. On the entire sampling, 1.8% fell in the range of 200 to 300 and this indicated the very poor water quality in Noyyal River.
5. The result from the data analysis show that, the water is certainly not fit for drinking purpose without any form of treatment, but may be considered for other purpose like washing, bathing of animals etc.

REFERENCES

Basanta KM, Chitta RP (2009). 'Assessment of water quality index in Mahanadi and Atharabanki Rivers and Taldanda Canal in paradip Area'. India J. Hum. Ecol. 26(3):153-161.

Dhage SS, Chandorkar AA, Rakesh K, Srivastava A, Gupta I (2006). Marine water quality assessment at Mumbai West Coast. Environ.

Int. 32:149-158.
Gupta DP, Sahara JP (2009). 'Physiochemical Analysis of Ground Water of Selected Area of Kaithal City (Haryana) India Res. 1(2). http://www.city-data.com/world-cities/Prague-Environment.html.
Indian Standard (IS) (1991). 10500, Indian standard drinking water – Specification.
Islam MR, Alam MDJB (2007). 'Water quality parameters along rivers'. J. Environ. Sci. Technol. 4(1):159-167.
Jurgen P (1991). 'Chemical analysis for drinking water'.
Lenin Sundar M, Saseetharan MK (2008). 'Ground water quality in Coimbatore, Tamil Nadu along Noyyal River'. J. Environ. Sci. Eng. 50(3):187-190.
Merriam Webster online dictionary (2010). 'http://www.merriam-webster.com/dictionary/pollution', pp. 08-13.
Ramakrishnaiah CR, Sadashivaiah C (2009). 'Assessment of Water Quality Index for the Groundwater in Tumkur Taluk, Karnataka State'. India E-J. Chem. 6(2):523-530.
Rainwater harvesting (2007). 'http://www.rainwater harvesting.org/crisis/river-noyyal.html' 02-01.
Sajidu SMI, Henry EMT (2007). 'Water quality assessment in streams and wastewater treatment plants of Blantyre, Malawi'. J. Phys. Chem. Earth 32:1391-1398.
www.cpcb.nic.in.

Water resources management and sustainable development in Sudan

Abdeen Mustafa Omer

Energy Research Institute (ERI), UK. E-mail: abdeenomer2@yahoo.co.uk.

For the thirty-nine million who live in Sudan, environmental pollution is a major concern; therefore industry, communities, local authorities and central government, to deal with pollution issues, should adopt an integrated approach. Most polluters pay little or no attention to the control and proper management of polluting effluents. This may be due to a lack of enforceable legislation and/or the fear of spending money on the treatment of their effluent prior to discharge. Furthermore, the imposed fines are generally low and therefore do not deter potential offenders. The present problems related to water and sanitation in Sudan is many and varied, and the disparity between water supply and demand is growing with time due to the rapid population growth and aridity. The situation of the sewerage system in the cities is extremely critical, and there are no sewerage systems in the rural areas. There is an urgent need for substantial improvements and extensions to the sewerage systems treatment plants. Further development of water resources for agriculture and domestic use is one of the priorities to improve the agricultural yield of the country, and the domestic and industrial demands for water. This article discusses the overall problem and identifies possible solutions.

Key words: Sudan, water resources development, community water supply, effective water-supply management, environment.

INTRODUCTION

In Sudan, more than ten million people do not have adequate access to water supply; twenty million inhabitants are without access to sanitation, and a very low proportion of domestic sewage being treated. The investment, which is needed to fund the extension and improvement of these services, is substantial (Omer, 1995). Most governments in developing countries are ready to admit that they lack the financial resources for proper water and sanitation schemes. Moreover, historically, bilateral and multilateral funding accounts for less than 10% of total investment needed. Thus the need for private financing is imperative.

Many water utilities in developing countries need to work in earnest to improve the efficiency of operations. These improvements will not only lead to better services but also to enhanced net cash flows that can be re-invested to improve the quality of service. Staff productivity is another area where significant gains can be achieved. Investment and consumption subsidies have been predicated on the need to help the poor to have, access to basic services and to improve the environment. Failure of subsidies to reach intended objectives is due, in part, to lack of transparency in their allocation. A key element to successful private participation is the allocation of risks. How project risks are allocated and mitigated will determine the financial and operational performance and success of the project, under the basic principle that the risk should be allocated to the party, which is best able to bear it. Many developing countries (Sudan is not an exception) encourage the participation of the private sector as a means to improve productivity in the provision of water and wastewaters services. Private-sector involvement is also needed to increase financial flows to expand the coverage and quality of services. Many successful private-sector interventions have been under taken. Private operators are not responsible for the financing of works, nonetheless they can bring significant productivity gains, which would allow the utility to allocate more resources to improve and extend services. Redressing productivity, subsidy and cross-subsidy issues before the private sector is invited to participate, has proven to be less contentious. The author had previously thought to encourage more private-sector involvement (Omer, 1995).

Sudan is geo-politically well located, bridging the Arab world to Africa. Its large size and extension from south to north provides for several agro-ecological zones with a variety of climatic conditions, rainfall, soils and vegetation. Water resources available to Sudan from the Nile system, together with groundwater resources, provide a potential for thirty years increase in the irrigated sub-sector. There are also opportunities for increased hydropower generation. The strategy of Sudan at the national level aims at the multi-purpose use of water resources to ensure water security for attaining food security, drinking-water security, fibre-security, hydro-energy security, industrial security, navigation, waste disposal and the security at the regional levels within an environmentally sustainable development context and in harmony with the promotion of basin-wide integrated development of the shared water resources (Noureddine, 1997). The government has continued to pay for the development and operation of water systems, but attempts are being sought to make the user communities pay water charges. In order to ensure the sustainability of water supplies, an adequate institutional and legal framework is needed. Funds must be generated (a) for production; (b) for environmental protection to ensure water quality, and (c) to ensure that water abstraction from groundwater remains below the annual groundwater recharge. At present, there are private-sector providers who do not have an enabling environment to offer the services adequately. There is a need for the government to have a mechanism to assist in the regulation and harmonisation of the private-sector providers. Privatisation is part of a solution to improve services delivery in water and sanitation sector. At present, there is a transitional situation characterised by: (1) A resistance to water charge; (2) Insufficient suitable law/law enforcement; (3) Insufficient capacities; and (4) Inadequate interaction between actors.

In a country relatively sparsely populated, there are extreme pressures on water and waste systems, which can stunt the country's economic growth. However, Sudan has recognised the potential to alleviate some of these problems by promoting renewable water and utilising its vast and diverse climate, landscape, and resources, and by coupling its solutions for waste disposal with its solutions for water production. Thus, Sudan may stand at the forefront of the global renewable water community, and presents an example of how non-conventional water strategies may be implemented. In Sudan, more than ten million people do not have adequate access to water supply, twenty million inhabitants are without access to sanitation, and a very little domestic sewage is being treated. The investment needed to fund the extension and improvement of these services is great. Most governments in developing countries are ready to admit that they lack the financial resources for proper water and sanitation schemes. Moreover, historically, bilateral and multilateral funding accounts for less than 10% of total investment needed. Thus, the need for private financing is imperative. Water utilities in developing countries need to work in earnest to improve the efficiency of operations. These improvements would not only lead to better services but also to enhanced net cash flows that can be re-invested to improve the quality of service. Staff productivity is another area where significant gains can be achieved. Investment and consumption subsidies have been predicated on the need to (a) help the poor, which have not an access to basic services, and (b) improve the environment. Failure of subsidies to reach intended objectives is in part, from lack of transparency in their allocation. Subsidies are often indiscriminately assigned to support investment programmes that benefit more middle and high-income families that already receive acceptable service. Consumption subsidies often benefit upper-income domestic consumers much more than low-income ones. Many developing countries (Sudan is not an exception) are encouraging the participation of the private-sector as a means to improve productivity in the provision of water and wastewaters services. Private sector involvement is also needed to increase financial flows to expand the coverage and quality of services. A key element to successful private participation is the allocation of risks. How project risks are allocated and mitigated determines the financial and operational performance and success of the project, under the basic principle that the risk should be allocated to the party, which is best able to bear it. Many successful private-sector interventions have been undertaken. Private operators are not responsible for the financing of works, nonetheless they can bring significant gains in productivity, which would allow the utility to allocate more resources to improve and extend services. Redressing productivity, subsidy and cross-subsidy issues before the private-sector is invited to participate, has proven to be less contentious. The author had previously sought to encourage more private-sector involvement (Omer, 1995).

This paper comprises a comprehensive review of water sources, the environment and sustainable development. It includes the renewable water resources, water conservation scenarios and other mitigation measures necessary to reduce climate change.

WATER RESOURCES

Sudan is rich in water (from the Nile system, rainfall and groundwater) and lands resources (Table 1). Surface water resources are estimated at 84 billion m^3 and the annual rainfall varies from almost nil in the arid hot north to more than 1600 mm in the tropical zone of the south.

Table 1. Land use, land-resource zones and water resources (Omer, 2002).

(a) Land use (millions of ha)

Geographical area (total Sudan area)	250.6
Land area	237.6
Cultivable area	8.4
Pastures	29.9
Forests and woodland	108.3
Uncultivable land	81.0
Area under crop (irrigated, rain-fed, mechanised, and rain-fed traditional)	10.0

(b) Land-resource zones

Zone	Area as % to total area of Sudan	Persons per km^2	Mean average rainfall range (mm)
Desert	44	2	0-200
QOS sands	10	11	200-800
Central clay plains	14	19	200-800
Southern clay plains	12	8	800-900
Ironstone plateau	12	7	800-1400
Hill area and others	8	16	Variable

(c) Water resources

	Available number	Static water level (m)	Number
Haffirs	824	0-0	824
Slow sand filters	128	0-0	128
Open shallow wells	3000	0-10	3000
Boreholes deep wells	2259	0-25	1248
		26-50	478
		51-75	287
		76-100	246

(d) Geological formations

Basins	Amount of water recharged (10^6 m^3)	Water level below land (m)	Aquifer thickness (m)	Velocity (m/year)	Abstraction (10^6 m^3/year)
Sahara Nile	136	30-100	300-500	1-2.5	7.3
Sahara Nubian	20.6	10-50	300-500	0.8-1.5	1.5
Central Darfur	47.6	25-100	250-550	0.3-6.0	5.5
Nuhui	15.4	75-120	200-400	1.0-2.75	1.6
Sag El Na'am	13.5	50-1000	300-500	1.0-25.0	2.5
River Atbara	150	100-150	250-300	0.3-5.0	2.3
Sudd	341	10-25	200-400	0.1-1.8	1.8
Western Kordofan	15	50-70	300-500	0.1-0.3	1.7
Baggara	155	10-75	300-500	0.1-2.4	11.9
Blue Nile	70.9	10-50	250-500	0.1-2.5	10.2
The Alluvial	N.A	Shallow	N.A	N.A	N.A
Gedaref	41.7	50-75	200-500	0.1-2.0	1.2
Shagara	1.1	25-30	200-300	0.1-2.5	0.7

The total quantity of groundwater is estimated to be 260 billion m^3, but only 1% of this amount is being utilised. Water-resources assessment in Sudan is not an easy task because of uncertainty of parameters, numerous degrees of freedom of variables, lack of information and inaccurate measurements. However, according to seasonal water availability, Sudan could be globally divided into three zones: (a) Areas with water availability throughout the year are the rainy regions (equatorial tropical zones); (b) Areas with seasonal water availability, and (c) areas with water deficit throughout the year, which occupy more than half the area of Sudan.

The most important research and development policies which have been adopted in different fields of water resources are: (1) The water resource; (2) Irrigation development; (3) The re-use of drainage water and groundwater; (4) Preventive and canal maintenance; (5) Aquatic weed control and river channel development, and (6) protection plans. The physical and human resources base can provide for sustainable agriculture growth and food security for itself and for others in the region. Failure to do so in the past derives from several causes and constraints, which are manageable. These include misguided policies, poor infrastructure, low level of technology use, recurring droughts and political instability. Perhaps the biggest challenge is that of finding resources for capital improvements in the light of changing water-quality regulations and ageing systems (James, 1994).

The desert environment is fragile and highly affected by human activities. Disturbances in the balanced ecosystems are apt to take place causing serious problems to the environment, and consequently, initiating geotechnical hazards. Urbanisation, climatic conditions, and geomorphic and geologic setting are usually the controlling factors influencing the types of these hazards. One of the potential geotechnical hazards that may occur under desert conditions is sand drifting and dune movement. The problem of sand drifting and dune migration is of special interest in Sudan as moving sand covers approximately one-third of the country. Because sand poses natural erosional-depositional hazards on the existing structures, such as roads and urbanised areas, it become necessary to study the behaviour of the sand forms in the different parts of the country.

Although deserts are known to be simply barren areas, they are scientifically defined in terms of water shortage or aridity, soil type, topography and vegetation. Anon (1979) presented a map showing the distribution of deserts in the world. Accordingly to this map, most of the Middle Eastern countries lie within the semi-arid, arid, and hyper-arid desert zones, with an aridity index (ratio between annual precipitation and mean annual potential evapotranspiration) ranging between 0.03 and 0.02. Most of the geotechnical hazards are associated with desert environments. The desert environment, being a fragile ecosystem, needs to be treated with care. Intercommunications between different national and international agencies and education of the layman should help to keep the system balanced and reduce the resulting environmental hazards. In addition, any suggested remedial measures should be planned with nature and be engineered with natural materials.

WATER AND SANITATION MANAGEMENT

Community water supply and sanitation management is a new form of cooperation between support agencies in the water and sanitation sector and communities. It involves a common search to identify problems with the local water supply and sanitation systems, to establish the possibilities for, and constraints on, management by communities, and to find possible solutions that may be tested. Some fundamental principles of community water and sanitation management are: (1) Increased management capacities are the basis for improved water and sanitation systems, and each community must develop its own specific management systems, and (2) Communities own the process of water charge; facilitators and local researchers participate in the community's projects, not the other way round.

Through this approach, the support agency is no longer the provider of technical goods or solutions, but the facilitator of process to enhance the capacity of the community to manage its own water and sanitation systems. Constraints include: (1) A lack of funds or substantial delays in allocating funds for essential requirements such as operation and maintenance of irrigation and drainage projects; (2) Deterioration in data-collection activities; (3) A lack of appropriate and consistent policies for water development for both large- and small-scale projects; (4) Serious delays in completing water projects after major investments such as dams and other hydraulic structures, and main secondary canals not being completed; (5) An absence or inadequacy of monitoring, evaluation, and feedback at both national and international levels; (6) A lack of proper policies on cost recovery, and water pricing or, if policies exist, absence of their implementation; (7) A shortage of professional and technical manpower, and training facilities; (8) A lack of beneficiary participation in planning, implementation, and operation of projects; (9) Inadequacy of knowledge, and absence of appropriate research to develop new technologies and approaches, and an absence of incentives to adopt them; (10) General institutional weaknesses and a lack of coordination between irrigation, agriculture, energy, healthy, environment, and planning; (11) Inappropriate project development by donor agencies, e.g. irrigation development with drainage, supporting projects which should not have been supported; and (12) A lack of donor coordination resulting in differing approaches and methodologies, and thus conflicting advice.

As developing nations strive to provide a safe and reliable drinking-water supply to their growing and increasingly urbanised population, is becoming more evident that new approaches to this problem will be needed. To meet this challenge, new methods of reclaiming and re-using water had been developed in cost-effective and environmentally sound ways (ODA, 1987; Seckler, 1992; Salih, 1992). Despite the constraints, over the last decade, the rate of implementation of rural and peri-urban water supply and sanitation programmes had increased considerably, and many people are now being served more adequately. The following are Sudan experience in water supply and sanitation projects:

At community level:

1. Participatory approaches in planning, implementation and monitoring.
2. Establishment and training of water tap committees.
3. Clear ownership of improved water supply and sanitation systems.
3. Technology and service level selection by consumers.
4. Sensitive timing of hygiene and sanitation education.
5. Establishment and training of reliable financial and maintenance management.

At district and national level:

1. Integrated multi-sectoral approach development.
2. Training approach and material development for district and extension staff.
3. Continuing support from integrated multi-sectoral extension team.
4. Establishment of technical support system.
5. Multi-sectoral advisory group including training and research institutions.
6. Development and dissemination of relevant information for district and extension staff.

WATER RESOURCE MANAGEMENT SYSTEMS

Water is a substance of paramount ecological, economical, and social importance. Interrelationships inherent in water use should encourage integrated water management. Water resources are to be better managed to:

1. Ensure more reliable water availability and efficient water use in the agricultural sector.
2. Mitigate flood damage.
3. Control water pollution.
4. Prevent development of soil salinity and water logging.
5. Reduce the spread of water-borne diseases.

The emerging water crisis, in terms of both water quantity and quality, requires new approaches and actions. Priority areas needing concerted action in various sectors are:

(a) Water use efficiency; (b) Flood control; (c) Management of scarce water resources; (d) Water quality management and provision of safe drinking water, and (e) Coordination and integration of various aspects of water management, and water management with other related resources and societal concern. The following are recommended:

1. Community must be the focus of benefits accruing from restructures, legislature to protect community interest on the basis of equity and distribution, handover the assets to the community should be examined; and communities shall encourage the transfer the management of water schemes to a professional entity.
2. The private sector should be used to mobilise, and strengthen the technical and financial resources, from within and without the country to implement the services, with particular emphasis on utilisation of local resources.
3. The government should provide the necessary financial resources to guide the process of community management of water supplies. The government should divert from provision of services and be a facilitator through setting up standards, specifications and rules to help harmonise the private sector and establish a legal independent body by an act of parliament to monitor and control the providers. Government to assist the poor communities who cannot afford service cost, and alleviate social-economic negative aspects of privatisation.
4. The sector actors should create awareness to the community of the roles of the private sector and government in the provision of water and sanitation services.
5. Support agencies assist with the financial and technical support, the training facilities, coordination, development and dissemination of water projects, and then evaluation of projects.

The development of new, modern, and complete water-resources-information systems is one of the basic needs for the implementation of the water-resources-management system. The decision process in drought or flood conditions, and also in over-exploitation cases, can only be correct if based on a reliable information system. A complete and comprehensive database on water availability, users, water quality monitoring, current technologies (like geographical information systems), is certainly the way to produce an efficient framework for decision-making. Lack of information is one of the most critical points regarding the development and implementation of the new management system (FAO, 1999). The types of data related to flood management include:

1. Topographic data (elevations, land use, soils, vegetation, hydrography).
2. Administrative data (political boundaries, jurisdictional boundaries).
3. Infrastructure data (roads, wells, utilities, bridges and culverts, hydraulic structure, properties, facilities) and imagery (satellite images, aerial photographs).
4. Environmental data (threatened and endangered species, critical aquatic and wildlife habitat, archaeological sites, water quality).

5. Hydrometeorology data (stream flows, precipitation, temperature, wind, solar radiation, soil water, discharge rating curves, flood frequency, flood plain delineation).
6. Economic data (stage-damage relationships, insured values, industries), and
7. Emergency management data (emergency plans, census data, organisational charts).

THE POLICY REGIME IN WATER QUALITY MANAGEMENT

Apart from effluent regulations, and sometimes, national water quality guidelines, a common observation is that few developing countries (Sudan is not an exception) include a water-quality-policy context. Whereas water supply is seen as a national issue, pollution is mainly felt at, and dealt with at, the local level. With few exceptions, national governments have little information on the relative importance of various types of pollution (agriculture, municipal, industrial, animal husbandry, aquaculture), and therefore, have no notion of which is of greatest economic or public health significance. Usually freshwater quality management is completely divorced from coastal management even though these are intimately linked. Consequently, it is difficult to develop a strategic water quality management plan or to efficiently focus domestic and donor funds on priority issues. A national water-quality-policy should include the following water quality components:

1. A policy framework that provides broad strategic and political directions for future water-quality management.
2. A strategic action plan for water-quality management based on priorities that reflect an understanding of economic and social costs of impaired water.

This plan should include the following components:

1. A mechanism for identifying national priorities for water-quality management that will guide domestic and donor investment.
2. A plan for developing a focused and cost-effective data programme for water quality and related uses, as a basis for economic and social planning.
3. A consideration of options for financial sustainability including donor support, public-private sector partnerships, regional self-support initiatives.
4. A regulatory framework that includes a combination of appropriate water-quality objectives (appropriate to that country and not necessarily based on Western standards) and effluent controls. This includes both surface and groundwater.
5. A methodology for public input into goals and priorities.
6. A process for tasking specific agencies with implementation so that accountability is firmly established and inter-agency competition is eliminated.
7. Specific mechanisms for providing drinking water monitoring capabilities, at the community level if necessary.
8. National data standards that must realistically reflect national needs and capabilities. Nevertheless, the objective is to ensure reliable data from those organisations that provide information for national water management purposes and at the community level for drinking water monitoring.

The design criteria in any water-quality programme are to determine the management issues which water quality data are required. Generally, there are four categories of data objectives:

1. Descriptive data that are typically used for government policy and planning, meeting international obligations, and for public information.
2. Data specific to public health.
3. Regulatory concerns, and
4. Aquatic ecosystem health.

The last category is not normally included in many developing countries for reasons of cost and complexity. In most developing countries, countries with transitional economies, and some developed countries, the technology of monitoring has changed little since 1970s, yet some of the largest advances in monitoring in recent years involve technical innovation that serve to reduce costs and increase efficiency. Admittedly, not all of these are inexpensive; however when deployed appropriately, they may eliminate traditional monitoring, or reduce costs by increasing the efficiency of more traditional approaches to chemical monitoring. Types of innovation include: Biological assessment, use of surrogates, use of enzymatic indicators, miniaturisation, automation, and simplification of laboratory analytical methods.

SUSTAINABLE DEVELOPMENTS

In the past decade, sustainability has increasingly become a key concept and ultimate global for socio-economic development in the modern world. Without a doubt, the sustainable development and management of natural resources fundamentally control the survival and welfare of human society. Water is an indispensable component and resource for life and essentially all human activities rely on water in a direct or in direct way. Yet supplying water of sufficient quantity and safe quality has seldom been an easy task. Although sustainability is still a loosely defined and evolving concept, researchers and policy-makers have made tremendous efforts to develop a working paradigm and measurement system for applying this concept in the exploitation, utilisation and management of various natural resources. In water resources arena, recent development had been

Table 2. Water and sustainable environment.

Technological criteria	Water and environment criteria	Social and economic criteria
Primary water saving in regional scale	Sustainability according to greenhouse gas pollutant emissions	Labour impact
Technical maturity, reliability	Sustainable according to other pollutant emissions	Market maturity
Consistence of installation and maintenance requirements with local technical known-how	Land requirement	Compatibility with political, legislative and administrative situation
Continuity and predictability of performance	Sustainability according to other environmental impacts	Cost of saved primary water

Table 3. Classification of key variables defining facility sustainability.

Criteria	Intra-system impacts	Extra-system impacts
Stakeholder satisfaction	Standard expectations met. Relative importance of standard expectations	Covered by attending to extra-system resource base and ecosystem impacts
Resource base impacts	Change in intra-system resource bases Significance of change	Resource flow into/out of facility system. Unit impact exerted by flow on source/sink system. Significance of unit impact
Ecosystem impacts	Change in intra-system ecosystems. Significance of change	Resource flows into/out of facility system Unit impact exerted by how on source/sink system Significance of unit impact

Table 4. Positive impact of durability, adaptability and energy conservation on economic, social and environment systems.

Economic system	Social system	Environmental system
Durability	Preservation of cultural values	Preservation of resources
Meeting changing needs of economic development	Meeting changing needs of individuals and society	Reuse, recycling and preservation of resources
Energy conservation and saving	Savings directed to meet other social needs	Preservation of resources, reduction of pollution and global warming

synthesised and presented in two important documents published by ASCE (1998) and UNESCO (1999), which attempt to give a specific definition and a set of criteria for sustainable water resource systems. When considering the long-term future as well as the present, sustainability is concept and goal that can only be specified and implemented over a range of spatial scales, of which urban water supply is a local problem with great reliance on the characteristics and availability of regional water resources.

Cleaner, leaner production processes-pursuing improvements and savings in waste minimisation, energy and water consumption, transport and distribution, as well as reduced emissions. Tables 2 to 4 indicate water conservation, sustainable development and environment. With the debate on climate change, the preference for real measured data had been changed. The analyses of climate scenarios need an hourly weather data series

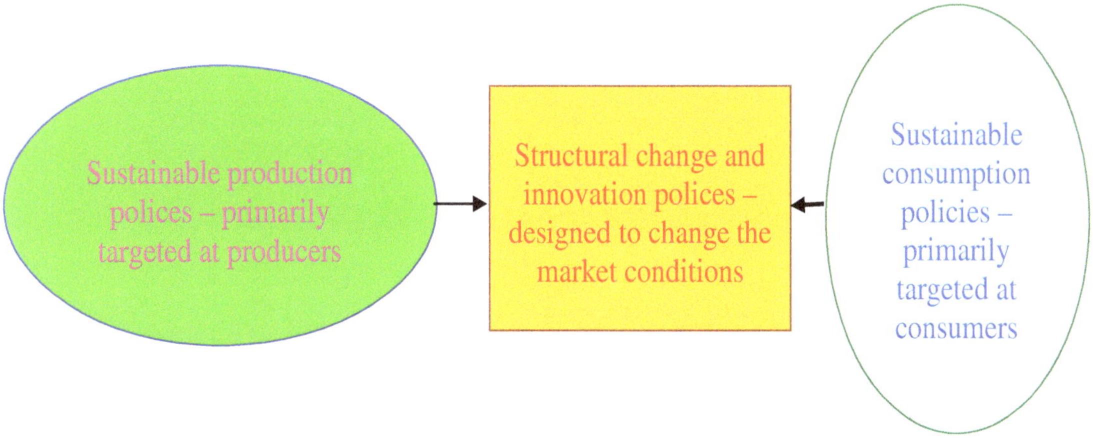

Figure 1. Link between resources and productivity.

that allows for realistic changes in various weather parameters. By adapting parameters in a proper way, data series can be generated for the site. Weather generators should be useful for:

1. Calculation of energy consumption (no extreme conditions are required),
2. Design purposes (extremes are essential), and
3. Predicting the effect of climate change such as increasing annually average of temperature.

This results in the following requirements:

1. Relevant climate variables should be generated (solar radiation: Global, diffuse, direct solar direction, temperature, humidity, wind speed and direction) according to the statistics of the real climate.
2. The average behaviour should be in accordance with the real climate.
3. Extremes should occur in the generated series in the way it will happen in a real warm period. This means that the generated series should be long enough to assure these extremes, and series based on average values from nearby stations.

Growing concerns about social and environmental sustainability have led to increased interest in planning for the energy utility sector because of its large resource requirements and production of emissions. A number of conflicting trends combine to make the energy sector a major concern, even though a clear definition of how to measure progress toward sustainability is lacking. These trends include imminent competition in the electricity industry, global climate change, expected long-term growth in population and pressure to balance living standards (including per capital energy consumption). Designing and implementing a sustainable energy sector will be a key element of defining and creating a sustainable society. In the electricity industry, the question of strategic planning for sustainability seems to conflict with the shorter time horizons associated with market forces as deregulation replaces vertical integration. Sustainable low-carbon energy scenarios for the new century emphasise the untapped potential of renewable resources. Rural areas can benefit from this transition. The increased availability of reliable and efficient energy services stimulates new development alternatives. It is concluded that renewable environmentally friendly energy must be encouraged, promoted, implemented, and demonstrated by full-scale plant especially for use in remote rural areas (Figure 1).

This is the step in a long journey to encourage a progressive economy, which continues to provide us with high living standards, but at the same time helps reduce pollution, waste mountains, other environmental degradation, and environmental rationale for future policy-making and intervention to improve market mechanisms. This vision will be accomplished by:

1. Decoupling' economic growth and environmental degradation. The basket of indicators illustrated shows the progress being made (Table 5). Decoupling air and water pollution from growth, making good headway with CO_2 emissions from energy, and transport. The environmental impact of our own individual behaviour is more closely linked to consumption expenditure than the economy as a whole.
2. Focusing policy on the most important environmental impacts associated with the use of particular resources, rather than on the total level of all resource use.
3. Increasing the productivity of material and energy use that are economically efficient by encouraging patterns of supply and demand, which are more efficient in the use of natural resources. The aim is to promote innovation and competitiveness. Investment in areas like energy efficiency, water efficiency and waste minimisation.
4. Encouraging and enabling active and informed individual and corporate consumers.

Table 5. The basket of indicators for sustainable consumption and production.

Economy-wide decoupling indicators
Greenhouse gas emissions
Air pollution
Water pollution (river water quality)
Commercial and industrial waste arisings and household waste not cycled
Resource use indicators
Material use
Water abstraction
Homes built on land not previously developed, and number of households
Decoupling indicators for specific sectors
Emissions from electricity generation
Motor vehicle kilometres and related emissions
Agricultural output, fertiliser use, methane emissions and farmland bird populations
Manufacturing output, energy consumption and related emissions
Household consumption, expenditure energy, water consumption and waste generated

GOALS AND CHALLENGES

Sudan needs assistance in developing and implementing (a) river-basin management, (b) diffuse source pollution, (c) environmental restoration, and (d) urban storm drainage. At present, the international, bilateral donor agencies, and relevant United Nations bodies provide such assistance. The international associations constitute an additional, but as yet untapped, source of assistance. The solution, which should be seriously explored, is the forging of partnerships with bodies such as the World Bank and the appropriate United Nations agencies. Advanced research and technology contribute to resolving water shortage and sanitation problems, and non-conventional reliable water supplies cannot be provided unless the environmental impacts are taken into consideration. Looking to the future, Sudan has set the following priorities for water-resource research and development until the year 2020:

1. Increase overall water-use efficiency to the maximum limit. This could be achieved by (a) improving the irrigation system and assure its flexibility to cope with modern farm irrigation system, (b) developing the farm system, (c) drawing up a proper mechanism for water charges;
2. Modify the cropping pattern; for example (a) planning the different cropping pattern according to water quality, (b) gradually replacing sugar cane by sugar beet, (c) introducing genetic engineering and tissue culture to develop salt tolerance crops, and (d) reducing the area of clover (*Berseem*);
3. Re-use all the possible agricultural drainage water using proper technological means to deal with its quality, especially after implementing the irrigation development programme;
4. Plan properly the re-use of sewage effluent after drawing up guidelines for its use;
5. Research agreements of losses and suggest conservation projects;
6. The conjunctive use and management of reservoirs and groundwater sources in the Nile valley, giving special consideration to drought conditions;
7. Develop non-renewable groundwater resources in the deserts on a sustainable basis;
8. Water harvest rainfall in desert areas and make full use of torrential streams and flash floods;
9. Use new economical technology of seawater desalination;
10. Raise public awareness about water resource scarcity and government management plans;
11. Consider laws to match with the required development and existing scarcity;
12. Establishment of efficient operation, maintenance and repair procedures;
13. Community participation in operation and maintenance;
14. The extent to which initial government investment can or should be recovered from water uses;
15. Domestic potable water supply should reach at least 25 L per day per person;
16. Water should be available for ten livestock units at 450 L/day;
17. Potable water must be available within two kilometres of individual residences.

The water quantity situation is highly variable in Sudan reflecting different levels of development and different

Table 6. Present water management of Sudan.

Using of resources	Sources	Institutions	Pricing principle	Price details
Urban	Surface and groundwater	National Water Corporation (NWC)	Full cost recovery	Progressive rate with increasing uses. Rates lower in the north
Major rural villages	Mostly groundwater	Rural Water Corporation (RWC)	Stand pipe free, recovery of recurrent costs, charges for yard and house connections	Progressive rates but less comparative to urban cities
Rural villages	Groundwater	District Councils	Stand pipe free, recovery of recurrent costs, charges for yard and house connections	Not available
Livestock	Surface and groundwater	Rural Water Corporation (RWC)	All investments and recurrent costs	Regressive, no charges on relatively small use
Mines	Surface and groundwater	National Water Corporation (NWC)	Full cost recovery	Progressive rates
Wildlife	Mostly surface	Rural Water Corporation (RWC)	Full cost of boreholes	Regressive

Table 7. Wastes in River Nile water.

Materials	(%)
Paper, wood	50.0
Ferrous residues	12.5
Glasses	11.0
Organic wastes	10.0
Plastics	5.0
Non-ferrous residues	1.5
Other	10.0

needs for water quality programmes in Table 6. The conventional paradigm of water quality monitoring is not suitable for the Sudan being too expensive, inefficient, and ineffective. Financial and sustainability issues include cost avoidance and cost reduction, local and accountability frameworks that encourage good business practices by senior programme managers, the use of new cost-effective technologies for monitoring, and a variety of donor/public/private sector linkages that focus on commercial benefits that permit the transfer of certain parts of water quality programmes to the private sector.

From a visual investigation of the River Nile in Table 7, the major sources are industrial effluents, crude sewage from blocked, broken or overloaded sewers, sewage effluents, surface runoff, and solid wastes which have been dumped into the river. Therefore remedial and improvement measures must be taken before the environment becomes further polluted and the natural resources are completely over-exploited (Omer, 2000). The challenges facing and enhancing the ecology in the twenty-first century are as follows: (a) Drinking-water sources should be treated with chemicals; (b) Suitable toilet facilities should be provided along the main roads to minimise pollution; (c) Proper arrangements should be made for litter dumping and waste disposal; (d) Local people should be fully educated about environment matters and hygiene; (e) Previous damage should not be allowed to continue while planning for a balanced development in the future, and (f) The concept of the ecosystem (involving education and interpretation of the natural environment) must be promoted.

Environmental pollution is a major problem facing all nations of the world. People have caused air pollution since they learned how to use fire, but man-made air pollution (anthropogenic air pollution) has rapidly increased since industrialisation began. Many volatile organic compounds and trace metals are emitted into the atmosphere by human activities. The pollutants emitted

Table 8. Main water resource issue in region.

Region	Water resource issues
South	Abundant water resources Localised scarcity of water and untapped water supplies High hydropower potential Water conflicts arising from immigration of Bagara Arabs (nomadic) from north to south Water-borne diseases International water conflicts (upstream and downstream countries)
Central	Water quality problems from untreated sewage and other pollution Water-borne diseases Potential use of rivers for navigation and recreational purposes Intensive erosion and sedimentation from agriculture High hydropower potential Excessive use in large urban and industrialised areas Frequent urban floods
North	Good water quality Scarcity of water resources Intensive erosion and sedimentation from agriculture Frequent urban floods
Northeast	Scarcity of water resources Water quality problems from untreated sewage In mining areas, water quality problems from effluent
West	Scarcity of water resources Water conflicts between nomadic and non-nomadic tribes Water-borne diseases Soil erosion and degradation caused by agriculture

into the atmosphere do not remain confined to the area near the source of emission or to the local environment, and can be transported over long distances, and create regional and global environmental problems. The privatisation and price liberalisation in energy fields has to some secured (but not fully). Availability and adequate energy supplies to the major productive sectors is needed. The result is that, the present situation of energy supplies is far better than ten years ago.

The challenge of overcoming the country's diversity

Sudan is a federal republic of 2.5 million km^2 located in the Eastern Africa. The country is divided into 26 states and a federal district, in which the capital, Khartoum is located. Sudan is known as a country of plentiful water, with highest total renewable fresh water supply in the region. Table 8 shows some of the most significant regional diversities concerning water issues.

Adequate water management is essential to sustain development. Competing needs for this beneficial resource include municipal supply, industry, and agriculture, among others. The National Water Act of 1994 (Law No. 1155) defines the objectives, principles, and instruments of the National Water Resources Policy and the National Water Resources Management system. The law establishes the institutional arrangement under which the country's water policies are to be implemented. The National Water Resources Policy was proposed to achieve:

1. Sustainability: To ensure that the present and future generations have an adequate availability of water with suitable quality.
2. Integrated management: To ensure the integration among uses in order to guarantee continuing development.
3. Security: To prevent and protect against critical events, due either to natural causes or inappropriate uses.

To achieve such objectives, water management must be implemented according to the following principles:

1. Water is a public good, and it is a finite resource that has economic value.
2. The use of water required to meet people's basic needs shall have priority, especially in critical periods.
3. Water management shall comprise and induce multiple uses.
4. The river basins are the appropriate unit for water management, and water management shall decentralise, with the participation of government, stakeholders and society.

Water resources plans are developed to guide future decisions and are to be developed for each river basin and state, as well as the country. The objective is to coordinate efforts and establish guidelines and priorities for water allocation and water pricing. The priorities established for water allocation will be used in critical drought conditions. Water pricing is the single most controversial instrument of the law. The pricing system is also the most difficult step to implement. The pricing system recognizes the economic value of water, as stated in the principles of the policy. The development of a new, modern, and complete water resources information system is one of the basic needs for the implementation of the water resources management system. The decision process in drought or flood conditions, and also in overexploitation cases, can only be correct if based on a reliable information system. A complete and comprehensive database on water availability, users, water quality monitoring, current technologies (like geographical information systems), is certainly the way to produce an efficient framework for decision-making. Lack of information is one of the most critical points regarding the development and implementation of the new management system. The institutional framework provides the basis by which all actions are taken, and an assessment of its functional character helps determine the collaborative potential. The resulting criteria for measuring a given community's institutional capacity can be found in Table 9.

WATER SCARCITY IMPACTS AND POTENTIAL CONFLICTS

The failure of water resources to meet the basic requirements of society has a host of social, economic, environmental, and political impacts. Water scarcity is man-made phenomenon brought about by the increasing demands of the population for water. The imbalance in the population- water resources equation strains society and has an adverse impact on domestic hygiene, public health, and cost of domestic water, and could impart political problems as a serious as bringing down government. On the social side, water scarcity adversely impacts job opportunities, farm incomes, credibility and reliability of agricultural exports, and ability of the vulnerable to meet the cost of domestic water. Economically, the adverse impact is displayed in the loss of production of goods, especially agricultural goods, the loss of working hours because of the hardships society faces as a result of water scarcity. The impacts of water scarcity on regional stability are addressed with reference to water in the Middle East Peace Process, taking into account the serious impacts of conflicts and potential water war.

Conditions of scarcity propel an increase in competition among the different sectors of water use with results, invariably, at the expense of irrigated agriculture. Pure market forces create a gradient under which water flows from the poor to the rich. Tough decisions await politicians, and the consequences are expected to displease one or more parties, and please others. The scene of domestic politics becomes as fluid as water itself, with politicians shifting positions continuously in response to domestic pressures. The political fallout from water resources scarcity on the domestic scene is parallel to the impact the scarcity has on domestic households in terms of basic needs for drinking and food preparation, on domestic hygiene, and on public health. Other important factors have a delayed response to water scarcity, and these pertain to the integrity of the environment, and deterrence it imparts on development investment and economic credibility of the country. The cost of mitigating these problems and of the provision of services to the increased urbanization could very well be beyond the ability of government to bear. The political consequences resulting from this will not be in favour of domestic stability, and social explosions can be anticipated.

A bilateral agreement was reached between Egypt and the Sudan in 1959 by which the two countries share the Nile flow: 55.5 billion cubic meters to Egypt, 18.5 billion to Sudan, and 10 billion were allocated to evaporation. Hopes are high for achieving a more extensive participation by the other riparian parties in what could be a multilateral treaty on the Nile encompassing the other riparian states in addition to Egypt and Sudan. The above agreement is not complete; it lacks the entry of other legitimate riparian states, lacks water quality components, and tends to focus on quantity measures, and miss important management issues. It is to be noted that regional relations, including those among the riparian parties, are connected to the political, economic, and trade network of international relations. Water is not the only determinant factor in shaping the nature of bilateral, regional, or international relations.

Water relations can be transformed into a positive sum game by which all parties can be made to win. One common gain to all is the environmental protection of the common watercourse or water body. Lack of cooperation

Table 9. Capacity assessment for flood management: institutional factors.

High capacity (plans etc. in places)	1. Basin-wide management plan has been drafted. 2. Natural mitigation strategy in place. 3. Basin-wide coordination and communications strategy instituted. 4. Trained emergency management staff coordinating at the regional level. 5. Effective regulatory policies that address floodplain occupancy. 6. Decentralised decision-making with a high degree of local autonomy. 7. Evidence of an updated national response plan. 8. Bilateral response agreements. 9. Evidence of regional preparedness and response training. 10. Some trained emergency management staff at the local and/or national level.
Medium capacity (evidence of activity on- going)	1. Evidence of some regulatory policies designed to address floodplain occupancy. 2. Attempts to decentralise decision-making, moderate local discretion. 3. No existing flood response plan. 4. No evidence of mitigation-related activities.
Low capacity (no formalisation in place nor apparently evolving)	1. Poor local-and national- level coordination and communications. 2. Little or no evidence of flood preparedness and response training. 3. No regulatory policies addressing floodplain occupancy. 4. Centralised decision-making, no evidence of local autonomy.

and agreement will most likely lead to environmental neglect and water quality degradation, which is loss to all. International encouragement to attain cooperation can, therefore, be brought to bear on the regional parties, and efforts of international lending agencies can be called upon to pool with the regional and international efforts to achieve this objective. It has been stipulated by many that under conditions of scarcity, water conflicts can lead to hostile actions between riparian parties. Experience in the region indicates that water, in its own right, has not been the cause of any of the wars that have broken out in the region.

Today's advanced societies heavily depend on energy. The principal sources of energy and electricity generation today are solar, wind, biomass, hydropower, and fossil fuel. Energy from hydropower is short of meeting the current or future energy requirements, and the fossil fuel resources, being depleted with time, will eventually run out. For human civilization to continue at its natural pace, new forms of affordable and clean energy will have to come on line. Failure of human civilisation to introduce new forms of energy will render that civilization doomed, and the quality of life will deteriorate. If this unlikely scenario actually takes place, the requirements will decrease because the mechanism of making it available for use (pumping) diminishes. The more likely scenario is more optimistic one, and it is that a new form of energy generation will be introduced in which case water desalination becomes affordable and its pumping from the coastal desalination plants become possible at reasonable cost. The way out of the looming water crisis rests, therefore, in the invention of new forms of energy generation that will make possible the reliance on desalination and in the recycling of wastewater for reuse in agricultural production and for environmental reasons. Integrated management of the three resources of water, energy, and the environment, will result in better results with a positive sum for society.

Table 10. Summary of the situation relating to data and information exchange in Nile basin.

River basin	Nile basin
Basin states or territories	Burundi, Democratic Republic of Congo, Egypt, Eritrea, Ethiopia, Kenya, Rwanda, Sudan, Tanzania, Uganda
Cooperative frameworks in place	Nine of the countries of basin are pursuing the development of a cooperative framework
Major languages spoken	More than 6 official languages and numerous unofficial languages
Major water issue facing the basin	Rapid population growth, environmental degradation, under development
External funding of cooperative basin initiatives	Extensive external funding of cooperative initiative
Range of GDP 'per capita' of the basin	\$550-\$3000
Extent of data/information exchange	Information exchange through the cooperative framework being developed is beginning to occur

COMMON LANGUAGE AND CULTURE

A common language and similar culture simplify communication and reduce the potential for misunderstandings. In Nile basin where several languages are spoken, an international language, English, is used with some success by multi-jurisdictional basin management authorities.

Primary factors promoting data and information exchange

Data and information exchange is more probable when needs are compatible and when there is potential for mutual benefit from cooperation in Table 10. Where countries are working on developments that are beneficial to both countries as well as other riparians, there is little incentive to hide project impacts. This means that since data and information exchange is unlikely to lead to pressure from surrounding countries that might restrict developments, countries have less reason to restrict access to their data and information resources. It is important, therefore to be no perceived clash of interests in development plans and needs. An example of this might be in developing their part of the basin primarily for hydroelectric development, while the lower riparians are more interested in developing the irrigation potential of their portion of the basin. By constructing large storage dams in the upper part of the basin, the river Nile seasonal flow might be evened out, reducing flooding downstream while increasing irrigation water supplies and even making downstream run-of-the-river hydroelectric projects more profitable. Ecosystem effects would have to be considered.

Sufficient levels of economic development

Sufficient levels of economic development across a basin are needed to permit joint funding of cooperative processes, particularly data collection and dissemination. Although countries with differing levels and forms of economic development may, at times, have more complementary needs than countries with similarly structured economies, the overall level of economic development is still significant. A wealthier country in a river basin may be able to assist with the funding of data collection activities in the neighbouring country with much needed data and helping to build confidence between the two countries.

Increasing water resources stress

As per capita water resources availability decreases as shown in Table 11, tensions between riparian nations may rise and make cooperation difficult. Stress may, therefore, reduce cooperation and data sharing rather than strife.

The historical background of the basin may have a lasting effect on current negotiations. Past conflicts can have a deleterious effect on the prospects for establishing cooperative practices, such as data sharing. Where there is a history of conflict between two nations, both nations may view the present situation primarily as competitive and focus on conflicting rather than common interests. Democracies may find it easier to negotiate cooperative arrangements with other democracies. Political differences can lead to legacies of mistrust developing between countries.

DISCUSSION

Water stress in Sudan

Water stress refers to economic, social, or environmental problems caused by unmet water needs. Lack of supply is often caused by contamination, drought, or a disruption in distribution. In an extreme example, when Sudan split

Table 11. Diverse water challenge (WRI, 2002).

Country	Egypt	Sudan
Per capita annual water resources 2000 (m^3)	34	1187
Per capita annual withdrawal (m^3)	921	666
Per capita annual withdrawal for agriculture (m^3)	86	94

four years ago between the rebel-led west and government-ruled north, the conflict led to unpaid water bills, which precipitated a dangerous health threat in the region, increasing the risk of water-born diseases such as cholera. Some analysts believe the disruption of distribution was a political ploy to put pressure on the rebel-led west.

While water stress occurs throughout the world, no region has been more afflicted than sub-Saharan Africa. The crisis in Darfur stems in part from disputes over water: The conflict that led to the crisis arose from tensions between nomadic farming groups who were competing for water and grazing land-both increasingly scarce due to the expanding Sahara Desert. As Mark Giordano of the International Water Management Institute in Colombo Sri Lanka says, "Most water extracted for development in sub-Saharan Africa-drinking water, livestock watering, and irrigation-is at least in some sense 'transboundary'". Because water sources are often cross-border, conflict emerges.

Improving water and sanitation programmes is crucial to spurring growth and sustaining economic development. Because it takes time to develop these programmes, a paradox emerges: Poor economies are unable to develop because of water stress, and economic instability prohibits the development of programmes to abate water stress. Developments in water storage could have prevented that drought from significantly affecting Sudan's economy. Hydropower can also spark economic development. Accordingly, some transboundary water agreements also play a clear role in fostering development, for example, by facilitating investment in hydropower and irrigation.

The role of agriculture in water stress

Agricultural development has the potential to improve African economies but requires extensive water supplies. These statics from the Water Systems Analysis Group at the Institute for the Study of Earth, Oceans, and Space at the University of New Hampshire reveal the urgent need for sustainable agricultural development:

1. About 64% of Africans rely on water that is limited and highly variable;
2. Croplands inhabit the driest regions of Africa where some 40% of the irrigated land is unsustainable;
3. Roughly 25% of Africa's population suffers from water stress;
4. Nearly 13% of the population in Africa experiences drought-related stress once each generation.

Another aspect of water-related stress is the relationship between water, soil and agriculture. Improved access to quality water is a long-term goal that requires more than humanitarian funds:

1. Because sub-Saharan Africa is subject to more extreme climate variability than other regions, it needs improved water storage capacity. Some experts say that large dam projects would create a more sustainable reserve of water resources to combat the burden of climate fluctuations, but other disagrees, stating the harmful environmental impact of large dams.
2. Many experts say more water treaties are needed. The transboundary water agreements have cultivated international cooperation and reduced the "probability of conflict and its intensity".
3. Better donor emphasis on water development is needed. Small-scale agricultural improvements also offer a solution to water stress, including the harvest of water in shallow wells, drip irrigation for crops, the use of pumps, and other technological innovations.

Farmers can access green water through drip irrigation systems that slowly and consistently deliver water to plant's root system, supplemental irrigation (supplementary to natural rainfall rather than the primary source of moisture during periods of drought) and rainwater harvesting (the collection of rainwater for crops, which reduces reliance on irrigation). Crops can grow poorly even during periods of rainfall, and most farms in Africa suffer from nitrogen and phosphorus depletion in soil. One way to assuage water stress in terms of food scarcity is to increase water-holding capacity with organic fertilisers that would increase availability and efficacy of green water.

Water supply problems in the Butana region - Central Sudan with special emphasis on Jebel Qeili area

The Butana region of Central Sudan is famous for its animal wealth and extensive pastures. Yet scarcity of water resources in the area especially during the dry seasons handicaps the proper utilisation of these pastures. The area is occupied by non-water-bearing basement rocks and the only source of water is from direct run-off. Thus large numbers of small-size water reservoirs, "haffirs", were constructed, but these are inadequate to provide enough water for the growing human and animal population. An all-year lake is here proposed to be constructed utilising the ring-structure the Jebel Qeili igneous complex, central Butana. This lake is expected to solve the present water problem and meet the future demand of Central Butana at the present rate of human and animal growth (Omer, 2001).

Southern Sudan

World Vision began its work in Sudan in 1972 through a partnership with the African Committee for Rehabilitation of the Southern Sudan (ACROSS) to provide emergency relief aid to war-effected families. Efforts included the reconstruction of the Rumbek community hospital and surrounding buildings, the provision of medicine and supplies, and education in preventative health care. Other projects during this period focused on training health and social workers in general medical aid and child welfare and instruction in water development, agriculture, handcrafts, and literacy (Omer, 2004).

The 1980s brought constant turmoil to the Sudanese people as the civil war raged on and severe drought parched the country. In 1983, approximately 1,500 refugees entered Sudan daily from violence-torn neighbouring countries, straining the already limited food supply. World Vision, through the ACROSS Refugee Settlement Project, responded by distributing blankets, grain, cooking oil, medical kits, and shelter to more than 50,000 people. Supplemental feeding for children also was provided. Numerous development projects were initiated during this time that assisted communities in improved crop production, animal husbandry, health care, clean water collection, infrastructure repair, and literacy. In 1989, World Vision became a founding member of Operation Lifeline Sudan (OLS), a partnership of non governmental organisations (NGOs) and UN agencies designated to coordinate the southern relief efforts.

During the 1990s World Vision conducted operations in all major regions of Southern Sudan. Project objectives included primary health care, water provision, agriculture, local grain purchase, enterprise development, and emergency relief efforts. World Vision focused on an integrated work approach that involved peace and advocacy, gender development, church support, and environment and natural resource initiatives. Some specific projects included:

1. The Kapoeta Medical Supplies Project provided health and educational assistance to more than 200,000 people to help reduce incidences of disease and suffering.
2. The Agriculture/Livestock Rehabilitation Project assisted the aforementioned families with food, seed, vaccinations, and agricultural consults.
3. The South Sudan Relief/Church Support Project coordinated a pastors' conference for 150 pastors and religious leaders from the Western Equatoria province.
4. The South Sudan Relief and Rehabilitation Project, a 10-year programme, provided 450,000 Sudanese with agriculture and economic development, food and water, health care, enterprise development opportunities, and emergency relief.

Beja people's problems

The Beja, a semi-nomadic group of people, who live in rebel-held areas of Eastern Sudan, need a huge amount of humanitarian assistance, a representative from the International Rescue Committee (IRC). Although Beja can be found throughout Northeast Africa, tens of thousands are currently trapped in an area of Eastern Sudan near the Eritrean border, held by Sudanese rebels since the late 1990s. Only two NGOs, both based in Eritrea, are able to access the 15,000 km^2 area at the moment, one of which is the IRC. The organisation estimates the Beja population in the area to be between 45,000 and 186,000 people (Omer, 2008).

Although it did rain in the area in 2004, a shortage of water had also posed serious problems. "Fresh drinking water is incredibly hard to come by. All the settlements have just focused around dry river beds, in which people dig hand-dug wells". Locusts would eat the foliage that usually sustains the Beja's goats and camels—upon which the Beja utterly depend for survival. A few immature locust swarms have formed in northeast Sudan near the Red Sea and the border of Egypt, the UN Food and Agriculture Organisation said in March 2004. Moreover, Beja grazing areas have been severely restricted by a front line between rebel forces and Khartoum government soldiers, the second to be opened by Southern rebels during Sudan's 21-year-old civil war.

Sudan is an example that projects the environmental plight of Africa, south of the Sahara – drought and desertification, floods, deforestation, loss of biodiversity, tribal and ethnic conflict and poverty are only too common. As a result, interest and commitment to environmental impact assessment practices have become mandatory by donors when executing new development projects. The ecological zones of Sudan in 1998 as:

1. Deserts: Cover almost 30% of the northern parts. Annual precipitation is less than 50 mm; soils are sandy. Sparse vegetation grows on seasonal 'Wadis' and the banks of the Nile.
2. Semi deserts: Cover above 20% south of the desert belt. Rainfall ranges from 50 to 300 mm. It is speckled with few Acacia trees and thorny bushes and zerophytes.

Figure 2. Water collection as a daily need for some people.

Figure 3. A typical donkey-drawn water tank used by water vendors.

3. Low rainfall woodland Savannah: Covers about 27% of the area of Sudan with rainfall less than 900 mm, with a nine-month dry period. Annual grasses are dominant. Heavy clay soils lie on the east of the Nile and the west is sandy. Most of the 36 million feddans of rain-fed agriculture and the 4 million irrigated lands fall within this heavily populated belt.
4. High rainfall woodland Savannah: 13% of the area with rainfall more than 900 mm and with broad-leafed trees in the southern parts of Sudan.
5. Swamps: Are probably the largest in the world and cover about l0% and fall in three main areas around the tributaries of the White Nile.
6. Highlands: Are less than 0.3% of the areas of Sudan and are scattered along the Red Sea coast, the south and the west of the country.
7. The Red Sea Coast-Marine ecosystem, mangrove swamps, coral reefs and associated fauna.

Environmental problems include:

1. Horizontal expansion in rain-fed and irrigated agriculture;
2. The complete absence of the environmental dimensions in policies, strategies, plans and programmes of management of resources;
3. Development is random and environmental evaluation does exist before or after execution of projects;
4. The economy and society, in spite of the century-long attempts at 'modernisation' are still dominated by subsistence way of living;
5. The economy is still affected seriously by the yearly, seasonal and geographical variability of rainfall for crop and livestock production;
6. Dependence on imported seeds and agricultural chemicals has increased cost of production;
7. Loss of land productivity and marketing policies decreased cash surplus;
8. The civil war in the south has grave economic and social costs;
9. Population distribution and rural-urban migration due to desertification and civil strife has led to deterioration of natural resources, indigenous knowledge and loss of local culture and dignity;
10. Problems of poor sanitation, limited industrial pollution and food hygiene have become more complex;
11. The energy crisis is aggravating desertification and affecting climate charge;
12. Vast water resources are badly managed;
13. Environmental education has only been recently incorporated in school curricula, and
14. Laws and legislation concerning the environment are not effective and law enforcement measures are not integrated.

Western Sudan

El Fasher, Darfur region, Sudan, 24 August, 2005 – Torrential rains have caused severe flooding in this city of 400,000 people and in nearby Abu Shook, a camp for people forced to flee their homes as a result of the ongoing Darfur conflict (Figures 2 to 3). The floods have destroyed hundreds of homes and have made El Fasher's water supply largely unsafe. UNICEF is mounting a concerted effort to restore basic services to those affected by the flood, and to prevent the outbreak of disease. Since the flood, UNICEF has assisted with the following:

1. Reinstalling pipes in Abu Shook and restoring the water supply by linking boreholes with pumps.
2. Testing the water quality each day. No bacterial contamination had been found.
3. Rebuilding 156 latrines and 88 bath stations.
4. Renting five tankers to deliver more water.
5. Repairing damaged schools and child-friendly spaces.
6. Providing daily door-to-door hygiene-promotion trainings.

7. Distributing jerry cans, soap, tarps, and mosquito nets.

CONCLUSIONS

A booming economy, high population, land-locked location, vast area, remote separated and poorly accessible rural areas, large reserves of oil, excellent sunshine, large mining sector and cattle farming on a large-scale, are factors which are most influential to the total water scene in Sudan. It is expected that the pace of implementation of water infrastructure will increase and the quality of work will improve in addition to building the capacity of the private and district staff in contracting procedures. The financial accountability is also easier and more transparent. The communities should be fully utilised in any attempts to promote the local management of water supply and sanitation systems. There is little notion of 'service, invoice and move on'. As a result, there are major problems looming with sustainability of completed projects. A change in water and sanitation sector approach from supply-driven approach to demand-responsive approach call for full community participation. The community should be defined in terms of their primary role as user/clients. Private-sector services are necessary because there are gaps, which exist as a result of the government not being able to provide water services due to limited financial resources and increase in population. The factors affecting the eco-environmental changes are complex, interrelated, and interactive. The deterioration problems of water and sanitation have attracted some attention in recent years. There is an urgent need to study possible rehabilitation measures to ensure a sustainable and excellent water quality and improved sanitation. Water resources plans are developed to guide future decisions and are to be developed for each river basin and state, as well as for the country. The overall objective is to coordinate efforts and establish guidelines and priorities for water allocation and water pricing. The priorities established for water allocation would be used in critical drought conditions. The water quality classification of water bodies by different classes of use is the basis for truly integrating the quality and quality of water management. Water pricing is the single most controversial instrument of the law. The pricing system recognises the economic value of water, as stated in the principles of the policy, but is also the most difficult step to implement. It is expected that the pace of implementation will increase and the quality of work will improve in addition to building the capacity of the private and district staff in contracting procedures. The financial accountability is also easier and more transparent. The communities should be fully utilised in any attempts to promote the local management of water supply and sanitation systems. A change in water and sanitation sector approach from supply-driven approach to demand-responsive approach call for full community participation. The community should be defined in terms of their primary role as user/clients. Private-sector services are necessary because there are gaps, which exist as a result of the government not being able to provide water services due to limited financial resources and increase in population. There is little notion of 'service, invoice and move on'. As a result, there are major problems looming with sustainability of completed projects. The factors affecting the eco-environmental changes are complex. There are interrelated and interact. The deterioration problems of water and sanitation have attracted some attention in recent years. There is an urgent need to study possible rehabilitation measures to ensure a sustainable and excellent water quality and improved sanitation.

REFERENCES

Anon (1979). Map of the world distribution regions', MAB Tech Note 7.

ASCE (1998). ASCE Task Committee on Sustainability Criteria, Sustainability Criteria for Water Resource Systems', Reston, Virginia, USA.

FAO United Nations Food and Agriculture Organisation (1999). The State of Food in Security in the World', Rome: Italy.

James W (1994). Managing water as economic resources', Overseas Development Institute (ODI), UK.

Noureddine RM (1997). Conservation planning and management of limited water resources in arid and semi-arid areas', Proceedings of the 9th Session of the Regional Commission on Land and Water Use in the Near East, Rabat: Morocco, pp. 15-21.

Omer AM (1995). Water resources in Sudan', NETWAS 2, Nairobi, pp. 7-8.

Omer AM (2000). 'Water and environment in Sudan: the challenges of the new millennium', NETWAS, 7(2): 1-3.

Omer AM (2001). Water development in Sudan: Present and future challenges. Arab Organisation for Agriculture Development (AOAD). Khartoum: Sudan. Arabic J. Irrig. Water Manage., 2:.48-58.

Omer AM (2002). 'Focus on groundwater in Sudan', Int. J. Geosci. Environ. Geol., 41(8): 972-976.

Omer AM (2004). Water resources development and management in the Republic of the Sudan. Water Energy Int., 61(4): 27-39.

Omer AM (2008). Water resources in the Sudan. Water Int., 32 (5): 894-903.

Overseas Development Administration (ODA) (1987). Sudan profile of agricultural potential', Survey, UK, 1987.

Salih AMA, Ali AAG (1992). Water scarcity and sustainable development', Nat. Resour., 28: 1.

Seckler D (1992). Private sector irrigation in Africa-Water resources and irrigation policy studies', Winrock International Institute for Agricultural Development.

UNESCO (1999). UNESCO Working Group M.IV, 'Sustainability criteria for water resource systems, Cambridge', United Kingdom: Cambridge University Press.

World Resources Institute (WRI) (2002). 'World Resources 2000-2001', USA.

Policy interventions in watershed management: The case of the Inchaban Watershed, Ghana

O. D. Mireku[1] and A. K. Mensah[2]

[1]Department of Geography and Regional Planning, University of Cape Coast, Cape Coast, Ghana.
[2]Department of Geography, Kenyatta University, Nairobi, Kenya.

Before the twentieth century, watershed management in most Ghanaian communities relied solely on religious-based restrictions such as the use of taboos and sacred groves, to deter people from encroaching on watersheds. However, the advents of Christianity, western education, and urbanization, and the desire to develop the resources of the country have reduced the effectiveness and respect for traditional restrictions for the protection of the environment. Also, customary administration over watersheds has had a lot of challenges in evaluation and assessment of environmental damages, enforcement of laws to bring illegal land users to book, and integrating the rights of land users with policies on regulation and management. In the face of these numerous problems in the reliance on customary laws and practices, several watershed management policies have been consolidated with other key water sector policies such as the Water Use Regulation LI 1962, the Integrated Water Resource Management Policy (IWRMP) of 1996 and the National Water Vision Policy of 1997to streamline the administration over local watersheds. Despite these policies, many watersheds are still under the threat of degradation. This study identifies the reasons why the government has become unsuccessful to manage watersheds in the country. It used the Inchaban Watershed as a case, and solicited information using in-depth interviews and focus group discussions from 41 stakeholders or respondents who were selected purposively. The results of the study revealed that capacity of the management institutions to enforce the laws and policies set out by government was challenged due to problems of legal pluralism posed by local chiefs, some management and other institutions as well as individual local users of the Inchaban Watershed.

Key words: Watershed management, management institutions, user institutions, policies, Inchaban Watershed, Ghana.

INTRODUCTION

Watershed is an elevated land that separates the headwater and tributaries of one river, or drainage basin, from those tributaries flowing into another river or drainage basin (Acheampong, 2009). World Bank (2001) defines watershed as the area that supplies water by surface or subsurface flow to a given drainage system or body of water, such as a stream, river, wetland, lake, or the ocean. According to Pradhan (2003), watershed is that area of land within which all living things such as human beings are inextricably linked or connected to a bounded hydrological system. Watersheds are, therefore, part of human communities, and become very important for communities to properly plan their use (Pradhan, 2003). To properly plan for watershed use calls for the establishment of sound management systems.

Darghouth (2008) defines watershed management as the integrated use of land, vegetation and water in a geographically discrete drainage area with the aim of protecting and conserving the watershed resources. According to the Global Water Partnership (2000), watershed management is a process which promotes coordinated development and management of watershed resources in order to maximize economic and social welfare in an equitable manner that sustains vital ecosystems and promote conservation. Watershed management is being supported by governments of developed and developing countries including Ghana for a number of reasons. Among them are the facts that watersheds support supply of food, water and medicinal products; provide habitat for plants and animals, and regulates climate (World Natural Resource Conservation (WNRC), 1996). Manuel (2007) discovered that to meet the objectives of watershed management, there must be supportive institutional arrangements. Ostrom (1999) has subsequently suggested that institutional arrangements for watershed management should consist of organizational roles, established national laws and regulations (policies) that shape structures for human actions in order to prevent future environmental problems.

History of policy interventions in watershed management in Ghana

Before the twentieth century, watershed management in most Ghanaian communities relied solely on religious-based restrictions (Bullock, 2008) and the use of taboos and sacred groves, to deter people from encroaching on watersheds (Opoku-Agyemang, 2008).These restrictions were, to a large extent, dependent on the respect for religious, local and cultural structures for the protection of the environment (Odame, 2010).

Unfortunately, the advents of Christianity, western education, and urbanization, and the desire to develop the resources of the country have reduced the effectiveness and respect for traditional restrictions for the protection of the environment (Opoku-Agyemang, 2008). Moreover, customary administration over watersheds has had numerous challenges in evaluation and assessment of environmental damages, enforcement of laws to bring illegal land users to book, and integrating the rights of land users with policies on regulation and management (Gibson, 2001).

In the face of these numerous problems in the reliance on customary laws and practices, the government of Ghana resorted to the enactment of state laws and policies to strengthen mandates of institutions in charge of managing water resources. The first comprehensive attempt to regulate the use of water resources, other than for industrial production activities, was the enactment of the *Rivers Ordinance Act (CAP 226) of 1903*. Section 10 of this Ordinance states that it shall be unlawful to pump, divert or by any means cause water to flow from any river, for purposes of large scale irrigation, mining or to generate power without a license from the appropriate quarters.

Unfortunately, there was no follow-up to the Rivers Ordinance Act. Consequently the ordinance was overtaken by time and other enactments which contained specific provisions that enabled agencies to perform certain functions, some of which were watershed-related (Bossman, 1998). For example, the *Forestry Ordinance of 1927* made provisions for catchment protection and control of water abstraction in forest reserves. The *Land Planning and Soil Conservation Ordinance of 1953* contained sections for checking soil erosion and the control of watercourses. State laws were very beneficial to some extent since they stressed the need to establish institutions and agencies to support watershed management (Odame, 2010).

The onset of the post-independence era opened the way for the establishment modern policies with specific legal enactments for water supply and drawing of economic products in watersheds. Table 1 shows some relevant watershed management policies and their associated legal enactments. As observed by Opoku-Agyemang et al. (1998), Table 1 demonstrates the attempts made by previous governments to improve watershed management by some agencies in Ghana. For example, to promote sound cooperation among water resource users and managers, *the Integrated Water Resource Management Policy (IWRMP)* was formulated and implemented in 1996. Five years after, the Water Use Regulation LI 1962 Policy was then implemented to streamline the administration and governance over local water bodies which were under serious threats of degradation. Realizing the weaknesses in the earlier policies, the National Water Vision Policy was formulated and consolidated with other key water sector policies to comprehensively manage the nation's water resources.

The attempts by government of Ghana was to emulate successful watershed management policies such as that of Restoration of the Alps of the United States, Integrated Soil and Water Management (ISWM) in Brazil, which were used for rehabilitation, restoration and conservation of watershed resources (Maarleveld and Dangbegnon, 1998).

However, in spite of all these policies and regulations, the Inchaban Watershed in the Western Region of Ghana is still under the threat of degradation. The Inchaban

Table 1. Watershed management policies formulated by the Ministry of Water and Housing (MWH).

Date formulated	Policies formulated	Goals to be achieved
1996	Integrated Water Resource Management Policy (IWRMP)	To promote sound cooperation among water resource users and managers
2001	Water Use Regulation LI 1962 Policy	To streamline the administration and governance over local water bodies
2006	Drilling for Water and Groundwater Development Regulation Policy LI 1827	To license drilling companies and ensure safe development of watershed resources
2007	The National Water Vision Policy	Being consolidated with other key water sector policies to comprehensively manage the nation's water resources
2012	The National Buffer Zone Policy	To initiate the development of programmes to safeguard watershed resource.

Source: Ghana Ministry of Works and Housing, 2010.

Watershed is degraded due to a number of anthropogenic factors such as expansion of cultivated areas, unsustainable fuel wood and timber harvesting, bushfires, and the development of settlements and other infrastructures (Carson, 1992). For instance, according to Button (2010), the rate in which the Inchaban watershed is built-up is 7.6% per year. The increasing encroachment of the watershed has adversely affected economic activities such as fishing and farming. For example, the quantity of charcoal production, a major economic activity of the inhabitants, dropped by 30% between 2007 and 2009 (Biney, 2010). This raises a number of questions about the adherence to the policies set out by government (Table 1) to support the institutions responsible for managing the Inchaban Watershed. It is against this background that this study seeks to identify the approaches and the challenges against the effective use of policies set out by government to effectively manage the Inchaban watershed. The subsequent sections look at the conceptual framework of the study followed by the methodology and analysis of results.

Conceptual framework

The integrated watershed management framework: IWM-framework

The Institutional Framework for Integrated Watershed Management developed by the Water Resource Commission of Ghana (WRC) has been adopted for this study.The choice of this conceptual framework for this study is informed by the definition of integrated watershed management as a comprehensive multi-resource management planning process involving all stakeholders within the management processes. Stakeholders in watershed management are expected to work cooperatively to identify an approach that is environmentally friendly and economically sustainable (Botero, 1986; United Nation Environment Programme (UNEP), 2004). In the framework, the WRC is the superior body empowered by an *Act of Parliament (Act 522 of 1996)* to grant rights to watershed users to allocate watershed resources, and to implement policies. The strength of this framework (Appendix Figure 1) is that it spells out clearly the processes to govern watersheds in terms of policy formulation and implementation. The IWM-framework designed by Water Resources Commission (WRC) was tailored to address the diffused functions and authority of institutions concerned with water resources management with the aim of integrating their roles, objectives, mandate, policies and laws. Going by this framework, the National Development Planning Commission (NDPC) and the Ministry of Works and Housing (MWH) are expected to work in collaboration with the WRC to specifically coordinate national development plans and also formulate national water policies. The WRC supervises the activities of key representative groups that are involved in water services delivery and utilization. These actors are the Water Research Planning Input Providers, Water Users, Regulatory Agencies and Civil Society Representatives. The roles and activities of the representative groups are summarized in Appendix Figure 1.

METHODOLOGY

Study area

The Inchaban Watershed is located in the Shama District of the Western Region of Ghana (Figure 1). The size of the watershed is 13,553.80 acres. The climate is dry-humid tropical (Acheampong, 2009), and has a double maximum rainfall; the main rainfall season lasts from June to early August, and the minor from September to November. The average annual precipitation is 1195 mm (Acheampong, 2009). The dry season is short, occurring from

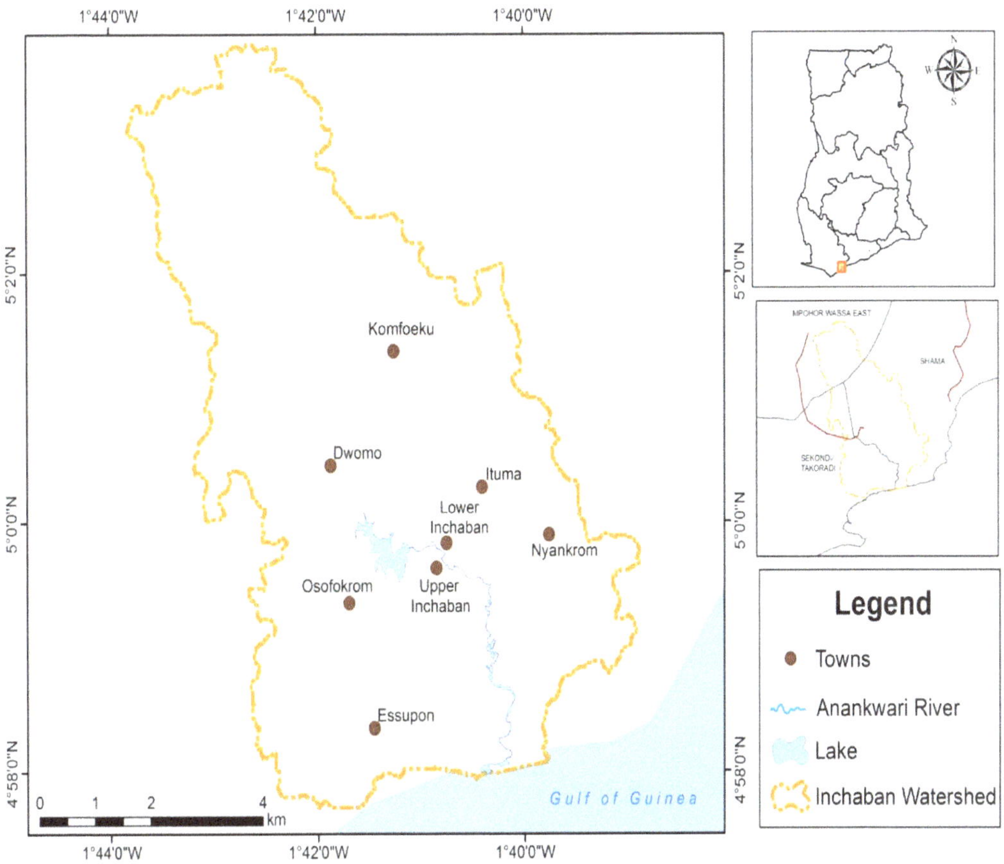

Figure 1. Map of the Inchaban Watershed. Source: Cartography Unit of the Department of Geography and Regional Planning (UCC), 2011.

December to February. The average annual humidity of the area is high (over 94%) and the mean annual temperatures are 29°C. The main vegetation in the watershed consists of woodland savannah near the coast, while a semi deciduous forest occupies the upper courses of streams. Mangroves occur along the southern portion of the watershed. The nature of the climate and vegetation of the district has limited the growth of most local food crops but rather sugar is extensively cultivated. Consequently, mining and charcoal burning activities have absorbed about 45% of the active labour force. Many pockets of farming activities occur in communities such as Dwomo, Nyankrom and Ituma whereas charcoal burning activities have been intensified in portions of the watershed at Inchaban.

The relief of the study area is undulating, gently sloping towards the coast, and is interspersed with plains in the west. The landscape is characterized by muddy lagoons and marshlands as a result of the undulating topography. The district is drained by River Anakwari. River Anakwari is dammed at Inchaban to supply potable water to Takoradi and its surrounding settlements that include Dwomo, Nyankrom, Ituma, Shama and Yabiw. Drainage in the district is very poor; the area is prone to flooding.

Research design

The descriptive research design was used in the study. This research design incorporated scientific methods which involved observing and describing the behavior of respondents without influencing them in any way (Creswell, 2003). The method helped to describe the social systems, social events and background information related to the study. It also helped to stimulate explanations (Sarantakos, 1998). This method as observed by Hakim (2000) is the best because it helps to give a thorough picture of a phenomenon, and the changes that have occurred in a phenomenon over time.

The study population

The target population comprised all the state user and management institutions, non-governmental organisations (NGOs), local authorities and individuals who engage in economic activities in the watershed. The relevant state institutions identified were the Water Resource Commission, the Environmental Protection Agency (EPA), the Forestry Commission (FC), the Mining Commission (MC), the Hydrological Service Department (HSD) and the Water Resource Institute (WRI) of Ghana. The non-governmental organisations were the Coastal Resource Centre (CRC) and the Friends of the Nation (FON). The state user institutions are the Irrigation Development Authority (IDA), the Ghana Water Company Limited (GWCL), the Community Water and Sanitation Agency (CWSA), the Mining Commission (MC), the Ministry of Food and Agriculture (MoFA) and Hydrological Service Department (HSD). The individual respondents comprised residents from three selected communities which were Ituma, Inchaban and Dwomo (Figure 1).

Table 2. Total sample size for the study.

Sample units	Sample size
Watershed management organizations	4
Watershed user organizations	4
Community chiefs	3
Individual users	
Crop farmers	9
Charcoal producers	12
Fishermen	9
Total	41

Sampling procedure

Two non-probability sampling methods were used to select the respondents: The purposive and convenience sampling methods. The purposive sampling technique was first used to select the chief of three settlements (Inchaban, Ituma and Dwomo) from a total of eight settlements (Komfoeku, Dwomo, Osofokrom, Essupon, Yabiw, Ituma, Nyankrom and Inchaban, within the watershed. The chiefs were selected because their subjects were engaged in economic activities in the watershed, and can therefore be in the position to provide relevant information to the study. Four organisations (EPA, FC, WRC and CRC) were selected from a total of seven watershed management organisations that were the Environmental Protection Agency (EPA), the Forestry Commission (FC), the Mining Commission (MC), the Hydrological Service Department (HSD), the Water Resource Commission, the Water Resource Institute (WRI) and the Coastal Resource Centre (CRC). The four selected management institutions were identified to be active, and each entrusted with different duties in the management of the watershed. Hence, each of them could provide unique information relevant to the study. Four watershed user institutions (IDA, GWCL, MOFA and CWSA) were also selected from a total of six user institutions that were the Irrigation Development Authority (IDA), the Ghana Water Company Limited (GWCL), the Community Water and Sanitation Agency (CWSA), the Mining Commission (MC), the Ministry of Food and Agriculture (MoFA) and Hydrological Service Department (HSD). The two institutions: MC and HSD that were not selected were observed to be dormant. The selected user institutions were identified to be active and engaged in several different economic activities in the watershed. As such these selected user institutions could provide useful information to the study. Additionally, the convenience sampling technique was used to select nine crop farmers out of 15 in Ituma, 12 charcoal producers out of 16 in Inchaban and nine fishermen out of 15 in Dwomo. Focus group discussions (FGDs) were conducted for these individual users in their homogeneous groups from the above-mentioned villages (Ituma, Inchaban and Dwomo). Therefore, in all, a total sample population of 41 was used in the study (Table 2)

Data collection instruments and techniques

The data collection instruments employed consisted of three (3) sets of in-depth interview guides, three sets of focus group discussion guides and an observation checklist. The in-depth interview guides containing a number of unstructured questions were each used to collect data from the chiefs in the three settlements, from the management institutions and from the user organisations respectively. The set of questions in the in-depth interview guides elicited responses on issues regarding the background characteristics of respondents, effectiveness of the approaches and the challenges confronting the management institutions responsible for managing the Inchaban Watershed.

Three sets of focus group discussion guides (Appendix 2) were used to collect data from individual users (farmers, fishermen and charcoal producers) of the watershed (Table 2).

Data processing and analysis

The data collected was conceptually organized into themes which were based on the objectives of the study such as assessment of the effectiveness of the approach used to manage the Inchaban Watershed and the challenges of the management institutions to implement policies set out by government and analyzed manually, while the data collection was still on-going. This method was employed so as to intensify probing into issues on more specific aspects of the objectives of the study. Again, the analysis during the data collection was important because it guided the study to facilitate a more effective treatment and coverage of the research topic (Benini, 2000). Conclusions drawn after the analysis were tested against notes and records from the field.

RESULTS AND DISCUSSION

The results of the study under the following sub-headings:

1. Assessment of the effectiveness of the approach used to manage the Inchaban Watershed
2. The challenges of the management institutions to implement policies set out by government.

Assessment of the approach used to manage the Inchaban Watershed

Focus group discussions (FGDs) to assess the management approach with the key societal groups in the local communities surrounding the watershed revealed that there have been several discussions with the individual users on best farming practices. Farmers, for example, had understood very well the concerns of the government about the conservation of the watershed. Indeed the chiefs of the settlements admitted that they had been involved in the discussions on planning and implementation of projects such as contour farming, vegetative barriers and earth bunds to erosion control. Interviews with the chiefs of settlements in the watershed agree with the bottom-up approach used in managing the Inchaban Watershed. This falls in line with the findings of Thoruw and Juo (1995), who made the observations that the first among the local authorities to confirm the practice of the bottom-up approach was a chief of one of the communities who recalled that:

"Officials from almost all the management institutions come to me to solicit my views on a number of projects to conserve the watershed. I quickly invite the leaders of the Crop Growers Association, and schedule time to meet the other farmers in the field. Agreeably, we meet the

farmers about a week or two, and have discussions on a few projects such as contour farming, vegetative barriers, and earth-bunds projects. The officials also educate the farmers and take some of their concerns to government for policy formulations." (Village chief, Inchaban Watershed)

Even though the bottom-up approach was started, for it to be successful, according to Leach and Sabatier (2002), local people are expected to be involved in important stages of the project cycle such as monitoring and evaluation. Leach and Sabatier (2002) further made the observations that participatory watershed planning must go beyond initial implementation of policies to yield good results. However, the focus group discussions with the individual users and chiefs indicated that the watershed institutions never made any special arrangements to involve the local people in monitoring projects. By implication, as identified by Martin (2008), the management institutions find it difficult to arrest free-rider behaviour among the individual watershed users. As a result, the management institutions such as EPA and FC could not evaluate projects such as tree planting, protection of river banks, protection of land from erosion, soil fertility projects, contour farming projects, vegetative barriers and earth bunds projects, land and water management projects, among others to reshape subsequent plans based on future impacts of the projects in the watershed. Instead, the management institutions resorted to a simple method of tasking some untrained influential members to provide security for implemented projects. With this approach, according to the charcoal producers, the institutions never hadgood feedback to assist them to redesign future projects.It therefore, became difficult to discourage negative practices on the watershed. One of these charcoal producers had this to say:

"The EPA and the FC came to plant some trees in the watershed in 2008. The officers in the management institutions had made fruitless attempts, over the years, to arrest illegal users because they relied on people from this community to provide security for their projects. It will be very advisable for some of the officers to come and live with us here to make their work effective" (Charcoal producer, Inchaban Watershed).

More importantly, the suggestions of the local communities were expected to reach top management authorities on time for quick implementation of projects and granting of watershed use rights. However, the user institutions and individual users complained that it took them about half a yearbefore their concerns were sent to the WRC. Additionally, the users of the watershed who had been convinced to accept the bottom-up approach complained about the bureaucracies involved to obtain permission from the top officials. For example, the district head of the Irrigation Development Authority (IDA), who had had several confrontations with the heads of the National Development Planning Commission (NDPC) and Water Resources Commission (WRC), in the course of obtaining permission for the local users for crop farming irrigation, poured out his sentiments about the management system as follows:

"We expect our higher authorities (NDPC and MWH) to submit our development plans to the WRC on time for quick feedback. Unfortunately, since 2005 we have not been given authority to grant permission on any project plan sent to the NDPC especially" (District Head, Irrigation Development Authority, Inchaban Watershed).

According to Sabatier et al. (2002), for successful watershed management, top managers are supposed to, in addition to existing policies; spell out other management policies without delay to allow needful economic activities. Findings from the study indicated that this was not the case. It was found out that the top managers (MWH, NDPC and WRC) of the Inchaban Watershed did not have policies to allow local users to conduct economic activities. Hence, the bottom-up approach was not attractive to local users.

Sectoral approach of watershed management

In other interviews, the heads of the selected institutions admitted that most of them had implemented programmes with different priorities regarding the management of the Inchaban Watershed. It also followed that the user and management institutions had, in addition to bottom-up approach, started using the sectoral management approach. These results fall in line with the findings of Pretty and Shah (2000), where watershed management agencies implement programmes according to their own targets and priorities.

Following the sectoral management approach, it was discovered that the Ghana Water Company Limited (GWCL) confined its operations to the improvement of the quality of water from the watershed. Whereas the Environmental Protection Agency (EPA) implemented projects such as the Invasive Aquatic Water Project, Water Hyacinth and Lettuce Projects towards protecting aquatic life in River Anakwari. The interviews again revealed that the Ministry of Food and Agriculture (MoFA) and the Irrigation Development Authority (IDA) had been focusing on soil rehabilitation projects. Incidentally, the management institutions had followed the national development plan that had a specific goal of conserving the watershed but took operations with different objectives, making them unsuccessful. With the fast depletion of the Inchaban Watershed, the heads of the management institutions expressed displeasure in the sectoral approach, and therefore proposed for the establishment of joint projects that will involve all the stakeholders who matter in managing the watershed. This will help take into consideration the views and opinions of all affected groups, users and individuals who use the Inchaban Watershed. When this is done, the concept of integration and better sustainability of watershed resources will be enhanced (Ozyuvaci, 1997).

On the other hand, the individual users of the Inchaban Watershed rather embraced the sectoral management style because they realized that when the institutions worked independently, they had a leeway to take permission from any of them (the institutions). For example, the local charcoal producers commended the Forestry Commission (FC) for supporting them with ideas and strategies to increase their production using the Acacia trees from the watershed. A fourty-year old charcoal producer had this to say:

"We are highly indebted to the Forestry Commission; the institution has, over the years, supported us with permission right to use the Acacia trees in the watershed. It is good that the Ghana Water Company Limited had concentrated on the management of water in rivers and left the forest trees for the Forestry Commission to manage. Taking separate functions like this will help to conserve the watershed" (Charcoal producer, Inchaban watershed).

In another instance, the crop farmers in a typical farming community (Ituma) in the watershed commended the Irrigation Development Authority (IDA) for educating them on soil conservation strategies. One of them at Ituma frankly said:

"Since 2000, we have had several supports from the IDA. Most at times, the IDA supplies us with equipment such as water pumps and cutlasses for cultivation to prevent soil erosion. The IDA also organizes workshops on projects for soil rehabilitation to support our work" (Crop Farmer, Ituma Village, Inchaban Watershed).

In summary, the FC, the GWCL and the IDA concentrated on the management of the forest trees, rivers and the soils in the Inchaban Watershed respectively. These findings conform to the results by Economic and Social Commission for Asia and the Pacific (ESCAP) (1997) when they assessed the use of sectoral management approach in the success of the Mahaveli Project Schemes worked on by state institutions in Sri Lanka.

Challenges to the implementation of watershed policies

Challenges related to coordination among the management institutions

The WRC, MWH and the NDPC operate at the top level of decision-making and are expected to coordinate the activities of the FC, EPA and the CRC to prevent conflicts with the user institutions (IDA, CWSA, GWCL and MoFA) (Amakye, 2002).

Here, we assess the level of coordination among the institutions at the top level of decision making and allocation of watershed resources. Interviews with the head of the WRC showed some level of collaboration among the institutions (WRC, MWH, and NDPC). For example, the head of the WRC had policies (Table 1) that the MWH and the NDPC respectively, had designed to support its (the WRC) work. Better still, we also assessed the extent to which this collaboration had gone. Further discussions on management duties with the head of the WRC revealed that the WRC had failed to take up subsequent follow-up duties with the NDPC and the MWH, to ensure that policies and laws instituted were being obeyed by the local users of the watershed. This shows a poor collaboration between these institutions (NDPC, MWH and WRC).

Social challenges

One realizes that as population grows, societies become dynamic and it calls for a review of laws and policies governing resources use (Heckathorn and Maser, 2001). In interviews with the local people concerning resource use in the watershed, it came out that there were illegal building construction, illegal agricultural encroachment and large-scale illegal logging in the Inchaban watershed (Plate 1). Such practices have been vehemently fought against by the management institutions such as the EPA, FC and WRC offices in the area by the use of prohibitive laws such as the use of land guards in protecting the forests, use of taboos, vegetation cannot be cleared along a strip of 30 m at both banks of streams and rivers and days and periods when fishing, farming and hunting are prohibited or forbidden. Yet, since 1996, the Ghana WRC had approved the 'Integrated Watershed Management Policy' which allows diverse uses of all watersheds in every region. Thus, the prohibitive laws contradict the policy of the government to promote the concept of integrated watershed management.

This meant that, with increasing demand for land for settlements and other private uses such as farming and construction of buildings, the work of the NDPC and MWH was to collaborate with the WRC to re-formulate policies that could permit urgent uses of the Inchaban Watershed in order to follow the approved management approach. Without any options for land for survival such as for farming, fishing and charcoal, the individual users buy watershed usage rights (rights to enter, right to use, right to take wood or selling the land out) from the government institutions. In an in-depth interview with the head of the WRC, some explanations were given about the institution's reluctance to allocate the watershed resources to the local users. This is what the head of the Water Resource Commission had to say:

Plate 1. Acacia trees cut down in the Inchaban Watershed. Source: Fieldwork, June 2012.

"We cannot permit all users including individuals to use the watershed because the watershed vegetation is fast depleting. We normally give priority to state institutions like the IDA, GWCL and CWSA. We had entrusted the use of the watershed to the state institutions whose activities are environmentally sustainable and in the interest of the entire society. Unfortunately, individuals rather get the chance to work on the lands demarcated. We have the information that some of the state institutions have sold their usage rights to these individuals but we find it difficult to prosecute them" (Head, WRC, Inchaban Watershed).

Legal capacity challenges

The *Water Use Regulation Act of 1962*, supports the integrated water resource management policy of Ghana, and gives WRC the mandate to regulate the use of all water bodies in the country. To make the work of WRC easier, the EPA, FC, NGOs and local authorities have also been assigned specific and separate roles to support the WRC to manage water bodies. Serious conflicts were detected among the local management and user institutions. The local chiefs in the first place, said that the state management institutions did not give them room to exercise their powers. From the *Statutory Land Administrative Act 125 of 1962* of Ghana, power is only given to the WRC to plan and manage watersheds and the subordinate institutions and civil society organizations such as the local chiefs are to conform, not exercise power. The chiefs recalled instances where they had given permission to some of the farmers and fisher folks to use the watershed, and have been chased away by the management institutions. For example, the queen mother of one of the communities made it clear that it was time they claimed portions of the watershed to support local economic activities. This is what she said:

"In colonial times, portions of the Inchaban Watershed were demarcated for our forefathers to use. It is just about time we reclaimed the lands that belong to us to support the local people here. We will not sit down and watch other people to use the watershed illegally. I have written several letters to take permission from the management institutions for the local users of the watershed but have not had any good feedback. Personally, I grant some of the local people usage right when they ask for help. I know they receive threats from the government authorities but we still support them in every way" (Queen Mother, Inchaban Watershed).

The *Statutory Land Administrative Act 125 of 1962* of Ghana supports the state watershed management institutions to regulate and control the use of all lands such as mineral sites, forest lands and water bodies that fall under the areas of interest of the state (Opoku-Agyemang, 2001). For state watershed management institutions to work effectively, government of Ghana has established institutions as the Lands Commission, the Survey, Town and Country Planning Department and the judicial courts to support the state's claim for lands for social development. However, the reports obtained from the management institutions showed that the state judicial system, unfortunately, is weak to support governance over the Inchaban Watershed. In most cases, the reports were that certain institutions took bribes from individual users and overlooked the illegal activities in the watershed. Others have also supported some political leaders to erect structures for self-owned businesses in the watershed. To make the situation worse, as was indicated by an official of the Environmental Protection Agency, the judicial courts went on adjourning cases involving illegal users of the Inchaban Watershed. He said frankly that:

"Since the last four years we have sent three major cases to the courts requesting the support of the Takoradi Court to stop the construction of buildings in the watershed. As I speak to you, there are two additional cases of illegal construction in the watershed. The courts kept on adjourning the hearing of these cases. We have persistently referred the cases to the local chiefs for support but, to our dismay, the chiefs go behind us to encourage the illegal users of the watershed. Some of the heads in the other sister watershed management institutions pay bribes to the court officials for the cases to be adjourned" (EPA official, Inchaban Watershed).

The poor cooperation among the management institutions posed legal challenges for the institutions to battle with. Once the management institutions neglected collaborative project building and thus followed, to a large extent, the sectoral management approach, there were always conflicting interests that resulted in serious legal challenges.

CONCLUSIONS AND RECOMMENDATIONS

Two main approaches (the bottom-up and the sectoral approaches) were discovered to be in use for the management of the Inchaban Watershed. The bottom-up approach was least practiced due to challenges of getting community support after the project implementation (to conserve the Inchaban Watershed). Also, the results obtained in this study indicated thatthere was not regular monitoring and evaluation by the management institutions to ensure successful project implementation. Again, there was limited involvement of the local community in important stages of the project cycle such as the monitoring and evaluation phase. These were challenges to the smooth and successful practice and adoption of the bottom-up approach in watershed management of the Inchaban watershed. Also, the sectoral management approach was adopted due to the fact that management institutions concentrated on different aspects of the watershed in order to conserve it.
Assessment was also made of the institutional linkages among the top watershed management institutions to see how best the policy of integrated watershed management has been implemented. Findings from the study indicated that the NDPC and the MWH collaborated poorly to support the work of the WRC. The NDPC and the MWH had only formulated watershed use policies and coordination of projects respectively but have neglected important follow-up duties to ensure that the policies are being implemented and enforced. Little has been done about monitoring, evaluation and assessment of the achievements over the years. The linkages between traditional authorities such as chiefs and management organizations (EPA, FC, WRC and CRC) were also poor and this poses legal challenges. The mandate of watershed management institutions supported by the *Watershed Resource Act 522 of 1996* though has been stated clearly in the Customary Land Administrative Policydocument (this document supports Ghanaian local chiefs to manage local and indigenous resources). It appears this document is not known to many chiefs and other traditional leaders.

In connection with the key findings and conclusions drawn, the following recommendations have been made:

1. Government should ensure better implementation and fixing the problems in the current policies or there should be system to promote sound institutional arrangements among the watershed management institutions responsible for the Inchaban Watershed.
2. It should be the policy of the government to educate local authorities to constantly support the operations of watershed management institutions.
3. To attract the participation of the communities surrounding the watershed, there is the need for building the capacity of government institutions in implementing more participatory, multi-stakeholder approaches to watershed management".

Conflict of Interest

The authors have not declared any conflict of interests.

REFERENCES

Acheampong PK (2009). The earth themes and variations.Cape Coast.University of Cape Coast, Ghana.

Amakye AK (2002). The changing focus in the development and the management of Ghana's water resources. Libreville, Gabon: Raider Books.

Benini A (2000). Construction of knowledge. Rome, Italy: Sergio Bonelli.

Biney P (2010).Policy to Improve Watershed Resources for Rural Communities in Ghana: A study of water bodies in Western Region. Accra, Ghana: Aabok Press.

Bossman C (1998). Institutions, institutional changes in Land administrative policies.Accra, Ghana: Digibooks.

Botero A (1986). Riparian forest effects on lateral stream channel migration in the glacial till plains. Amer.J. Wat. Res. Assoc. 34(25): 179-184.

Bullock GF (2008). A perspective of river basin development: Respect for Religious,Local and Cultural Structures.22: 157-184.

Button B (2010). Water Resources Management in Ghana. J. Agric. 28(1): 65-75.

Carson B (1992).The land, the farmer, and the future: A soil fertility management strategy for Nepal. Kathmandu, Nepal: McGraw Hill.

Creswell JW (2003). Research design in social research. Thousand Oaks. CA: Sage.

Darghouth RA (2008). Evaluating water institutions and water sector performance. World Bank Technical Report on watershed management. Retrieved from siteresources.worldbank.org/.../Watershed Experience on 3/01/2012.

Department of Geography and Regional Planning (2011). Map showing Shama District.University of Cape Coast,Cape Coast, Ghana.
Economic and Social Commission for Asia and the Pacific (1997). Guidelines and manual on land-use planning and practices in watershed management and disaster reduction. Bangkok, Thailand.
Gibson JL (2001). Social networks, civil society, and the prospects for consolidating Russia's Democratic Transition. Amer. J. Pol. Sci. 45: 51-69.
Global Water Partnership (2000). The world fact book: Population growth rate estimate. Geneva, Switzerland.
Government of Ghana (1996).Water Resource Management Act 522.Accra: Government of Ghana.
Government of Ghana (2006). Drilling for Water and Groundwater Development Regulation Policy LI 1827, Accra: Government of Ghana.
Government of Ghana (2007).National Water Vision Policy. Accra: Government of Ghana.
Government of Ghana (2012).National Buffer Zone Policy. Accra: Government of Ghana.
Government of Gold Coast (1903).River Ordinance Act 210. Gold Coast: Government of Gold Coast.
Government of Gold Coast (1927).Forestry Ordinance Act 212.Gold Coast: Government of Gold Coast.
Government of Gold Coast (1953).Land Planning and Soil Conservation Ordinance Act, CAP 210. Gold Coast: Government of Gold Coast.
Hakim C (2000). Research design: Successful design for social and economic research. Manchester. England: Routledge Press.
Heckathorn D, Maser M (2001). Bargaining and the sources of transaction costs: The case of government regulation. J. Law Econ. Org. 3:69-98.
Leach WD, Sabatier PA (2002). Approaches in Watershed Management. Rome, Italy: Packer.
Maarleveld M, Dangbegnon C (1998). Managing natural resources: A social learning perspective. Agric. Hum. Val.16 (3): 267-280.
Manuel DK (2007). Integrated watershed management: Principles and practices. Rochester, NY: Lark Books.
Martin G (2008). Community Involvement in Watershed Management.Buffalo NY: Oxford University Press.
Ministry of Works and Housing (2010). Watershed Management Policies. Accra: Government of Ghana.
Odame AK (2010). Integrated Water Resource Management, Environment and Equity in National Water Laws. Accra, Ghana. West Africa Press.
Opoku-Agyemang KB, Micheal N, Manu D, Bossman J (1998). Addressing Diffused State of Water Resource Management Institutions. Lund, Sweden: University of Lund Press.
Opoku-Agyemang M (2001). Shifting Paradigms: Towards the Integration of Customary Practices into the Environmental Law and Policy in Ghana. Unpublished Master's thesis. University of Ghana, Accra, Ghana.
Opoku-Agyemang SA (2008). Customary laws in land Administration in Ghana. Accra, Ghana. Anansesem Press.
Ostrom G (1999). Crafting institutions for self-governing irrigation systems.San Francisco, CA: Contemporary Press.
Ozyuvaci N (1997). Integrated Watershed Management for Sustainable Development of Renewable Natural Resources. Retrieved from www.nepjol.info/index.php/AEJ/article/viewFileon 1/04/2012
Pradhan BB (2003). Emerging options in watershed management.Retrieved from indiaenvironmentportal.org.in/.../icimod-policy on 10/04/2012.
Pretty J, Shah P (2000). Soil and water conservation: A brief history of coercion and control. London, England: Sage.
Sabatier PA, Pelkey NW, Leach WD (2002). Policy to Improve Watershed Resources for Rural Communities in Kano, Nigeria.Illinois Press.
Sarantakos S (1998). Social Research. Buffalo, NY: Macmillan Books.
Thoruw T, Juo A (1995). The Rationale for Using a Watershed as a Basis For Planning and Development. Amer. Soc. Agron.8:93-116.
United Nations Environment Programme (2004).Assessing progress in linking the management of catchments and coasts. Nairobi, Kenya.
Water Resources Commission (1996). Integrated Water Resource Management Policy, Ghana. Government of Ghana.
World Bank (2001). Issues in Watershed Management: A Pilot Project in Colombia, Kent, WA.
World Natural Resource Conservation (1996). Natural resource management and livelihoods, Geneva, Switzerland.

APPENDIX

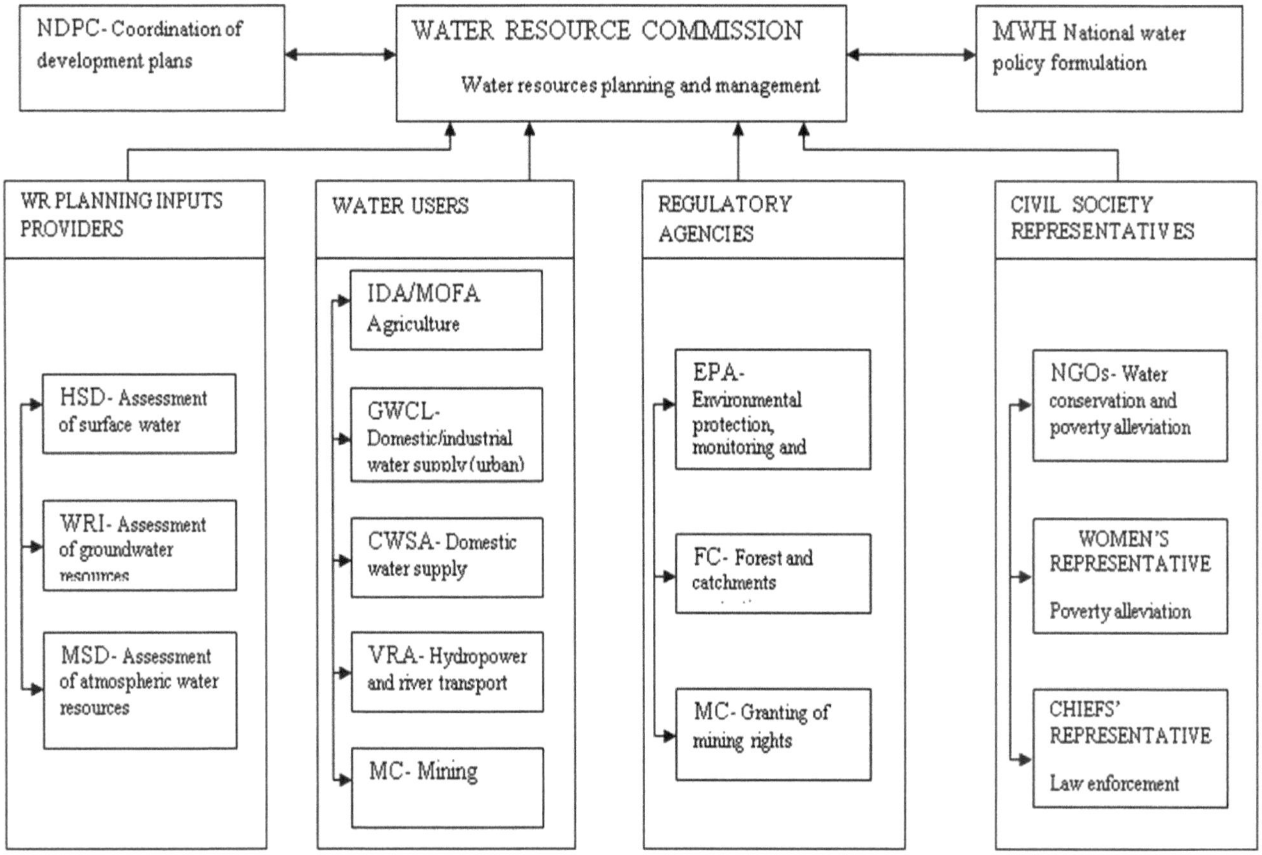

Appendix Figure 1. The Integrated Watershed Management Conceptual Framework. Source: Ghana Water Resources Commission, 2000.

Appendix 1: The Integrated Watershed Management Framework. Source: Ghana Water Resources Commission, 2000

Appendix 2: Focus group discussion guide for crop farmers, fish farmers and charcoal producers

Date of interview:
Place of interview:
Interviewer name:
Time of interview:

The main objective of the study is to examine the institutional arrangements in managing the Inchaban Watershed. The study is primary for academic work, and therefore you are assured of full confidentiality, privacy and anonymity of all the information that you will give. Please or kindly answer each question to the best of your ability by providing responses that best reflect your opinion.

Section A: Watershed management approach

1. Mention some of the organisations in charge of managing the watershed. Probe: What roles do they play?
2. Do the government authorities involve you in managing the watershed? If Yes, Probe: Where is involvement done? Indecision making, formulation of management policies, implementation of project policies, monitoring and evaluation.
3. What are some of the projects initiated in your community to manage the watershed? Probe: The nature of the projects (e.g. terracing, tree planting etc), when the projects were started, which organisation initiated the projects, challenges encountered in starting projects, success stories of projects. Are the projects done jointly by individuals and government organisations, or are the projects done jointly by NGOs and government organisations?
4. Do the organisations have different management activities which concentrate on different aspects of the watershed (use the sectoral approach)? If Yes, Probe: Type of activities (e.g. contour-bunding, landslide fencing, gully control, plantation growing, etc.), times that the departments do such activities, aspects of the watershed (e.g. forest, land or water) that the activities are aimed at, challenges encountered in this approach, advantage of this approach.

Section B: Economic activities of user organisations and their effect on the economic activities of local inhabitants

1. What are some of the economic activities that user organisations undertake on the watershed? Probe; farming, fishing, sand winning, (any other?, Please specify).
2. What products do they obtained from the watershed? Probe: (a) Food; which kind (cereals, grains, tubers, etc.), how do the activities of user organisation affect your output level (increase or decrease by how much). What is the quantity? (b) Employment; what type of employment, how does the activity of user organisations affect the security of your job, are there any other sources of employment?
3 Do the user organisations have any alternative source instead of using the watershed?

Section C: Activities of watershed management organisations and their effects on the economic activities of the people

1. What are some of the conservatory activities of the management organisations? Probe; tree planting, terracing or contour "bunding", (any other?)
2. How does the activity of the management organisation affect the following. Probe: (a) Food, which kind (cereals, grains, tubers, fish etc.) how does the activity of user organisation affect your output level (increase or decrease by how much); what is the quantity. (b) Employment, what type of employment, how does the activities of user organisations affect the security of your job, are there any other sources of employment?
3. Do the user organisations have any alternative source instead of using the watershed?

Section D: Assessment of the drainage basin management institutions on the principles of transparency, accountability and participation

1. Transparency (openness of governance processes and free access to official information): Are there available information about who, how and what decisions are made at the district assembly level? Probe (If yes, is the information available in a format and language that is easily understood by non-experts, accurate and up-to date? Is the information timely (that is, was it available prior to key decision-making processes such as planning, a town-hall meeting or available websites or on public notice boards); How long does it take to get the information?
2. Accountability (a set of controls, counterweights and supervision modes that make officials and institutions in the public and private sector answerable for their actions): Are there internal control mechanisms, checks and balances within the drainage basin management institutions to ensure internal accountability. Probe: Are there monitoring and evaluation institutions that ensure to check the services provided by the drainage basin management institutions? Is there an independent body that oversees and monitors state institutions to ensure that established norms and standards are met? Has the state oversight institutions got the legitimate power to demand accountability on both fiscal management and performance of management institutions? Are there laws, rules and regulations that govern the accountability relationship between oversight institutions?
3. Participation (the possibility for citizens to provide informed, timely and meaningful input and influence decisions at various levels): Are there mechanisms for citizens to express themselves and influence decisions and processes in the management of drainage basin? Probe: Are your decisions being heard and taken into consideration during town hall meetings? If Yes to Question 3 at which stage are you given the opportunity to make suggestions to the management of the drainage basin?

Formulation of management policies [ ] Often [ ] very often [ ] not often [ ]
Implementation of project policies [ ] Often [ ] very often [ ] not often [ ]
Monitoring and evaluation [ ] Often [ ] very often [ ] not often [ ]

In each case indicate how often (the opportunity is given decision making (Often, Very often, Not Often).

Rainwater harvesting: An option for dry land agriculture in arid and semi-arid Ethiopia

Binyam Alemu Yosef[1] and Desale Kidane Asmamaw[2]

[1]Department of Soil and Water Resources Management, College of Agriculture, Wollo University, Ethiopia.
[2]Department of Geography, College of Social Science, Samara University, Ethiopia.

Subsistence rain-fed agriculture has been widely practiced for many centuries in Ethiopia and this sector has been highly dependent on rainfall. Thus, rainfall remains the crucial component of the weather elements for improving agricultural productivity. Extreme climatic conditions and high inter-annual or seasonal variability of this weather element could adversely affect productivity, because rainfall controls the crop yields and determines the choice of crops that can be grown. One of the reasons for low crop production in semi arid areas is marginal and erratic rainfall exacerbated by high runoff and evapotranspiration losses. Rainfall in terms of amount and frequency in a growing season is essential for planning and management of agricultural practices. To avert this problem, the successive Ethiopian government (from the time of Aksumite kingdom to the present) and the local community practiced different water harvesting techniques. The *in-situ* and *ex-situ* rainwater harvesting techniques have shown significant impact on improved soil moisture, runoff and ground water recharge; and increased agricultural production, which in turn reduces risks and deliver positive impacts on other ecosystems. Besides, rainwater harvesting has a potential of addressing spatial and temporal water scarcity for domestic consumption, agricultural development and overall water resources management. High water loss through seepage, lack of awareness and being very labor intensive to irrigate the whole fields by pumping the water manually from the pond and applying directly to the crop has been the main challenges of adopting harvested water technologies.

Key words: Rainfall, runoff, semi-arid Ethiopia, soil moisture, water harvesting, yield.

INTRODUCTION

Agriculture is the main economic activity in sub-Saharan Africa (SSA) supporting over 67% of the population, out of which 60% depends on rainfed agricultural practices; generating 30-40% of the country's Gross Domestic Product (Rockström, 2002). However, rainfall is poorly distributed in these countries (Ngigi, 2003). High losses occur due to high surface runoff during high intensity rains, poor crop rooting conditions, past and present soil erosion and evaporation losses from soil, and crop canopy in particular during pre-planting and early crop stages (Rockström, 2003). Uncontrolled runoff can cause damaging flash floods; severe erosion; increased water turbidity and substantial water loss. These losses of valuable water resources increase food insecurity and societal poverty, which are the greatest threats to sustainable development in the region (Ngigi, 2003; Temesgen et al., 2007). This implies that erratic rainfall patterns has been the serious challenge of food production in these areas (Fischer et al. 2004), and this will be further worsened by climate change which is expected to increase rainfall variability in many African countries that are already at least partly semi-arid and arid (Ngigi, 2003).

The primary limiting factor for crop-yield stabilization in semi-arid regions is the amount of water available in the crop rooting zone (Lal, 2001). Rainfall intensity in Sub-Saharan Africa can often be greater than the infiltration rate and the water holding capacity of the soil, which can trigger an excess of runoff (Rockström, 2002). In Sub Saharan Africa where rainfall is low, unpredictable and also expected to decline due to climate change, rainwater storage in farm ponds, water pans, subsurface dams, and earth dams is gaining importance as a supplement to irrigation and livestock watering (Ngigi, 2009). It is also an effective strategy to manage floods *in-situ* and *ex-situ*, particularly in high rainfall areas like the Ethiopian highlands. Hence, rainwater harvesting (RWH) could be used to satisfy water demands during dry spells and to create opportunities for multiple uses. De Fraiture and Wichelns (2010) estimated that 78% of water consumed by crops comes directly from rainfall. It is believed that rainfall, if managed properly, is the remaining potential source for feeding the current huge population of the world as there are limitations to further increase of the area under irrigation (Rockström, 2003). This calls for increased attention towards rain fed agriculture or a green revolution (Rockström et al., 2009).

Subsistence rain-fed agriculture has been widely practiced for many centuries in Ethiopia and over 80% of the population's livelihood is contingent upon this sector which contributes ~45% to the national GDP, on average (Bewket, 2009). The Ethiopian agriculture is characterized by extreme dependence on rainfall, low use of modern agricultural inputs and low output levels. The sector is highly vulnerable to drought (Bewket, 2009), which is the single most important climate related natural hazard impacting the country from time to time for many decades. Ethiopia has significant rainfall and it is the ultimate source of water with surface water, ground water and other water sources. Based on grid-based average annual rainfall and the land area, the study estimates that Ethiopia receives about 980 billion (~1 trillion) m^3 of rainfall a year (Awulachew, 2010). While Ethiopia has abundant annual rainfall, the rainfall varies spatially, temporarily and inter-annually (yearly cycles). About 80 percent of rainfall occurs between June and September, while yearly variability can also be significant (e.g., about 30% average variation year over year). Consequently, increasing rainwater storage capacity and improving water control and rainwater management techniques, especially rain water harvesting, are critical to ensure that Ethiopia gets maximum use of its rainfall (Awulachew, 2010).

The term rain water harvesting (RWH) is used in different ways and, thus, no universal classification has been adopted (Ngigi, 2003). However, it can be defined as capturing and storing seasonal excess runoff and diverting it for household and agricultural uses (Hatibu and Mahoo, 1999). RWH refers to all technologies where rainwater is collected to make it available for agricultural production or domestic purposes (Liniger et al., 2006). Nigigi (2003) also defined rain water harvesting as a method for inducing, collecting, storing and conserving local surface runoff for agriculture in arid and semi-arid regions. It is considered as the single most important means to increase agricultural productivity and provide a source of domestic water supply in drought prone areas (Getaneh and Tsigae, 2013). This makes cultivation of crops twice or more a year possible, as well as the possibility for supplementary irrigation when rains stop early. Rain water harvesting includes roof water harvesting, *in-situ* water harvesting, run-off harvesting, flood water harvesting and subsurface water harvesting (Finkle and Sergerros, 1995).

The sustainability of RWH is based on reliable water supply and production, effectiveness of water use (increase rainwater productivity) and minimal negative impacts on natural resources (Pachpute et al., 2009). RWH systems are generally categorized into two; *in-situ* water conservation practices, small basins, pits, bunds/ridges; and runoff-based systems (catchment and/or storage) (Awlachew et al., 2005) (Figure 1). The storage system is usually used in supplemental irrigation. The *in-situ* systems, which enhance soil infiltration and water holding capacity, have dominated over storage schemes in Ethiopia until recently. Surface runoff from small catchments and roadside ditches is collected and stored in farm ponds holding an average of about $60m^3$ of water. *In-situ* rainwater harvesting techniques (IRWHT) that aims at maximizing benefits of rainfall where it falls are tremendously used and produce good results, especially in semi arid zones (Li et al., 2000; Tian et al., 2003; Li et al., 2006; Ito et al., 2007; Oloro et al., 2007; Rockström et al., 2009 and Vohland and Barry, 2009). Ngigi (2003) attributed the dominant use of IRWHT to its simplicity, low cost and can be practiced in all land use systems.

Improved rainwater management for agriculture has many potential benefits in efforts to reduce vulnerability and improve productivity (Awulachew et al., 2005). As a result of RWH and irrigation development, Ethiopia's agricultural sector has witnessed consistent growth since 2003. For instance maize production has expanded at 6% per annum and the aggregate export value across all commodities has grown at 9% per annum as well as underpinning an 8% annual growth rate in GDP

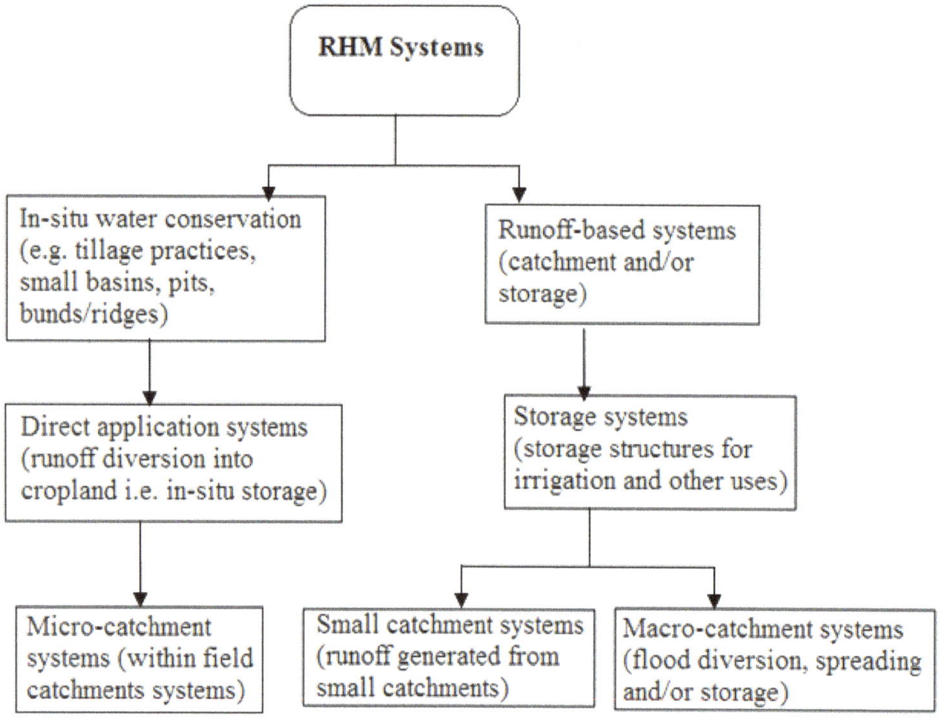

Figure 1. Adopted classification of RWH systems (Ngigi, 2003).

(Awulachew, 2010). In terms of increasing irrigation potential incrementally to the formal irrigation component, through RWH and better water management, the study estimates that RWH can provide an additional 0.5 M ha in irrigation. Thus, *in-situ* agricultural water management including RWH is crucial to improve smallholder livelihood and income in Ethiopia.

Frequent dry spells and droughts exacerbate the incidence of crop failure and hence food insecurity and poverty (Awlachew et al., 2005). The need to improve food production in Ethiopia cannot be overemphasized. In addressing the problems of rainfall variability, the recurrent droughts and food insecurity, the government of Ethiopia has given top priority to a variety of water harvesting programmes to supplement the rain-fed agriculture. Therefore, rainwater harvesting has been recognized as a promising way for improving the water availability for crop production, domestic use and water for livestock in the arid and semi-arid parts of the country. This paper therefore, reviews the present status of RWH researches in the light of an option for dry land agriculture in the arid and semi-arid Ethiopia.

RAINWATER HARVESTING TRENDS IN ETHIOPIA

Over centuries, generations of farming communities in Ethiopia have developed different farming technologies that can provide a basis on which to build improved land husbandry (Alamerew et al., 2002). The history of water harvesting in Ethiopia dated back as early as the pre Axumit period (560 BC) (Habtamu, 1999). It was a time when rainwater was harvested and stored in ponds for agricultural and water supply purposes. Anthropologist Fattovich (1990) have documented evidences of the remains of ponds that were once used for irrigation during this period. A roof water harvesting set-up is still visible in the remains of one of the oldest palaces in Axum; the palace of the legendary Queen of Sheba. Other evidences include the remains of one of the old castles in Gondar, constructed in the 15-16th century, which used to have a water harvesting set up and a pool that was used for religious rituals by the kings.

In south of the country, the Konso people have had a long and well established tradition of building level terraces to harvest rainwater to produce sorghum successfully under extremely harsh environment; low, erratic and unreliable rainfall conditions (Hailemichael, 2011). It is indeed one of the wonders of this country and it has been practiced for millennium; a symbol of struggle for survival by the Konso people against the adversaries of nature (Figure 2). Hailemichael (2011) shows, everyone in the farming system acquires the skill of terrace construction as part of routine farming practices. Thus, rainfall is simply redirected through carefully constructed walls and channels. Researchers noted that this is a country where indigenous knowledge and practices are acknowledged to the extent that the 'Konso

Figure 2. Indigenous hillside stone terracing and mixed cropping *in-situ* soil and water conservation practices at Konso southern Ethiopia (A) (Hailemicael, 2011) and Half-moon ponds, promoted for water resources conservation in Tigray, northern Ethiopia (B) (Mekonnen and Haile, 2010).

cultural landscape' has been recognized as a UNESCO World Heritage Site (Alamerew et al., 2002.). Efforts over the last decades to promote water harvesting have produced some bright spots particularly with micro-watershed interventions, which include water harvesting as part of an integrated participatory approach to sustainable land management. In addition to that, there are different half-moon ponds, promoted for water resources conservation in Tigray, northern Ethiopia (Figure 2B) (Mekonnen and Haile, 2010)

Ethiopia's agriculture is predominantly rainfed with a potential of nearly 3.5 million ha of land suitable for irrigated agriculture (Awulachew et al., 2005; Awulachew and Merrey, 2007). The population has grown dramatically over the last three decades, increasing from 25 million in the 1960s to nearly 90 million in 2014 leaving a pressure on the agricultural land, forests and the environment at large. Thus, rainwater harvesting and management (RHM) play paramount roles for increasing gain yields and support the growing population (Awulachew, 2010). The RWH techniques most commonly practiced in Ethiopia today are runoff irrigation (run-off farming), flood spreading (spate irrigation), *in-situ* water harvesting (ridges, micro basins, conservation tillage, etc.) and roof water harvesting.

The importance of these techniques, though dates back in the antiquity, has not been recognized until very recently, following the devastating drought and famine of the 1980s (Alem, 1999). RWH is a simple low-cost technique that requires minimum specific expertise or knowledge and offers many benefits. It has regained importance as a valuable alternative or supplementary water resource in Ethiopia (Desta, 2003). Utilisation of rainwater is now an option along with more 'conventional' water supply technologies, particularly in rural areas, but increasingly in urban areas as well. RWH has proven to be of great value for arid and semi-arid countries or regions like Ethiopia where more than 66% of the country is categorized as arid and semi-arid (Georgis, 2000; 2002; Rockström, 2000; Rockström, 2001).

The development of irrigation and agricultural water management holds significant potential to improve productivity and reduce vulnerability to climactic volatility in any country (Awulachew, 2010). Although Ethiopia has abundant rainfall and water resources, its agricultural system does not yet fully benefit from the technologies of water management and irrigation (Georgis, 2000; 2002). The majority of rural dwellers in Ethiopia are among the poorest in the country, with limited access to agricultural technology, limited possibilities to diversify agricultural production given underdeveloped rural infrastructure and little or no access to agricultural markets and technological innovations (Awulachew, 2010). A field experiment was conducted under natural rainfall conditions to investigate the effects of farmyard manure and straw mulch on runoff, soil loss, *in-situ* water conservation and yield and yield components of an improved bread wheat variety (HAR-1480) grown on vertisol of Sinana area, south- eastern Ethiopian highland. The results revealed that there was a highly significant difference ($P<0.0001$) between the treatments regarding their effect on runoff depth, soil loss and *in-situ* water conservation (Birru et al., 2012).

TYPES OF RAINWATER HARVESTING

In-situ rainwater harvesting

In-situ RWH, a technique which involves the use of methods that increase the amount of water stored in the soil profile by trapping or holding the rain where it falls (Alamerew et al., 2002; Ngigi, 2003). This may involve small movements of rainwater as surface runoff in order to concentrate the water where it is most wanted. It is sometimes called water conservation and is basically a prevention of net runoff from a given cropped area by holding rainwater and prolonging the time for infiltration. This system works better where the soil water holding capacity is large enough and the rainfall is equal or more than the crop water requirement, but moisture amount in the soil is restricted by the amount of infiltration and or deep percolation (Hatibu and Mahoo, 1999). *In-situ* water harvesting methods that concentrate soil water in the rhizosphere for more efficient use by plants are critically needed for moisture stress areas. It means rainwater is conserved where it falls, whereas *ex-situ* water harvesting systems involve transfer of runoff water from a "catchment" to the desired area or storage structure (Critchley and Siegert, 1991). Land and water conservation interventions on sloping lands include bench/fanya juu terraces, retention ditches, stone lines, vegetative buffer strips, contour bunds, contour farming and other activities that reduce loss of runoff water. They are primarily used to reduce soil erosion and to improve rainfall infiltration and conservation in the soil profile (Bossio et al., 2007).

In-situ rainwater conservation technologies are distinct from runoff farming systems in that they do not include a runoff generation area; instead it aims at conserving the rain where it falls in the cropped area or pasture (Alem, 1999). The most commonly implemented technology is conservation tillage which aims to maximize the amount of soil moisture within the root zone. A number of cultural moisture conservation practices such as mulching, ridging, addition of manure, etc could fall under this category. Small field/farm structures such as tied ridges/bunds within cropped area that conserve direct rainfall without 'external' outside cropland boundary, that is, no distinct catchment area, except overflow from upstream sections also falls under this category. *In-situ* rainwater conservation technology is one of the simplest and cheapest and can be practiced in almost all the land use systems. *In-situ* water conservation systems are by far the most common (Rockström, 2000) and are based on indigenous/traditional systems (Reij et al., 1996; Leisa, 1998). The primary objective has been to control soil erosion and hence, manage the negative side effects of runoff soil and water conservation measures, that is, to ensure minimal runoff is generated. The positive effect of *in-situ* water conservation techniques is to concentrate within-field rainfall to the cropped area (Rockström, 2002).

In a semi-arid context, especially with coarse-textured soil with high hydraulic conductivity, this means that *in-situ* conservation may offer little or no protection against the poor rainfall distribution (Rockström, 2000). In such cases, the farmers will continue to live at the mercy of the rain. In effect, the risk of crop failure is only slightly lower than that without any measures. However, soil improvements and management would enhance realization of better yields.

Agronomic practices such as the use of farm yard manure (FYM), timely weeding and mulching are used to enhance water availability in the soil by improving the water holding capacity and reducing soil water evaporation. Mulch and manure treatments had a highly significant effect ($P<0.0001$) on *in-situ* rainwater retention (Birru et al., 2012). The results demonstrated that *in-situ* soil moisture conservation increased significantly for 6 ton ha^{-1} mulch as compared to the control treatment. Soil moisture storage in the 6 ton ha^{-1} mulch treatment was 216.11 mm, which was 39.15 mm higher in comparison to the control. However, the mean soil moisture storage for 4 ton ha^{-1} straw mulch (STR) was 215.40 mm, which was statistically at par in comparison to that retained at 6 ton ha^{-1} straw mulch treatment.

The average rainwater depth retained for 6 ton ha^{-1} farmyard manure application (181.81 mm) was significantly higher than that retained under control treatment (176.96 mm). The results further indicated that soil moisture storage was increased by 22.1, 21.7, 19.0, 2.7, 0.74 and 2.53% for STR-6, STR-4, STR-2, FYM-6, FYM-4 and FYM-2, respectively, as compared to the control (Birru et al., 2012). From these results one can understand that leaving crop residues could have the potential to conserve much of the incoming rainfall and contribute towards sustainable crop production by alleviating the impacts of drought spells which frequently occur in the growing season. The main limitation of this technique is its high labour demand, especially on steep slopes where proper structural measures are required. Some amount of training and site-specific design/layout is also needed. In one example from the Anjenie watershed of Ethiopia (Akalu and Adgo, 2010), long-term terracing increased yields of teff, barley and maize significantly. In contrast, cultivation on steep un-terraced hillsides had negative gross margins.

Similarly, Vancampenhout et al. (2005) obtained positive yield results and increased soil water holding capacity using stone bunding on field crops in the Ethiopian highlands. Rockstrom (2003) reported that in-situ RWH had a significant effect on grain yield and by using this system in Burkina Faso they were able to increase the yield of sorghum from 715 kg ha^{-1} to 1,057 kg ha^{-1}. Micro-basin water harvesting structures (e.g., half-moons, eyebrow basins and trenches) have also been proven to be effective in improving tree survival and growth in degraded lands. Experiences from northern Ethiopia have shown that these structures improved tree survival and growth significantly compared to non-treated landscapes (Derib et al., 2009). The seedlings grown in micro-basins were thicker, taller and more productive than those grown in normal pits, suggesting a need to integrate tree planting with soil water management.

Tillage normally assists in increasing the soil moisture holding capacity through increased porosity, increasing the infiltration rates and reducing the surface runoff by providing surface micro-relief or roughness which helps in temporary storage of rain water. Previous research results have shown that the depth of tillage is the most important factor controlling or affecting soil moisture characteristics (Hudson, 1987). Deep tillage helps to increase porosity, reduce surface sealing of the soil and permits roots proliferation to exploit soil water and nutrients at deep horizons (Kidane et al., 2012). Significant reduction of surface runoff and increase in crop yields have been shown to occur with increased depth of tillage.

Temesgen et al. (2012) conducted an experiment on the impact of conservation tillage on hydrology and agronomy on famer's fields from the year 2009-2010. The soil moisture measurements had been taken continuously at the lower and upper sides of each plot, for a period of one month only (due to vandalism). Although the measurement period is short, the sampled results clearly revealed that soil moisture in traditional tillage (TT) (average 34.6%vol.) is significantly higher ($\alpha = 0.05$) than that of conservation tillage (CT) (average 31%vol.) at 0-15 cm depth while the reverse holds true at 15-30 cm layer (33.5 and 31.6%vol.), in CT and TT, respectively. Similar studies have been conducted in an on-farm experiment in the northwestern highlands of Ethiopia (Kidane et al., 2012). There were significant differences ($P< 0.0001$) in soil moisture content between tillage treatments as well as in the upper and lower sides of the plots. Significantly, ($P \leq 0.05$) different soil moisture contents between the upper and lower sides of the *fanya juus* were observed under traditional tillage practice 0.305±0.003 and 0.323±0.003 m^3, respectively.

Temesgen et al. (2007) observed field trial on conservation tillage compared with traditional tillage in a moisture deficit area in Ethiopia. Sub-soiling along the same lines (STS) resulted in the least surface runoff (Q_s =17 mm-season^{-1}), the highest transpiration (T =196 mm-season^{-1}) and the highest water productivity using total evaporation (W_{PET}=0.67 kg-m^{-3}) while TT resulted in highest surface runoff (Q_s=40 mm-season^{-1}), least transpiration (T =158 mm-season^{-1}) and low water productivity (W_{PET}=0.58 kg-m^{-3}), respectively. Araya et al. (2011) reported runoff values of 46.3, 76.3 and 98.1 mm from *derdero* (DER), *terwah* (TER) and conventional tillage, respectively. There were significant differences ($P < 0.0001$) in infiltration rates in the soils between winged subsoiler (WS) and traditional tillage (TT) treated plots (Kidane et al., 2012). The initial and steady state infiltration rates under WS plots were 0.84±0.005 and 0.1±0 cm m^{-1}, respectively.

On the other hand, 0.54±0.006 and 0.05±0.004 cm m^{-1} at 01 and 60 m were observed under the TT treated plots. The infiltration rate in the WS treated plots was twice higher than the TT tilled plots and this implies that implementation of conservation tillage is more important for *in-situ* moisture conservation. Temesgen et al. (2012) evaluated the hydrological and grain yield impact of conservation tillage on field experiment and they identified that more surface runoff occurred for traditional tillage compared to conservation tillage, and that the differences between the two was more in the wheat plot than in *tef*. The average reduction of surface runoff was 48% in the wheat plot due to the application of CT, with the daily averages of 4.8 and 2.5 mm d^{-1} in TT and CT, respectively. In *tef* crop, the surface runoff reduction was 15% with an average of 4.5 and 3.8 mm d^{-1} in TT and CT, respectively.

Ex-situ water harvesting

Water harvesting ponds

The occurrence of repeated drought and dependence on rainfall which has been erratic and uneven in Ethiopia over the past decades has resulted in a widespread crop failure which in turn has brought a growing awareness of the importance of small-scale water harvesting at both household and community levels (Haile, 2007). In order to alleviate the problem of recurrent drought and household food security, the government of Ethiopia has taken household level water harvesting ponds and shallow wells development as one strategy of the country's irrigation development (Haile, 2007). In Ethiopia, particularly in moisture deficit areas, irrigation development is considered. Therefore an absolute necessity and not an option has been adopted as the frontline strategy to bring about sustained food security due to its profound role in boosting agricultural productivity in highly rain dependent agricultural farming systems.

Moreover, since 2002, household level water harvesting irrigation development has attracted the attention of the policy makers due to the small initial investment, low government recurring cost, short development period, relative freedom of organization and freedom from management difficulties (Alem, 1999: Mekonnen and Haile, 2010). In an effort to address the problems of recurrent droughts and food insecurity, the government of Ethiopia has given top priority to a variety of water harvesting programs to supplement rain fed agriculture of which water harvesting ponds have been widely implemented (Alem, 1999).

Although advanced water harvesting systems has been introduced to Ethiopia, traditional ponds have been widely used for many centuries; some estimates it as early as 560 BC (Fattovich, 1990). Its inhabitants are used to harvesting rainwater for both human and livestock watering in most rural areas, particularly in the arid and semi-arid regions where annual rainfall is less than 600 mm (Alem, 1999). Ponds are simple to construct and can be managed by the community. Approximately 15 to 20 % of the people and over 80% of

Figure 3. Water harvesting ponds using Geomemmbrane and micro basins in Ethio-Somali Region (A) and Amhara region (B-D) Ethiopia, respectively (www.gcca.eu/2014).

the livestock in Ethiopia uses water from either rivers/streams and ponds (Alem, 1999). Currently, the Ethiopian government promotes the use of water harvesting technologies such as geomambrane in different parts of the country especially in the semi-arid and arid regions (Figure 3). However, the adoption and its successfulness have been disappointing due to lack of awareness of development agencies and farmers.

The distribution of these ponds generally, is in the arid and semi-arid areas where the Sahelian climatic condition prevails. Traditional ponds are the major sources of water in the Ethiopian rift valley where ground water is deep and other sources of water are not feasible. These days, the use and promotion of ponds even for livestock watering is increasingly becoming difficult and challenging by the spread of deadly child-hood malaria, and for this reason most NGOs are unable to promote and support pond construction due to environmental constraints (Mekonnen and Haile, 2010).

Roof water harvesting

Roof water harvesting is a system of collecting rainfall water from the roof of a building and storing it in some storage facilities for future use when there is shortage of water (Haile and Merga, 2002). Large scale and modern water supply schemes in rural Ethiopia remains a challenge owing to the unique and rugged terrain and the scattered settlement pattern of the rural people. One technique that appeals today that can be of significant importance in the development of the subsector is roof water harvesting at household level. This technique is so important in the rural highland areas where the terrain is rugged and the villages and hamlets are scattered. In such areas, it is difficult to think that communities can be served by centralized water supply schemes, at least it is expensive. Other sources require long walk and time for women and children to fetch water (Alem, 1999).

The roof water harvesting in Ethiopia has the advantage of being low cost, relatively simple in design (household technology), less laborious and time saving (Alem, 1999). It provides adequate water during the rainy season, a period when the rural people are busy with the farm activities and when there is shortage of labor. They are more appropriate in areas where there are no rivers, ground water sources and where rainwater is the only feasible means of water supply. The emergence of this technique these days is due to the increasing shortage of water from the conventional sources, shallow wells, perennial springs, rivers/streams. In earlier times, roof water harvesting practices were confined to urban areas only. However, its use in the rural areas are increasingly becoming important these days as more people in the rural areas are having corrugated roof houses.

EFFECTS OF RAINWATER HARVESTING

Soil moisture

Globally, the total volumes of water stored within the soil are huge, but at any given locality they are relatively small and quickly depleted through evapotranspiration. Because of this, in recent decades there has been increased interest in various *in-situ* rainwater management techniques that enhance infiltration and water retention in the soil profile (World Bank, 2006). It is important to note that RWH can be used to rehabilitate degraded land and retain moisture (FAO, 2001). Water harvesting retains moisture in-situ, through structures that reduce runoff from fields and hold water long enough to allow infiltration (Rockström, 2002). Improved in-field water harvesting can increase the time required for crop moisture stress to set in and thus can result in improved crop yields and have resulted in positive effects on soil fertility, moisture conservation and agricultural productivity (Alemu and Kidane, 2014, Kidane, 2014, Kidane et al., 2012).

These are techniques for improving the soil moisture by enhancing infiltration and reducing runoff and evaporation (Ngigi et al., 2005; SEI, 2009). An important potential of in -situ water harvesting is to limit nutrient leakage from the fields by controlling soil erosion (Herweg and Ludi, 1999; Gebremichael et al., 2005; McHugh et al., 2007; Gebreegziabher et al., 2009). Water can be stored in artificial constructions (e.g. water tanks, drums, jars, jerry cans, cisterns), in surface reservoirs (ponds, dug-outs, artificial reservoirs) and in the sub-surface as soil moisture or groundwater. Soil moisture conservation (recharge of shallow aquifers) is the key to high productivity (Herweg and Ludi, 1999). Hence, RWH is efficient in increasing the soil moisture for crops in water scarce areas (Ngigi, 2003). In addition, in the dry lands of the Tigray Regional State in northern Ethiopia, the regional government and the general population have been making efforts to control the degradation of natural resources since 1991. As a result, many soil and water conservation measures have been initiated, particularly for soil erosion control, including the construction of stone bunds to conserve *in-situ* soil moisture (Alemu and Kidane, 2014).

Surface runoff

During rainy season, rainwater is collected in large storage tanks which also help in reducing floods in some low lying areas. Apart from this, it also helps in reducing soil erosion and contamination of surface water with pesticides and fertilizers from rainwater run-off which results in cleaner lakes and ponds (Rockström, 2002). The main characteristic of flood water harvesting is a turbulent channel of water flow harvested either by diversion or spreading within a channel bed/valley floor where the runoff is stored in soil profile (Critchley and Siegert, 1991). Harvesting runoff water for supplemental irrigation is a risk-averting strategy, pre-empting situations where crops have to depend on rainfall whose variability is high both in distribution and amounts. By using underground spherical tanks in cascades and having a combined capacity of 60 m^3, seasonal water for supplemental irrigation for an area of about 400 m^2 is guaranteed (Mtisi and Nicol, 2013). Indirect benefits of RWH in terms of reduced incidence of downstream flood damage had been noted (Johnson et al., 2001; Gleick, 2002).

Run-off harvesting from a catchment using channels or diversion systems and storing it in a surface reservoir-water pans/ponds (Rockstrom, 2000) has shown that, the yields and reliability of agricultural production can be significantly improved with water harvesting. In this system, surface runoff from small catchments or adjacent road runoff is collected and stored in manually and/or mechanically dug farm ponds. Although, this technique requires relatively high investment costs compared to *in-situ* systems. Evaluation of RWH in a surface reservoir in the four Great Horn of African (GHA) countries (Ethiopia, Kenya, Tanzania and Uganda) revealed that, it was slowly being adopted with high degree of success (Kiggundu, 2002). In northern Shewa of Ethiopia, runoff from farm lands is stored down streams in large pits for later use, for irrigating plants using watering cans (Alem, 1999).

Ground water recharge

Methods for increasing groundwater recharge include pumping surface water directly into an aquifer and/or enhancing infiltration by spreading water in infiltration basins (World Bank, 2006). There are two main techniques of rain water harvesting namely storage of runoff on surface for future use and recharge to groundwater and shallow aquifer (MacDonald and Davies, 2000). Water harvesting can also have a positive impact on soil conservation, erosion prevention, groundwater replenishment and the restoration of ecosystems. Options for increasing groundwater recharge include constructing small dams or bunds, terracing, contour trenching, sub-surface dams and planting trees or planting vetiver grass (Gebremichael et al., 2005).

Soil storage has more advantage as it does not require lifting and water application to the root of each plant. It also contributes to the recharge of the ground water (Oweis et al., 2001; William et al., 1992). By deep ponding, moisture deficit for crops such as cereal grains, which are more needed by the farmers for their food security, is tackled. Soil storage has more traditional base and lower cost so that its adoption rate can be faster compared to deep ponding. Therefore, more emphasis needs to be given in the future (Desta, 2005). The same trend was also observed for available soil water for rain-fed crops. Water availability from rehabilitated gullies using check dams was the main source of surface irrigation water, which was supplemented by shallow and deep ground water wells. Groundwater levels in the wells increased up to 2.5 m, while the irrigated areas are increased and the number of hand-dug wells also significantly increased. Newly emerging springs and irrigated fields as well as increasing crop diversity and yields were some of the indicators of the improved water resources and supply (Mekonen and Brhane, 2011).

Yield increment

One of the reasons for low crop production in semi-arid areas is marginal and erratic rainfall exacerbated by high runoff and evapotranspiration losses. The in-field RWH techniques has been shown to improve the yield of maize and sun flower on some benchmark ecotypes in South Africa (Henslley et al., 2000). RWH can be used to re-establish vegetation cover to improve crop growth in order to alleviate poverty and enhance food security. Successful interventions in rain-fed

Figure 4. RWH pond picture to prove the point that RWH facility can promote yields (Lemma, 2005 (B); Woldearegay, 2012 (A)).

agriculture in Somali Region Ethiopia have transformed the livelihoods of many poor farmers (FAO, 2001). Improved RWH may result in improved crop yields, food security and livelihood among households (Rockström, 2002). Supplementary irrigation increased crop yields by 20%. With RWH, farmers have diversified to include horticultural cash crops and the keeping of dairy animals. For instance US$ 735 (per ha) compared to US$ 146 normally is now earned from rain-fed maize (Mtisi and Nicol, 2013). This has contributed to food security, better nutrition and family income.

Most of the future growth in crop production in developing countries is likely to come from intensification, with irrigation playing an increasingly strategic role (Postel, 2003; IAASTD, 2009; FAO, 2011). Access to irrigation water is critical to raising and stabilizing crop production (Postel, 2003). Irrigation has direct benefits in terms of production and income. However, there have also been associated impacts whose costs may at times outweigh the benefits of production (Johnson et al., 2001; Gleick, 2002). One of the promising breakthroughs for upgrading rain-fed agriculture in the semi-arid lands remains on how efficiently small scale farmers can utilize practices such as RWH. It is viable in areas with annual rainfall as low as 300 mm (Ngigi, 2003). Besides increased yields, Ngigi et al. (2005) reports that RHM is also aimed at stabilizing variations in crop yields and ensuring food security. Lemma (2005) noted that the survival rate of trees in the irrigated areas of home gardens (84%) had shown a better result when compared to home gardens without pond. This implied that the availability of supplementary irrigation in the backyard has significant contribution on the survival of trees (fruit trees) in the home garden. For instance, about 75% additional yield could be obtained for green pepper and onion crops as compared to tomato crop, which had up to 83% yield increment obtained by using the effective water application method (Lemma, 2005) (Figure 4).

According to the study by Rämi (2003) in Tigray region, Wukro district boasts 30 ponds, mainly clay and plastic-lined serving a total of 80 households. Small gardens with peppers, tomatoes, maize and root crops, which were planted during the rainy season and freshly planted fruit and coffee trees were found around most of the ponds. In the Amhara region (north-western part of the country) from the total completed water harvesting structures, reaching 242,000 in number, over 42,000 have started production, as a result 21,194 ha of land is under irrigation and 148, 244 farm households are benefiting. Of this 14%, 21,194 are women headed households. Irrigated area in the region is primarily aggregated from shallow well, river diversion and spring development (Desta, 2005). In addition to that, according to the participatory rural appraisal (PRA) findings, sorghum yield in Kobo District, Northern Ethiopia doubled with availability of flood water. Similarly, pepper yields increase by up to 400% with application of floodwater (Alamerew et al., 2002).

In Oromia region (central, eastern and western part of the country) the total irrigated land is 65,508 ha, when the plan was 68,565 ha (95.5% achievement). By this 343,953 (92%) households have become beneficiaries. Again 379 ha traditional irrigation through river diversion is under establishment on top of 31,311 ha that already exists. Of the total planned 216,290 ponds, 75% is in food insecure districts and the remaining 25% is in non-food insecure districts. The stored water apart from drinking and crop production, have been used for selling, making mud for house construction, making soil blocks and raising seedlings at the nursery (Desta, 2005). RWH has the potential to increase the productivity of cropland by increasing the yields and by reducing the risk of crop failure (Tesfay, 2011). Alemie et al. (2005) also reported 20,000, 10,000 and 12,000 kg/ha yield of tomato, pepper and onion respectively, due to efficient use of RWH technologies in Wukro and Mehoni Destricts, northern Ethiopia. The farmers managed RWH structures to increase the farm household's agricultural yield by improving the availability of water during the dry spell periods (Amha, 2006).

CHALLENGES AND OPPORTUNITIES OF RAINWATER HARVESTING

The main challenges of adopting RWH technologies are that much of the harvested water is lost through seepage and thus it is not sufficient and it is very labor intensive to irrigate the whole fields by pumping the water manually from ponds and applying directly to the crops (Moges, 2009). The other challenge is to find ways of selecting and promoting appropriate RWH interventions that are well matched to the site-specific, biophysical and socio-economic circumstances (Gowing et al., 2003). The use of on-farm storage reservoirs faces evaporation and seepage losses and silting (Thome, 2005). It is important to minimize the adverse effects of these problems in the design of a surface-water storage facility. Silting may be minimized by arresting the silt and sand on the catchment area itself, mainly through controlling catchment erosion but also by installing silt-traps (Thome, 2005).

Rainfall distribution is another natural challenge of RHM systems. During extreme drought years, very little can be done to bridge a dry spell occurring during the vegetative crop growth stage, if no runoff producing events occur during early growth stages (Rockström, 2000). Despite significant efforts by the Government of Ethiopia and other stakeholders, improving agricultural water management from RWH is hampered by constraints in policies, institutions, technologies, capacity, infrastructure, and markets. Addressing these constraints is vital to achieve sustainable growth and accelerated development of the sector in Ethiopia (Awulachew, 2010). The challenge for agriculture in Sub-Saharan Africa is the variability in rainfall, characterized by high intensity storms, and high frequency of dry spells and droughts. RWH systems can turn these inherent challenges into opportunities.

Where there is no surface water, or where groundwater is inaccessible due to hard ground conditions, or where it is being over exploited, recharge alternatives should be implemented. In a very simplified sense, if a society is threatened or subjected to water scarcity it should respond by attempting to get and store more water using simple, inexpensive and traditional concepts of rain water harvesting technologies (Mupenzi et al., 2011). RWH systems operate on different scales (plot, field and catchment) to modify the water balance in order to increase the rainfall use efficiency (Gowing et al., 2003). RWH is one option to irrigate and produce high value crops to reduce poverty and food insecurity. The assumption is that producing high value crops enables farmers to get returns from selling the product and thus increasing their ability of generating income. For example, in Aleaku Gubantaboke Kebele (Bulbula District) some farmers ability of generating

income. For example, in Aleaku Gubantaboke Kebele (Bulbula District) some farmers have clearly confirmed the importance of RWH as best option for pepper (seedling) production. Realizing the importance of pepper in terms of cash returns, some farmers have already shifted from maize, teff and wheat to pepper and they reported a net return of US$502.26 per quarter hectare (Moges, 2009).

The Ethiopian highland has a large potential for RWH implementation on open woodland land use types to improve the water availability for livestock which has a nexus effect on water productivity. Other benefits of rain-water harvesting are that it makes water available at the point of consumption and reduces the need to pump or haul it over long distances thus saves time and human labour. The experience of RWH activities indicates that they could be used as catalysts for development in a bid to alleviating poverty and promoting socio-economic well-being of the rural people (Mbugua and Nissen-Petersen, 1995).

CONCLUSION AND RECOMMENDATION

RWH has been implemented in many parts of Ethiopia since the time of Aksumite Civilization (560 BC) to the present. However, the adoption of the technology had been disappointing due to many challenges. The sustainability of RWH techniques largely depends on the livelihood improvements of farmers, especially those who practice traditional rain fed agriculture. The adopted RWH techniques had enhanced soil moisture, recharge ground water, reduced runoff and increased yield which in turn preserve the ecosystem.

The benefits of RWH technologies extend beyond rain fed farming to the whole ecological system. Therefore, external catchment based RWH had eminent potential of mitigating rainfall-related crop production risks in the arid and semi-arid regions in Ethiopia. Site specific well-managed RHM therefore, as well as irrigation application are keys in helping Ethiopia overcome major challenges including population pressure; food insecurity; soil and land degradation; high climate variability and low agricultural productivity.

Conflict of Interest

The authors have not declared any conflict of interest.

ACKNOWLEDGEMENTS

We acknowledge all the authors of the scientific papers used in this article. We are also indebted to our colleagues of the department who helped us during the course of the study.

REFERENCES

Akalu TF, Adgo ET (2010). Water harvesting with geo-membrane lined ponds: Impacts on household incomes and rural livelihoods in Minjar

Shenkora district of Ethiopia. In Mati, B.M. Agricultural water management interventions delivers returns on investment in Africa. A compendium of 18 case studies from six countries in eastern and southern Africa. VDM Verlag, pp. 42-52.
Alamerew E, Fentaw B, Ali S (2002). Traditional rainwater harvesting systems for food production: the case of kobo Wereda, Northern Ethiopia.
Alem G (1999). Rainwater harvesting in Ethiopia: An overview. Integrated development for water supply and sanitation. 25th WEDC Conference. Addis Ababa, Ethiopia. pp. 387- 390.
Alemie A, Gebrie A, Teka D (2005). Pond water productivity under the present use in Tigray region, Northern Ethiopia. MU-DCG project Technical Report Number 4, Mekelle University, Mekelle.
Alemu B, Kidane D (2014). The Implication of Integrated Watershed Management for Rehabilitation of Degraded Lands: Case Study of Ethiopian Highlands. J. Agric. Biodivers. Res. 3(6):78-90.
Amha R (2006). Impact Assessment of Rainwater Harvesting Ponds: The Case of Alaba Woreda, Ethiopia. MSc Thesis, Addis Ababa University.
Araya T, Cornelis WM, Nyssen J, Govaerts B, Gebregziabher T, Oicha T, Getnet F, Raes D, Haile M, Saire KD, Deckers J (2011). Effects of conservation agriculture on runoff, soil loss and crop yield under rainfed conditions in Tigray, Northern Ethiopia. Soil Use and Manage. 27:404-414.
Awulachew SB (2010). Irrigation potential in Ethiopia: Constraints and opportunities for enhancing the system. International Water Management Institute (IWMI).
Awulachew SB, Merrey DJ (2007). Assessment of Small Scale Irrigation and Water Harvesting in Ethiopian Agricultural Development. International Water Management Institute (IWMI).
Awulachew SB, Merrey DJ, Kamara AB, Van Koppen B, Penning de Vries F, Boelee E, Makombe G (2005). Experiences and opportunities for promoting small–scale/micro irrigation and rainwater harvesting for food security in Ethiopia. Colombo, Sri Lanka: IWMI. v. (Working paper 98). P. 86.
Bewket W (2009). Rainwater harvesting as a livelihood strategy in the drought- prone areas of the Amhara region of Ethiopia. Organization for Social Science Research in Eastern and Southern Africa (OSSREA).
Birru O, Quraishi Sh, Bedadi B (2012). Effects of straw mulch and farmyard manure on runoff, erosion, *in-situ* water conservation, and yield and yield components of wheat at the highlands of Bale, south eastern Ethiopia. African Journal of Agricultural Research 7(44):5879-5886.
Bossio D, Critchley W, Geheb K, van Lynden G, Mati B (2007). Conserving landprotecting water. In: Water for Food, Water for Life: A Comprehensive Assessment of Water Management in Agriculture, London: Earth scan and Colombo: International Water Management Institute, pp. 551-583.
Critchley W, Siegert K (1991). A Manual for the Design and Construction of Water Harvesting Schemes for Plant Production. FAO Corporate Document Repository. http://www.fao.org/docrep/U3160E/u3160e00.htm#Contents.
De Fraiture C, Wichelns D (2010). Satisfying future water demands for agriculture. Ag. Wat. Manag. J. 97:502-511.
Derib SD, Assefa T, Berhanu B, Zeleke G (2009). Impacts of micro-basin water harvesting structures in improving vegetative cover in degraded hillslope areas of north-east Ethiopia. The Rangeland J. 31(2):259-265.
Desta L (2003). Comparative advantage of soil storage (runoff farming) for crop production over structural storage in the current effort of rainwater harvesting in Ethiopia. Proceeding of 'Rainwater Harvesting Stakeholders' Workshop. Addis Ababa, 27th May, 2003.
Desta L (2005). Concepts of Rainwater Harvesting and its Role in Food Security – the Ethiopian Experience. Ministry of Agriculture and Rural Development, Department of Water Harvesting, Small Scale Irrigation and Rural Infrastructure Development. P. 5.
FAO (2001). Water Harvesting in Western and Central Africa. RAF/Publication, FAO, Rome, Italy.
FAO (2011). The State of the World's Land and Water Resources for Food and Agriculture Organization. Food and Agriculture Organization of the United Nations, Rome.
Fattovich R (1990). Remarks on the PreAxumite Period in Northern Ethiopia. J. Ethiopian Studies 23:1-33.
Finkle M, Sergerros M (1995). Water Harvesting. Proceedings of the SADC ELMS practical Workshop held in Windhock and Okakarara, Namibia, 10-28 May, 1993. Report series No. 33.
Fischer T, Turner N, Angus J, McIntyre L, Robertson M, Borrell A, Lloyd D (2004). "New Directions for a Diverse Planet: Proceedings of the 4th International Crop Science Congress," Brisbane, Australia, 26 September–1 October 2004.
Gebreegziabher T, Nyssen J, Govaerts B, Getnet F, Behailu M, Haile M, Deckers J (2009). Contour furrows for in situ soil and water conservation, Tigray, Northern Ethiopia. Soil, Tillage Res. 103:257-264.
Gebremichael D, Nyssen J, Poesen J, Deckers J, Haile M, Govers G, Moeyersons J (2005). Effectiveness of stone bunds in controlling soil erosion on cropland in the Tigray Highlands, northern Ethiopia. Soil Use Manag. 21:287-297.
Georgis KW (2000). Conservation tillage research findings: The Ethiopian experience. In Conservation tillage for dryland farming. Technological options and experiences in Eastern and Southern Africa. eds.E.K. Biamah, J. Rockstrom and G.E. Okwach. RELMA, Workshop Report No. 3:148-151.
Georgis KW (2002). Water harvesting, the key to successful crop production and resource management in water stress environments: Eastern African experience. In Workshop on the experiences of water harvesting in drylands of Ethiopia: Principals and practices, eds. M. Haile and S.N. Merga. Dryland Coordination Group. DCG Report No. 19:55-64.
Getaneh M, Tsigae A (2013). Comparative analysis of lining materials for reduction of seepage in water harvesting structures, Adet, Ethiopia. Int. J. Dev. Sust. 2(2):1623-1635.
Gleick PH (2002). Soft water paths. Nature 418: 373.
Gowing JW, Young MDB, Hatibu N, Mahoo HF, Rwehumbiza F, Mzirai OB (2003). Developing Improved Dry land Cropping Systems for Maize in Semi-Arid Tanzania. Part II. Use of a Model to Extrapolate and Add Value to Experimental Results. Expl. Agric. 39: 293-306.
Habtamu G (1999). Rainwater harvesting concepts and issues. Paper presented at the Founding Conference of the Ethiopia Rainwater Harvesting Association (ERHA). December 17, 1999. Addis Ababa.
Haile M, Merga SN (2002). Workshop on The Experiences of Water Harvesting in the Drylands of Ethiopia: Principles and Practices December 28-30 2001, DCG Report No. 19, Mekelle, Ethiopia.
Haile T (2007). Assessment of Institutional Setup and Effect of Household Level Water Harvesting in Ensuring Sustainable Livelihood. Dry lands Coordination Group Report No. 52, Oslo, Norway.
Hailemicael A (2011). Indigenous talents of Konso people to cope up climate change susceptibility, Ethiopia. Arba Minch University, Ethiopia.
Hatibu N, Mahoo H (1999). Rainwater harvesting technologies for agricultural production: A case for Dodoma, Tanzania.
Henslley M, Botha JJ, Van Staden PP, Du Tota A (2000). Optimizing Rainfall Use Efficiency For developing Farmers with Limited Acccess to Irrigation Water, WRC report No 878/1/100. Water Research Commission, Pretoria, South Africa.
Herweg K, Ludi E (1999). The performance of selected soil and water conservation measures-case studies from Ethiopia and Eritrea. Catena pp. 99-114.
Hudson N (1987). Soil and Water Conservation in semi-arid areas. FAO Soils Bulletin No. 57. FAO, Rome. P. 172.
IAASTD (2009). International Assessment of Agricultural Knowledge, Science and Technology for Development: Summary for Decision Makers of the Sub-Saharan Africa Report. Assessment. International Assessment of Agricultural Knowledge, Science and Technology for Development, Johannesburg. P. 24.
Ito M, Matsumoto T, Quinones M (2007). Conservation tillage practice in sub-Saharan Africa: The experience of Sasakawa Global 2000. Crop Prot. J. 26:417-423.
Johnson N, Revenga C, Echeverria J (2001). Managing water for people and nature. Science 292:1071-1072.
Kidane D, Temesgen M, Abdelkadir A (2012). Effect of Winged Subsoiler and Traditional Tillage Integrated with *Fanya Juu* on

Selected Soil Physico-Chemical and Soil Water Properties in the Northwestern Highlands of Ethiopia. E. Afr. J. Sci. 6(2):21-32.

Kidane DA (2014). Conservation tillage implementation under rainfed agriculture: Implication for soil fertility, green water management, soil loss and grain yield in the Ethiopian Highlands. Int. J. Agric. Sci.4(9):268-280.

Kiggundu N (2002). Evaluation of rainwater harvesting systems in Rakai and Mbarara Districts, Uganda. GHARP case study report. Greater Horn of Africa Rainwater Partnership (GHARP), Kenya Rainwater Association, Nairobi, Kenya.

Lal R (2001). World cropland soils as source or sink for atmospheric carbon. Adv. Agron. 71:145-191.

Leisa (1998). Challenging water scarcity [Editorial]. ILEIA Newslett. Low Ext. Input Sustain. Agric. 14:1.

Lemma G (2005). The role of Household Ponds on the expansion of Homegardens in Tigray, Ethiopia. MSc Thesis. Department of Tropical Land Resource Management, Mekelle University, Ethiopia.

Li X, Gong J, Wei X (2000). *In-situ* rainwater harvesting and gravel mulch combination for corn production in the dry semi-arid region of China. J. Arid Env. 46:371-382.

Li X, Shi P, Sun Y, Tang J, Yang Z (2006). Influence of various in situ rainwater harvesting methods on soil moisture and growth of *Tamarix ramosissima* in the semiarid loess region of China. J. Hydrol. 328:338-346.

Liniger H, Schwilch G, Hurni H (2006). Soil and Water Conservation, Global Change, and the Millennium Development Goals –An Evaluation by WOCAT. 14th International Soil Conservation Organization Conference. Water Management and Soil Conservation in Semi-Arid Environments. Marrakech, Morocco, May 14-19, 2006 (ISCO 2006).

MacDonald AM, Davies JK (2000). A brief review of groundwater for rural water supply in sub-Saharan Africa. British Geological Survey Technical Report WC/00/33pp. BGS Keyworth UK.

Mbugua JK, Nissen-Petersen E (1995). Rain water an underutilized resource-SIDA, Nairobi, Kenya.

McHugh OV, Steenhuis TS, Fernandes ECM (2007). Performance of in situ rainwater conservation tillage techniques on dry spell mitigation and erosion control in the drought-prone North Wello zone of the Ethiopian highlands. Soil and Tillage Research 97: 19-36.

Mekonen K, Brhane G (2011). Impact assessment of soil and water conservation measures at Medego watershed in Tigray, northern Ethiopia. Maejo Int. J. Sci. Technol. 5(03): 312-330.

Mekonnen Y, Haile M (2010). The Potential of in situRain Water Harvesting for Water Resources Conservation on Malaria Transmission in Tigray, Northern Ethiopia. MEJS 2(2):49-63.

Moges G (2009). Identification of potential rain water harvesting areas in the Central Rift Valley of Ethiopia using a GIS based approach. MSc Thesis, Wageningen University, Netherland.

Mtisi S, Nicol A (2013). Good Practices in Water Development for Dryland Regions of Ethiopia, Kenya, and Uganda. P. 99.

Mupenzi JP, Lahai L, Tafadzwa LM, Lie Y, Jiwen G (2011). Rain water harvesting and Artificial Recharge in Africa. Nature Sci. 9(4): 41-45.

Ngigi SN, Savenije HHG, Thome JN, Rockstro J, de Vriesd FWTP (2005). Agrohydrological evaluation of on-farm rainwater storage systems for supplemental irrigation in Laikipia district, Kenya. Agricultural Water Management 73:21-41.

Ngigi SN (2003). What is the limit of up-scaling rainwater harvesting in a river basin? Physics and Chemistry of the Earth 28:943-956.

Ngigi SN (2009). Climate Change Adaptation Strategies: Water Resources Management Options for Smallholder Farming Systems in Sub-Saharan Africa. The MDG Centre for East and Southern Africa, The Earth Institute at Columbia University, New York, P. 189.

Oloro V, Steenhuis T, Abebe B, Fernandes E (2007). Performance of in situ rainwater conservation tillage techniques on dry spell mitigation and erosion control in the drought-prone North Wello zone of the Ethiopian highlands. Soil Till. Res. J. 97:19-36.

Oweis T, Prinz D, Hachum A (2001). Water Harvesting: Indigenous knowledge for the Future of the Drier Environments. ICARDA, Aleppo, Syria. P. 40. ISBN: 92-9127-116-0.

Pachpute J, Tumbo S, Sally H, Mul M (2009). Sustainability of Rainwater Harvesting Systems in Rural Catchment of Sub-Saharan Africa. Wat. Reso. Manag. 23:2815-2839.

Postel SL (2003). Securing water for people, crops, and ecosystems: New mindset and new priorities. Natural Resources Forum 27:89-98.

Rämi H (2003). Ponds filled with challenges, Water harvesting experiences in Amhara and Tigray. UN OCHA, Ethiopia.

Reij C, Scoones I, Toulin C (1996). Sustaining The Soil: Indigenous Soil and Water Conservation in Africa. Earth Scan, London, UK. P. 228.

Rockström J (2000). Water resources management in smallholder farms in Eastern and Southern Africa: An overview. Physics and Chemistry of the Earth 25(3):278-288.

Rockström J (2001). Green water security for the food makers of tomorrow: Windows of opportunity in drought prone savannahs. Water Sci. Technol. 43(4):71-78.

Rockström J (2002). Potential of Rainwater Harvesting to Reduce Pressure on Freshwater Resources. Dialogue on water, food and environment. International Water Conference, Hanoi, Vietnam, October 14-16.

Rockström J (2003). Water for food and nature in the tropics: Vapour shift in rainfed agriculture. *Phil.* Trans. R. Soc. Land, 358:1997-2009.

Rockstrom J, Barron J, Fox P (2002). Rainwater management for increased productivity among smallholder farmers in drought-prone environments. Physics and Chemistry of the Earth. 27:949-959.

Rockström J, Kaumbutho P, Mwalley J, Temesgen M, Mawenya L, Barron J, Mutua J, Damgaard-Larsen S (2009). Conservation farming strategies in East and Southern Africa: Yields and rain water productivity from on-farm action research. Soil Till. Res. 103(1):23-32.

Stockholm Environment Institute (SEI) (2009). Rainwater harvesting: A lifeline for human well-being, York. P. 69.

Temesgen M, Rockstrom J, Savenije HHG, Hoogmoed WB (2007). Assessment of Strip Tillage Systems for Maize production in semi-arid Ethiopia: Effects on grain yield and water balance. UNESCO-IHE, Wageningen University, The Netherlands.

Temesgen M, Uhlenbrook S, Belay S, van der Zaag P, Mohamed Y, Wenninger J, Savenije HHG (2012). Impact of conservation tillage on the hydrological and agronomic performance of Fanya juus in the upper Blue Nile (Abbay) river basin. Hydrol. Earth Sys. Sci. 16:4725-4735.

Tesfay G (2011). On-farm water harvesting for rainfed agriculture development and food security in Tigray, Northern Ethiopia. Drylands Coordination Group Report No. 61, 07, 2011, Norway.

Thome JN (2005). Evaluation and design of runoff storage systems for supplemental irrigation in Laikipia District, Kenya. MSc Thesis. Department of Environmental and Biosystems Engineering, University of Nairobi, Kenya.

Tian Y, Su D, Li F, Li X (2003). Effect of rainwater harvesting with ridge and furrow on yield of potato in semiarid areas. Field Crop Res. 84: 385-391.

Vancampenhout K, Nyssen J, Gebremichael D, Deckers J, Poesen J, Haile M, Moeyersons J (2005). Stone bunds for soil conservation in the northern Ethiopian highlands: Impacts on soil fertility and crop yield. Soil Till. Res. 90(1-2):1-5.

Vohland K, Barry B (2009). A review of *in-situ* rainwater harvesting (RWH) practices modifying landscape functions in African dry lands. Agric. Ecosys. Environ. 131:19-127.

William C, Chris R, Alain S (1992). Water Harvesting for Plant Production, Volume II: Case Studies and Conclusions for Sub-Saharan Africa. World Bank Technical Paper No 157, Africa Technical Department Series.

World Bank (2006). Ethiopia: Managing Water Resources to Maximize Sustainable Growth. Washington, DC: World Bank.

22

Hydrological assessment and suspended sediment loading of the Chini Lake catchment, Pahang, Malaysia

Sujaul I. M.[1], Ismail B. S.[1], Muhammad Barzani G.[1], Sahibin A. R.[1] and Mohd Ekhwan T.[2]

[1]School of Environment and Natural Resource Sciences, Faculty of Science and Technology, Universiti Kebangsaan Malaysia, 43600 UKM Bangi, Selangor, Malaysia.
[2]School of Social Development and Environmental Studies, Faculty of Social Science and Humanities, Universiti Kebangsaan Malaysia, 43600 UKM Bangi, Selangor, Malaysia.

Hydrological evaluation and sediment loading of the Chini Lake catchment was investigated. Sediment loads and water discharge from seven selected feeder rivers were measured over a period of one year (January to December 2006). Suspended sediment concentration was measured by the standard method. The annual rainfall from 1984 to 2006 at the Chini Lake catchment ranged from 1487.70 mm (1997) to 3071.40 mm (1994), with the annual rainfall for 2006 being 2544.50 mm. Stream flow rate during the sampling periods was relatively slow, ranging from 0.001 to 1.31 m^3s^{-1} (or an average of 0.21 m^3s^{-1}). The highest and lowest stream flow discharges were recorded from the Gumum River and Cenahan River subcatchment. The amount of sediment load ranged from 0.49 to 166.02 $kg/km^2/day$ (or an average of 30.57 $kg/km^2/day$) in the study area. The highest sediment load was recorded in the wet season and the lowest in the dry season. Human activities have significantly affected the hydrological functions and deposition of sediment and thus influenced the variation in sediment loading.

Key words: Hydrological characteristics, feeder river, stream flow, sediment loading, Chini Lake.

INTRODUCTION

Precipitation is the main cause of variability in the water balance over space and time. Changes in precipitation have very important implications on hydrology and water resources. Hydrological variability over time in a catchment is influenced by variations in precipitation over diurnal, temperature, seasonal, soil covering and, annual and decadal time scales (Viesmann and Lewis, 2003). Sediment is recognized as a significant water pollutant. In addition, sediment transports chemicals, which compounds its effect on water quality. Nutrients and suspended sediment loads into rivers are increased by timber harvesting, changes in land use and river channelization (Slaymaker, 1982; Nakamura et al., 1997; 2004). The sediment yield is highly dependent on soil erosion and sediment transport stages. Large volumes of nutrient and suspended sediment inputs have caused degradation of aquatic habitats due to declining water quality (Nakamura et al., 2004; Takamura et al., 2003).

The hydrology of a lake is responsible for the stability of its ecosystem. The water level of the Chini Lake depends on the feeder rivers discharge that comes from the surrounding area. The Chini Lake formation is believed to be similar to that of the Bera Lake (Furtado and Mori, 1982). The average altitude of the lake area is 76 m below sea level. The main water sources contributing to the main lake body comprise: firstly, the natural tributaries of the Chini River which are the Paya Merapuk River to the southwest, the Melai River to the south, the Jemberau and Kura kura Rivers to the southeast, the Cenahan and Gumum Rivers to the north east and the Datang River to the north; secondly the backflow from the Pahang River that occurs during the flood season. Based on the topographical map, most of the lowland within the study area has been converted into agricultural land including rubber and oil palm plantations and mixed crops. Logging activities are still going on in the study area. Most of the sediment problems in and around the Chini Lake are associated with anthropogenic effects such as forestry, agriculture and mining activities. The main objective of the study was to evaluate the hydrological factors and identify the stream flow and discharge pattern of the feeder rivers for the dry and wet seasons, as well as assess the sediment loading into the lake.

MATERIALS AND METHODS

Study area

The Chini Lake (Tasik Chini) is located in the southeastern region of Pahang, Malaysia (Figure 1). It is located approximately 100 km from Kuantan, the capital of Pahang. The lake system lies between 3°22´30˝ to 3°28´00˝N and 102° 52´40˝ to 102°58´10˝E and comprises 12 open water bodies (that are referred to as "laut" by the local people) and is linked to the Pahang River by the Chini River. A few communities of the indigenous Jakun tribe live around the lake. The Chini Lake is the second largest natural fresh-water lake in Peninsula Malaysia, encompassing 202 ha of open water, as well as 700 ha of Riparian Peat and Lowland Dipterocarp forest (Wetlands International Asia Pacific, 1998). The Chini Lake is surrounded by various vegetated low hills and undulating land which constitute the watershed of the region. There are three hilly areas surrounding the lake: (1) Bt. Ketaya (209 m) located southeast; (2) Bt. Tebakang (210 m) in the north and (3) Bt. Chini (641 m) located southwest. The Chini Lake catchment is representative of the upstream site of the Pahang River. The study area has humid tropical climate with two monsoon periods, characterized by the following bimodal pattern. The southwest and northeast monsoons bring an annual rainfall of 1488 to 3071 mm. The mean annual rainfall for the past 20 years is 2,500 mm and the temperature range is from 21 to 32°C. Potential evapotranspiration (PE) is between 500 to 1000 mm. The open water area has expanded since 1994, as a result of increased water retention after the construction of a barrage downstream Chini River. The lake drains northwest into the Pahang River via the Chini River, which meanders for 4.8 km before reaching the Pahang River.

Sampling and stream flow measurement

Seven sampling stations were selected for the study based on the seven feeder rivers that flowed into the Chini Lake. These seven stations were established on each feeder river of the lake: the Datang River (Station 1), the Cenahan River (Station 2), the Gumum River (Station 3), the Kura kura River (Station 4), the Melai River (Station 5), the Paya Merapuk River (Station 6) and the Jemberau River (Station 7). The catchment characteristics such as area, velocity, and flow discharge were evaluated for each feeder river. The area of the catchment was measured from the topographical map of the Chini Lake (scale 1: 25,000). The river morphology such as width, depth and flow were measured at each sampling station. The stream flow and river width were measured using a Flow Meter (model FP101) and the Rangefinder (Bushnell Model 20-0001) was used to measure the distance. At each site, water samples were taken from the mid-channel approximately 10 cm below the surface with care being taken not to disturb the bed sediment. Water samples from the main feeder rivers were collected for measurement of the total sediments. The sediment yield was determined based on calculation of the total suspended solids (TSS) which includes inorganic (sand, silt and clay) and organic materials, and total dissolved solids (TDS) in the water samples. The standard laboratory methods (APHA, 1998) were used. The total rainfall during the study period was obtained from the nearby weather station at Chini Dua, while the rainfall data prior to 2006 was obtained from the Meteorological Department at Petaling Jaya, Selangor, Malaysia.

Suspended sediment load measurement

In the current study, suspended sediment load was calculated by the turbidity threshold sampling (TTS) method. The USDA Forest Service Redwood Sciences Laboratory in Arcata, California developed this methodology to improve the accuracy and efficiency of suspended sediment load estimations (Lewis and Ead, 2001). The discharge-weighted suspended sediment load (Qs) in kilograms (kg) per area (km^2) per day ($kg/km^2/day$) for the river cross-section was obtained by multiplying the concentration, C, in kilograms per liter (kgl^{-1}) with the discharge, Q, in liters per day ($lday^{-1}$), followed by division of the drainage area A (km^2). The discharge, drainage area and concentration values are given in Table 2. The equation used is as follows:

$$Qs = (C*Q)/A$$

RESULTS AND DISCUSSION

Subcatchment areas of the Chini Lake

On the basis of morphometrical characteristics, the Chini Lake catchment consists of thirteen subcatchments (Figure 2). The names and areas of the subcatchments are detailed in Table 1. The Gumum River subcatchment is the largest subcatchment (1306.24 ha, or 21.49% of the total study area), the Paya Merapuk River subcatchment is the second largest (20.90% of total catchment area) and the Datang River subcatchment the third largest (7.92% of the study area).

Rainfall

Hydrological analysis was carried out to evaluate the water level characteristics of the water body as well as the drainage systems (Gray, 1970; Ceballas and Schnabel, 1998). The annual rainfall from 1984 to 2006 at the Chini Lake Catchment ranged from 1487.70 mm (1997) to 3071.40 mm (1994) (Figure 3). The average rainfall was 2344.30 or 195.36 mm $month^{-1}$. The number of rainy days in the study area ranged annually from 130 to 197 with an average of 175 or 15 days/month. The highest number of rainy days per year was obtained in 1993 and 1994 (197 and 190 days respectively), while 2005 had the lowest number of rainy days (130 days) (Figure 4). The number of rainy days during 2006 was 134 days. The highest number of rainy days per month (16 days) was recorded during the wet season in 2006 (December), while January and March 2006 recorded the

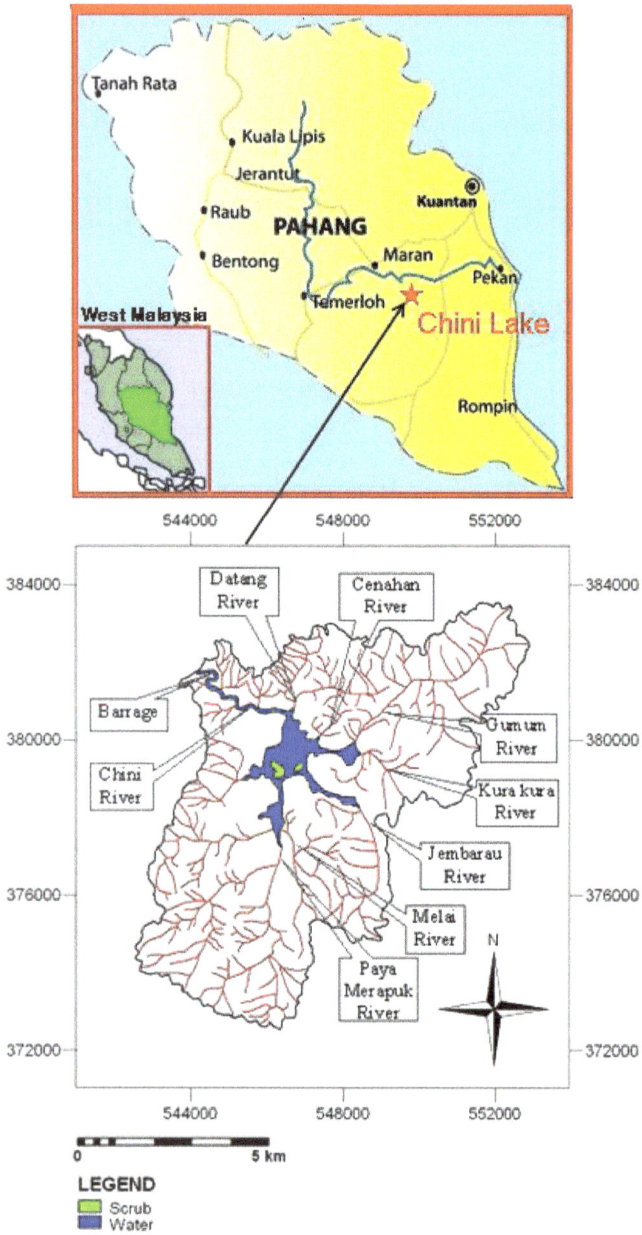

Figure 1. Study area and feeder rivers of the Chini Lake catchment.

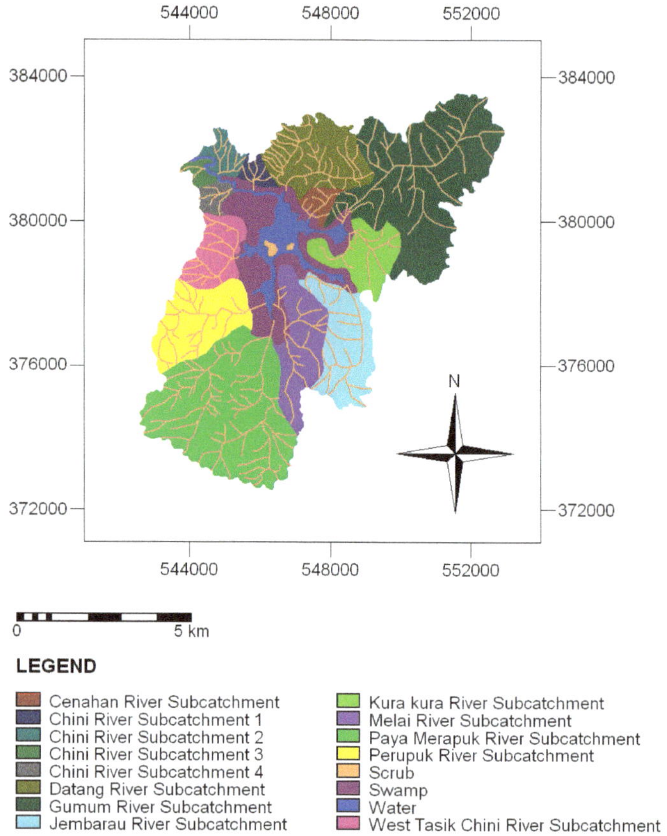

Figure 2. Subcatchment areas of the Chini Lake (DOA, 2006)

Table 1. Area (ha) of subcatchments in Chini Lake catchment area.

River subcatchment (ha)	Area (ha)	Area (%)
Perupuk River subcatchment	450.40	7.41
Paya Merapuk River subcatchment	1270.28	20.90
Melai River subcatchment	395.72	6.51
Jembarau River subcatchment	455.72	7.49
Kura kura River subcatchment	292.12	4.81
Gumum River subcatchment	1306.24	21.49
Cenahan River subcatchment	69.24	1.14
Datang River subcatchment	481.36	7.92
Chini River subcatchment 1	59.08	0.97
Chini River subcatchment 2	92.76	1.53
Chini River subcatchment 3	43.84	0.72
Chini River subcatchment 4	65.08	1.07
West Tasik Chini River subcatchment	247.08	4.06

lowest number of rainy days (7 days) (Figure 6). The highest rainfall recorded was 440.20 mm in December 2006 and the lowest, 76.50 mm in March 2006 (Figure 5). The annual rainfall for 2006 was 2544.50 mm.

Stream flow of the feeder rivers

Stream flow discharge from the seven feeder rivers to the Chini Lake was relatively low, ranging from 0.001 to 1.31 m^3s^{-1} or an average of 0.21 m^3s^{-1}. Daily discharge ranged seven feeder rivers are shown in Figure 7. The stream from 86.40 to 113,184 m^3, averaging 18,144 m^3/daily.

The seasonal (dry and wet) discharge variability of the flow rate of the Gumum River was higher (1.31 m^3s^{-1}) and the Cenahan River lower (0.01 m^3s^{-1}). The Paya Merapuk River was considered the second highest (1.26 m^3s^{-1}). The stream flow of these rivers depended mainly on the rainfall pattern. The fastest stream flow rate was observed during the wet season, especially in September,

Table 2. Suspended sediment load of the selected feeder rivers of the Chini Lake catchment.

Feeder river	Drainage area (km^2)	Water discharge (m^3s^{-1})	Sediment load (ton/km^2/year)
Melai	3.96	0.19	9.10
Paya Merapuk	12.7	0.4	7.58
Chenahan	0.69	0.04	14.91
Jemberau	4.55	0.1	5.77
Kura kura	2.92	0.07	6.53
Gumum	13.06	0.42	16.45
Datang	4.81	0.22	17.74

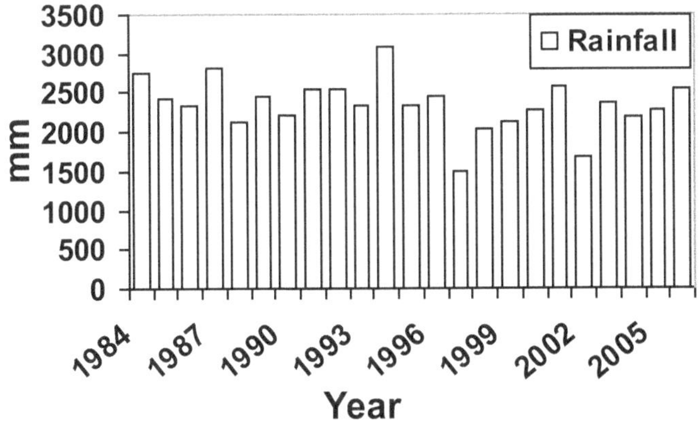

Figure 3. Distribution of rainfall from 1984 to 2006.

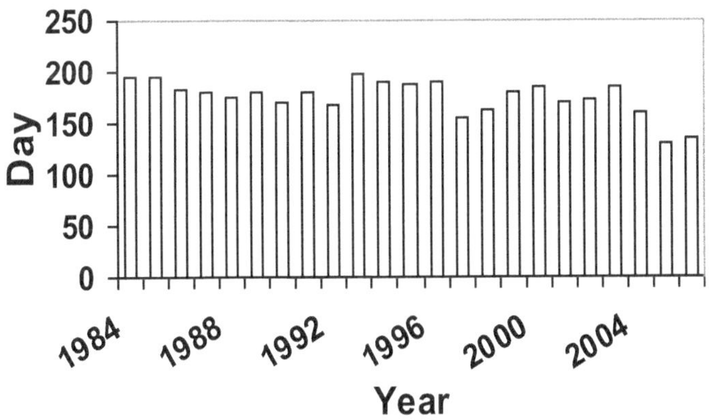

Figure 4. Distribution of the number of rainy days from 1984 to 2006.

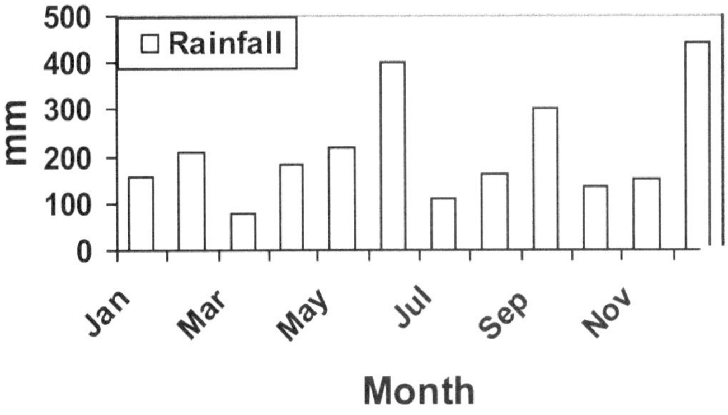

Figure 5. Distribution of rainfall from January to December 2006.

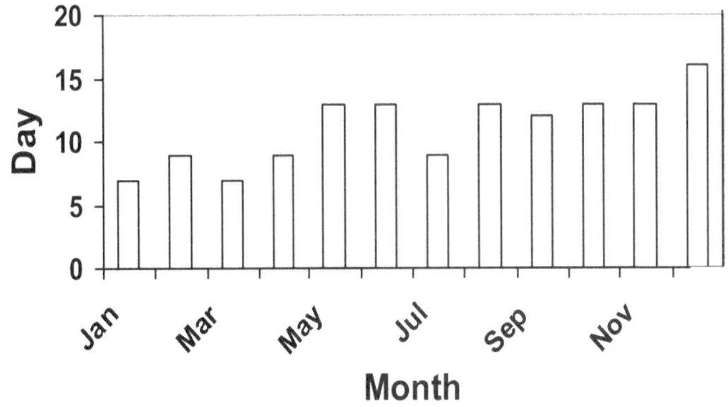

Figure 6. Distribution of the number of rainy days from January to December 2006.

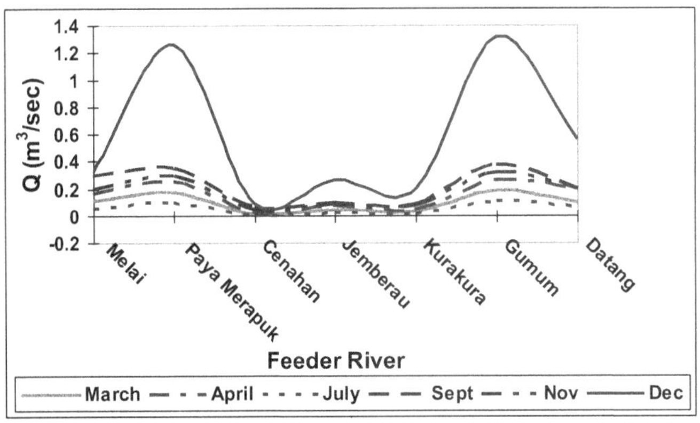

Figure 7. Stream flow distribution of the seven feeder rivers of the Chini Lake catchment.

November and December 2006, while in the dry season, especially during March, April and July 2006 a lower stream flow rate was recorded. Stream flow discharge from the seven subcatchments to the Chini Lake was relatively slow. The highest stream flow discharge was recorded from the Gumum River subcatchment as it was the largest subcatchment (1,306.24 ha) of the Chini Lake.

Lower stream flow rate was recorded at the Cenahan River subcatchment (69.24 ha) as it was a small subcatchment. The results of this study highlight that intense precipitation through the Chini Lake catchment controlled the discharge rate of the seven feeder rivers. The maximum discharge was recorded in the study area due to the monsoonal storm intensity especially from September to December. Discharges of all feeder rivers were low during the month of July due to the low intensity of rainfall.

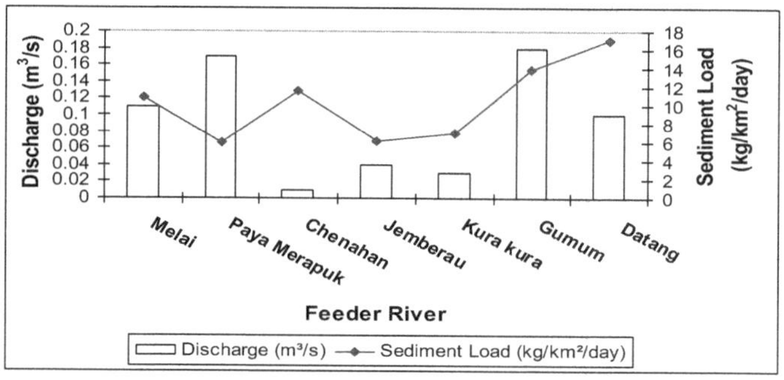

Figure 8. Distribution pattern of sediment loads and discharges of the seven feeder rivers in March 2006.

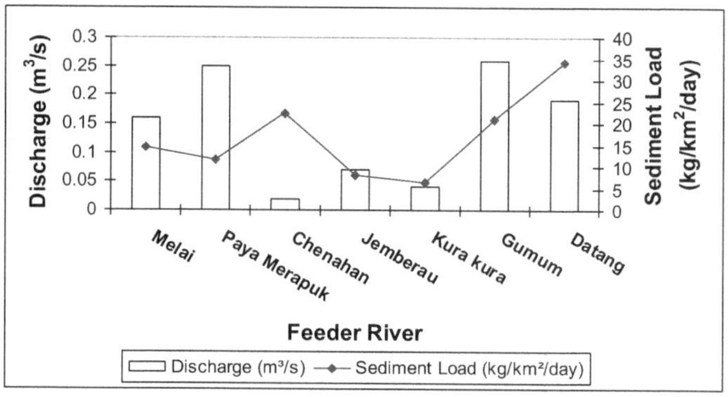

Figure 9. Distribution pattern of sediment loads and discharges of the seven feeder rivers in April 2006.

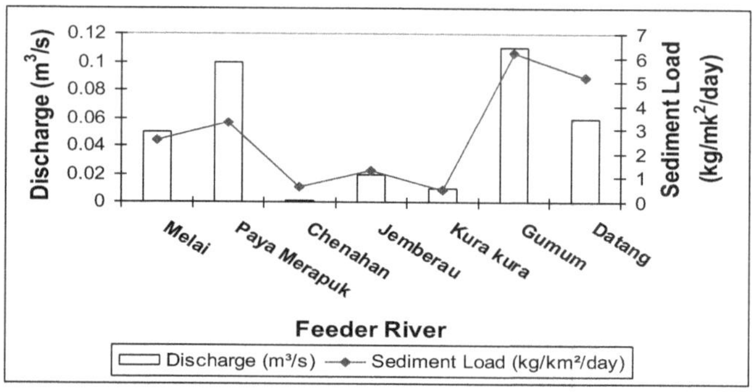

Figure 10. Distribution pattern of sediment loads and discharges of the seven feeder rivers in July 2006.

Sediment loading

The sediment concentrations varied from month to month reflecting the influence of the rainfall regime. Sediment loads varied depending on the discharge, sediment concentrations and the study area. The highest sediment concentration (19.16 mgL^{-1}) was observed in December 2006, the lowest (0.11 mgL^{-1}) in the dry month of July 2006 where water flow was also low. The average sediment load for the seven feeder rivers is shown in Table 2. The amount of sediment load was recorded from the seven feeder rivers, and they ranged from 5.98 to 17.05 $kg/km^2/day$ (or, an average of 10.36 $kg/km^2/day$) in March, 6.42 to 34.10 $kg/km^2/day$ (or, an average of 16.84 $kg/km^2/day$) in April, 0.49 to 6.18 $kg/km^2/day$ (or, an average of 2.81 $kg/km^2/day$) in July, 11.10 to 48.57 $kg/km^2/day$ (or, an average of 28.78 $kg/km^2/day$) in September, 5.53 to 39.22 $kg/km^2/day$ (or, an average of 19.64 $kg/km^2/day$) in November and 57.53 to 166.02 $kg/km^2/day$ (or, an average 104.96 $kg/km^2/day$) in December 2006. The details are in Figures 8 to 13. The mean sediment load in the Chini Lake was 30.56 $kg/km^2/day$ (11.16 $ton/km^2/year$). The calculated sediment loads deposited in the Chini River were: 9.38 $kg/km^2/day$ in March, 13.14 $kg/km^2/day$ in April, 5.28 $kg/km^2/day$ in July, 15.50 $kg/km^2/day$ in September, 13.50 $kg/km^2/day$ in November and 40.96 $kg/km^2/day$ in December 2006 (Figure 14a and 14b). The mean sediment load deposited in the Chini River was 16.29 $kg/km^2/day$ (5.95 $ton/km^2/year$). The highest sediment load was recorded in December and the lowest in July 2006. The highest load in December 2006 was due to rain and probably related to the high flow and concentration of sediment yield. The results also indicated that the total rainfall during December 2006 directly influenced the flow of the feeder rivers into the Chini Lake catchment. Mohd Ekhwan (2005) while carrying out a study at the Bebar River catchment (in Malaysia) found that sediment yield varied with the rainfall and factors such as channel slope, relief and basin size.

The study showed that the Chini Lake suffered high sediment loading from its feeder rivers. The highest suspended sediment discharge recorded was from the Gumum River subcatchment and it ranged from 6.18 to 166.02 $kg/km^2/day$ (or an average of 45.08 $kg/km^2/day$).

The Jemberau River subcatchment had the lowest sediment discharge ranging from 1.33 to 57.53 (or an average of 15.80). The Gumum River subcatchment was located within the settlements and agricultural areas. The highest suspended sediment concentration came from this subcatchment due to its land use pattern. The discharge from the Gumum River subcatchment was higher due to the size of the drainage area of all the feeder rivers. The Gumum River carried the largest amount of sediment into the catchment. The Datang River carried the second largest amount of sediment, discharging more than 41.01 $kg/km^2/day$ into the lake. Logging activities and shifting cultivation were the main land use patterns in the Datang River subcatchment area. The Jemberau River carried a low load of suspended sediment. The Jemberau River subcatchment was located under forested areas thus the

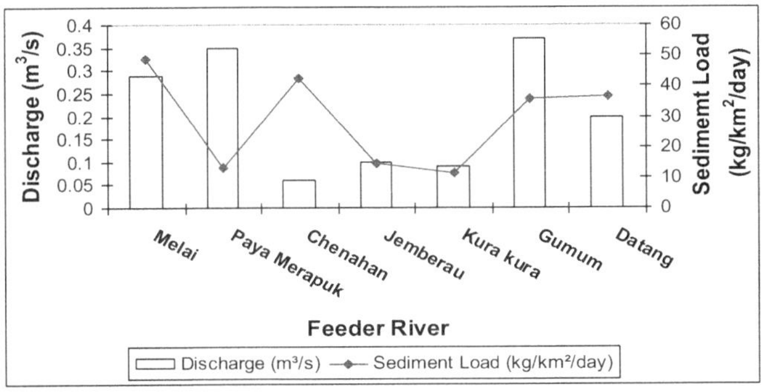

Figure 11. Distribution pattern of sediment loads and discharges of the seven the feeder rivers in September 2006.

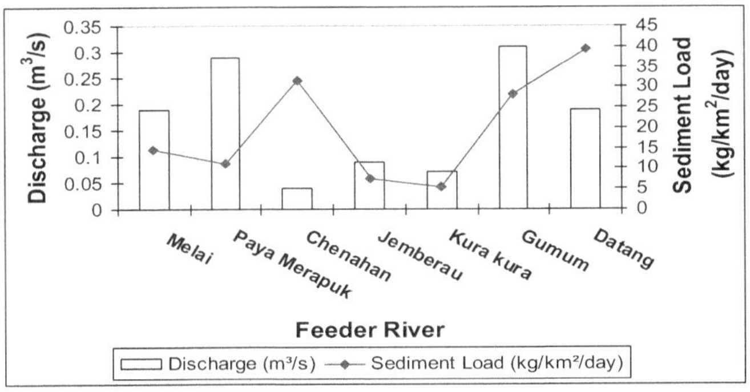

Figure 12. Distribution pattern of sediment loads and discharges of the seven feeder rivers in November 2006.

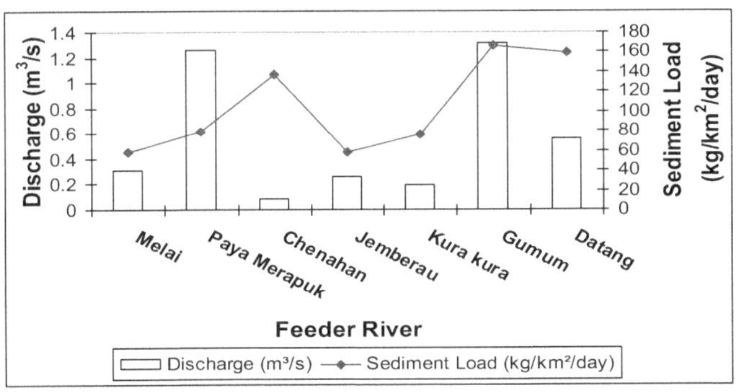

Figure 13. Distribution pattern of sediment loads and discharges of the seven feeder rivers in December 2006.

drainage area and the stream flow was also low. The study indicated that the sediment load varied from month to month reflecting the influence of the rainfall regime. All the feeder rivers had lower sediment loads during the dry season and higher in the wet season. There was a relationship between the water discharge and sediment loads during the study period (Figures 8 to 14). Suspended sediment concentration depended on hydrological behavior during the very event. The hydrological response was influenced by the land type

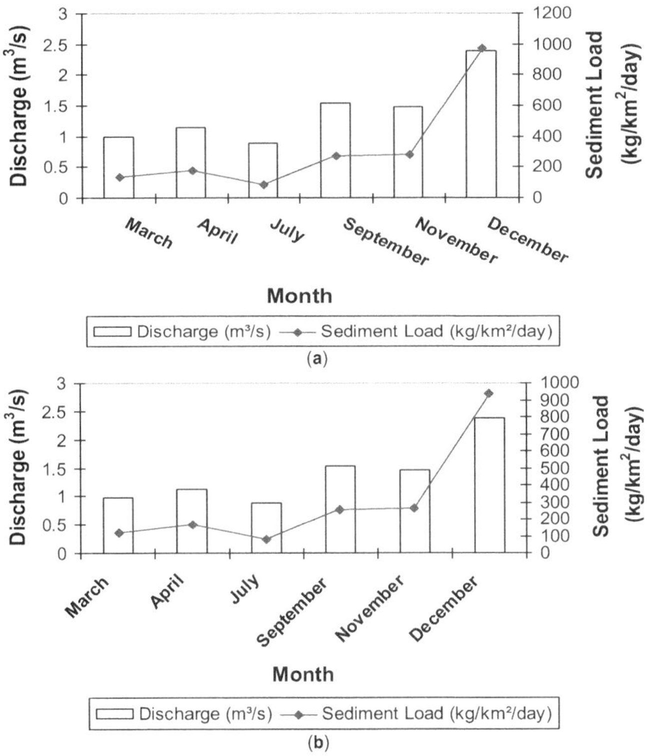

Figure 14. Distribution pattern of sediment loads and discharge variations of the Chini River (a) Up stream and (b) Down stream in 2006.

and study area. The study area was also subjected to intense erosion because of the land relief and the monsoon climate. Aha et al. (2008) stated that high concentration of sediment load was dependant on land use, stream channels, vegetation cover, climatic conditions and degree of channelization.

Conclusion

The Chini Lake is a shallow and small natural lake, which is considered a sensitive ecosystem as it responds to changes from its surroundings. In the study the hydrology and sediment dynamics in the Chini Lake catchment were examined in order to clarify the main factors degrading the lake ecosystem. The hydrology and sediment loading into the Chini Lake was dependant on the hydrometeorological conditions. Stream flow discharge from each feeder river into the Chini Lake was directly related to the rainfall pattern. In the dry season the discharge from the feeder rivers was lower than that in the wet season. A series of rain events in the study area favored erosion and mass movement, which also provided sediment sources for the studied catchment. Erosion and sediment yield from this catchment was higher than those from other forested catchment areas in Malaysia because some small scale disturbances were already present in the catchment plus there was the contribution of sediment from anthropogenic activities like progressive deforestation, cultivation and construction activities which led to soil exposure and degradation. Variability in sediment loading necessitates the need for long-term monitoring of sediment yield. As a long-term measure, environmental policies regulating land-use development and management practices should be formulated and implemented.

ACKNOWLEDGEMENT

This study was conducted and supported by the Ministry of Science and Technology, Malaysia through the IRPA grant, code: 09-02-02-0117-EA294, Zamalah Scheme and OUP fund (Code: UKM-OUP-FST-2008) UKM, Malaysia.

REFERENCES

Aha YS, Nakamura F, Mizugaki S (2008). Hydrology, suspended sediment dynamics and nutrient loading in Lake Takkobu, a degrading lake ecosystem in Kushiro Mire, northern Japan. Environ. Monit. Assess. 145:267-281.

APHA (1998). Standard Methods of Water and Waste Water Analysis (Eds.) 20th Edn. American Public Health Association, Washington.

Ceballas A, Schnabel S (1998). Hydrological behavior of a small catchment in the Dehesa land use system. J. Hydrol. 210:146-160.

DOA (2006). Land use map of Tasik Chini, Malaysia. Scale 1: 50,000. Director of National Mapping, Department of Agriculture, Putrajaya.

Furtado JJ, Mori S (1982). Tasik Bera: The Ecology of Freshwater Swamp. Dr. W. Junk Publication, Netherlands.

Gray DM (1970). Handbook on the Principles of Hydrology. National Research Council, Canada.

Lewis J, Eads R (2001).Turbidity threshold sampling for suspended sediment load estimation. In: Proceedings, 7th Federal Interagency Sedimentation Conference. 25-29 March 2001. Reno, Nevada. pp. 110-117.

Mohd ET (2005). Hydrometeorological Condition and Sediment Yield in the Upstream Reach of Sungai Bebar, Pahang, Malaysia. Biodiversity Expedition. Peat swamp forest project, Pahang Forestry Department, Forest Research Institute Malaysia, pp. 41-45.

Nakamura F, Kameyama S, Mizugaki S (2004). Rapid shrinkage of Kushiro Mire, the largest mire in Japan, due to increased sedimentation associated with land-use development in the catchment. Catena 55:213-229.

Nakamura F, Sudo T, Kameyama S, Mieko J (1997). Influences of channelization on discharge of suspended sediment and wetland vegetation in Kushiro Marsh, northern Japan. Geomorphology, 18:279-289.

Slaymaker O (1982). Land use effects on sediment yield and quality. Hydrobiologia, 91:93-109.

Takamura N, Kadono Y, Fukushima M, Nakagawa M, Kim BHO (2003). Effects of aquatic macrophytes on water quality and phytoplankton communities in shallow lakes. Ecol. Res. 18:381-395.

Viessman W, Lewis GL (2003). Introduction to Hydrology (Eds.). 5th Edn. Pearson Education Ltd., Upper Saddle River, USA.

Wetlands International Asia Pacific (1998). The ecological assessment of Tasik Chini Pahang, Peninsular Malaysia: An evaluation of its conservation value and environmental improvement requirements. WIAP, Kuala Lumpur.

Influence of Nambul River on the quality of fresh water in Loktak lake

N. K. Sharat Singh[1], Ch. Bino Devi[2], M. Sudarshan[3], N. Sanamacha Meetei[4], T. Brajakumar Singh[4] and N. Rajmuhon Singh[2]

1Department of Physics, Manipur University, Canchipur – 795 003, Manipur, India.
2Department of Chemistry, Manipur University, Canchipur – 795 003, Manipur, India.
3UGC-DAE Consortium for Scientific Research, Kolkata Centre, Bidhan Nagar, Kolkata – 700 098, India.
4Ecology and Enviroment Wing, Government of Manipur, Porompat – 795 005, India.

The major and trace elements in the water samples of Loktak lake and Nambul river, the most polluted river draining into the lake were determined by using Energy Dispersive X-ray Fluorescence (EDXRF) technique. The physico-chemical parameters as well as the concentration of trace elements so far detected were compared with that of WHO limits for both cases and a comparative study were done to investigate the influence of the river to the quality of water of the lake. The co-efficient of correlation among the elements so far detected from both water samples were observed and it was found to be highest between copper (Cu) and manganese (Mn) was found followed by that of Cu and zinc (Zn) leading to the conclusion that if Cu concentration was higher, then there would be higher concentration of Mn and Zn. Though the polluted water of the Nambul river is drained into the lake, the concentration of different elements in the lake water were found to be within the safe limit prescribed by WHO except for iron (Fe).

Key words: EDXRF, trace element, Loktak lake, Nambul River, Eutrofication.

INTRODUCTION

Some elements present in drinking water at ppm or ppb levels, can improve or affect the function of the human physiology as deficit or excess of certain elements may lead to many diseases. Since drinking water is one of the most essential pathways to uptake harmful heavy metals by people, the quality control is particularly important in water supply. Some of the trace elements known to be essential are Cr, As, Co, Cu, F, I, Fe, Mn, Mo, Ni, Se, Si, Sn, V, Zn, and the other essential major elements are C, H, O, N, S, Ca, P, K, Na, Cl and Mg totaling twenty six essential elements (Madan Lal et al., 1991). Excess or deficit of such elements may lead to metabolic disorder to our body hence we have to keep them in proper level as far as possible. The elements present in the water are mainly dependent upon its source and the environment. Elements entering water bodies are derived from industrial effluents, weathering of rocks, leaching of metals from garbage and solid waste dumps, animal and human excretions, insecticides, fungicides and pesticides etc. Loktak lake is considered as the lifeline of Manipur, due to its importance in the socioeconomic and cultural life of the people. It is the largest natural fresh water lake in the north-eastern region of India with an area of 236.21 km^2 and plays an important role in the ecological and economic security of the region. It is an intermonte lake surrounded by tertiary hills. These tertiary hills, dominantly consisting of sedimentary sequences belonging to Dishang and Barails Surma and Tipam groups with little amount of igneous and metamorphics are the integral part of Indo-Burmese Range now known as Indo-Myanmar Range. There are a number of islands in the lake namely: Thanga, Karang, Ithing Sendra etc.

The sewage and other waste products from industries in the Imphal town are drained into Nambul river and help directly to dump into the Loktak lake. The quality of water of this river is badly affected while it enters the city as the sewage, waste products from industries are dumped into it. Population of 0.28 million living within Nambul river catchments generates 31,207 m^3 of sewage daily. With rapid urbanization, industrialization and increasing population of the city the river gets polluted day by day leading to the increase of pollution levels of Loktak lake. The pollution of the lake is also being increased day by day due to the use of excessive amount of fertilizers, pesticides, insecticides and fungicides by the farmers, cultivators and people around the lake. Trace metals and their compounds are the non-degradable pollutants and they cannot be destroyed. Again, with the consumption of fishes and other edible plants collected from this lake (if it is severely polluted), the people may risk from many diseases due to food chain. So it is very necessary to analyse the quality of water of the lake. Adoption of suitable elemental analysis techniques for the treatment of polluted water and proper management of it has been a challenging interest in this modern world. Both the trace elements and major elements in the liquid samples can be analysed effectively by EDXRF, NAA and PIXE for their quick, highly sensitive multielemental analysis capability (Heiden et al., 2010; Joshi et al., 2006; Ekbatani et al., 2009; OH Jong-Seok et al., 1990).

MATERIALS AND METHODS

The water samples were collected from 10 different stations in and around the lake in post-monsoon (November) including the heart of it that is, Karang (L7), the island of this lake (N 24°32', E 93°49'; 2516 ft). The map of Loktak lake along with various feeder streams is shown in Figure 1. The collection were done from north western (L1), south-western (L2, L3, L4, L5), southern (Ll0), eastern (L8), northern (L9) and it was done in this season to know the pollution level when there is no raining in Manipur. Similarly, the polluted water samples were collected from 10 different stations about 2 km apart of Nambul river and named as S1, S2 .... S10. The collection was made from upper to lower course of the river within Imphal city area in the same season. Five drops of concentrated HCl were added to each of the samples having 500 ml capacity as preservative and kept it with tight seal. The physico-chemical parameters like pH, turbidity, conductivity and DO were measured on the spot. 1000 ppm of $PdCl_2$ solution was made by dissolving 5 mg of it in 5 ml of distilled water. 200 μl of this solution was added to 50 ml of the water sample. After this the pH of it was maintained at 9 by treating with ammonia solution. 2.5 ml of saturated NaDDTC (sodium di ethyl di thio carbamate, $C_5H_{10}NaS_2.H_2O$) was added to precipitate the metals contents as their carbamates while $PdCl_2$ is added as a co-precipitant. Filtration was done to collect the precipitate on a nucleopore membrane having 2.5 cm diameter and pore size 0.2 μm by using a millipore vacuum filtration unit. The uniform thin layer of precipitate was used for irradiation. The blank target was also prepared by following the same procedure with the NIST standard water (SRM–1643e) as external standard. "Trace element in water"-SRM 1643e was used in different concentrations and calibration was done using them. Water standard of a different concentration was then checked which matched very well with the given value. The samples were analysed by the Jordan Valley Ex-3600 Energy dispersive X-ray fluorescence (EDXRF) spectrometer, which works with an oil-cooled system Rh anode X-ray tube having maximum voltage 50 kV and current of 1 mA. The measurements were carried out in vacuum chamber with a 0.05 mm thick Ti filter which was inserted in between the sample and source so as to find the optimum detection of elements in the sample at anode voltage 14 kV and current 300 mA. These parameters of 14 keV and 300 mA with Ti filter were used mainly for elements between Cr and Zn. One could use higher voltage with different filter and current-voltage settings to detect higher Z elements. The characteristic X-rays emitted from the sample due to the bombardment of primary X-ray from the Rh anode, were detected by the 12.5 mm^2 Si (Li) detector of resolution 143 eV at Mn K_α 5.9 keV, cooled at liquid nitrogen temperature (77K). The spectra were analysed by the inbuilt software called ExWIN integrated with the system (Raychaudhuri et al., 2008). The NIST water standard, (SRM-1643e) was used as external standard reference material.

RESULTS AND DISCUSSION

The various physico-chemical parameters so far measured of the lake along with the mean value of Nambul river are shown in Table 1. The pH of Loktak water was in the range 6.59 to 7.35 which were well within the permissible range of WHO limit (6.5 to 9.2). The turbidity was within the WHO permissible limit (25 NTU) except for water in Moirang site (31.1 NTU) of the lake. Similarly, the conductivity and DO for the lake were within the WHO permissible limits. The comparative mean values of pH, turbidity, conductivity and DO are shown as bar chart in Figure 2. As evident from the figure, the pollution level of Nambul river is far greater than that of the lake. The high value of turbidity signifies the decrease of penetration of light due to suspended particulates including disease causing pathogens, thereby affecting the habitats of many aquatic species living in deep water. These organisms can cause symptoms such as nausea, cramps, diarrhea, and associated headache. Similarly, conductivity was ranging from 78.9 to 152.0 μS/cm, leading to the conclusion that there were many dissolved inorganic materials in the lake which was enhanced by the Nambul river water. This is so because when concentration of salts dissolved increases the conductivity also increases.

The elements detected in the water samples collected from both the lake and river along with their concentrations and WHO limits (Sudarshan et al., 2000) are shown in Table 2. The elements namely: vanadium, chromium, manganese, iron, cobalt, nickel, copper, zinc were detected altogether. Further elements namely: Cr, Cu, Zn, Pb, Cd, Mn, Hg had been determined by using Atomic Absorption Spectroscopy (AAS) in the different sediment samples of the lake and their average concentration were found to be 0.38, 0.10, 1.02, 0.10, 1.15, 1.26, 3.53 mg/L (ppm) respectively (Devi et al., 2006). These elements can be derived from rocks namely: (a) Tipam groups consisting of mottled clay, mottled sandy-clay, sandy-shale, clayey shale and sandstone.

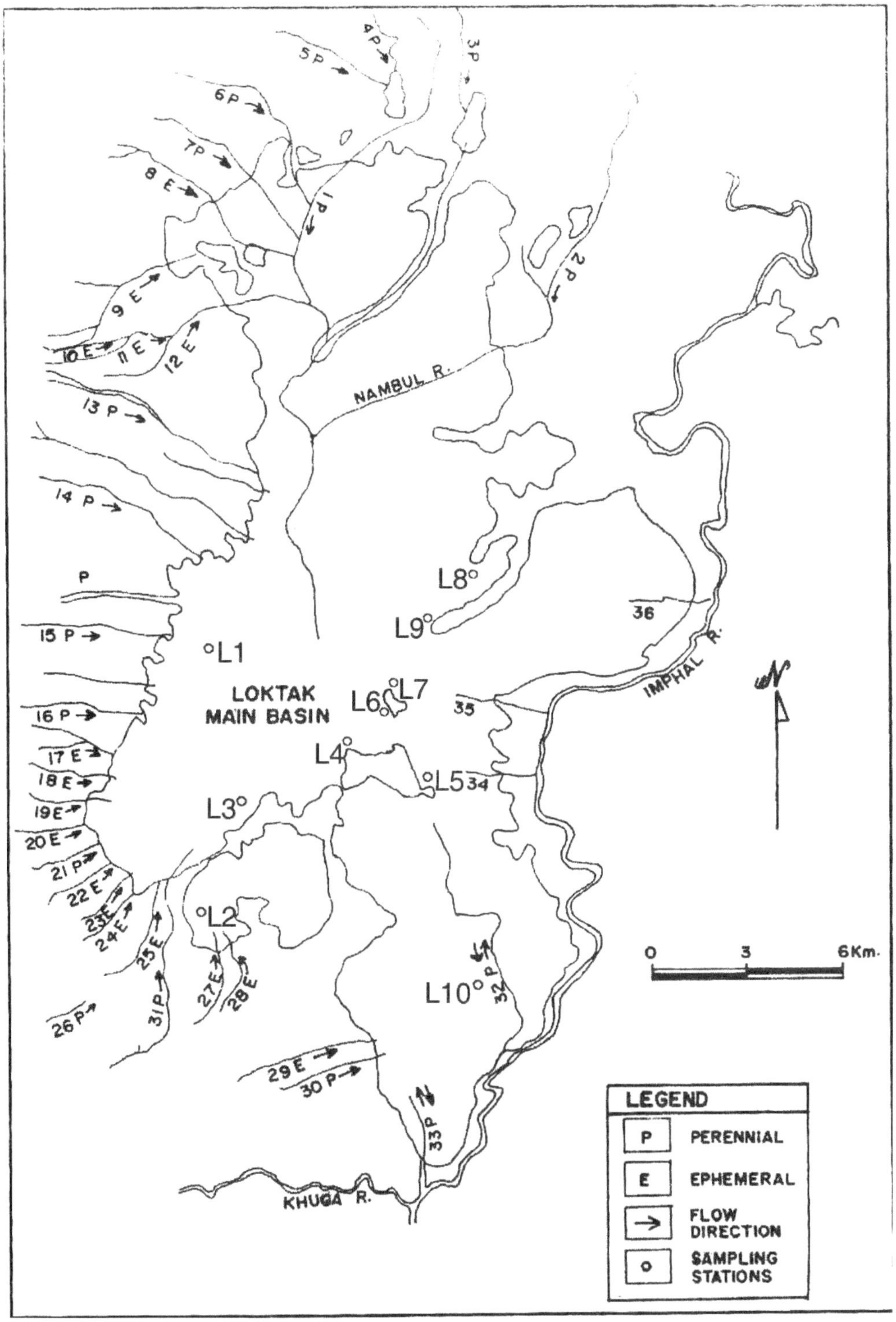

Figure 1. The map of Loktak lake along with its various feeder streams.

Greenish to blue, moderate to course ferruginous sanstone with sady shale, clay often brown to orange due to weathering molasses type of deposits; (b) Disang group consisting of Areneceous rocks, dark gray splintery shale interbedded sandy-shale and siltstone in upper part. Argillaceous gray shale interbedded with mudstone mudstone and minor amount of silt in lower part; (c) Barail group consisting of massive sandstone, alternation of shale and sandstone, and bedded structure, Flysch sediments of turbidite character; (d) Surma group

Table 1. Physico-chemical parameters of Loktak lake and Nambul river (mean value): temperature 25°C.

Sample	pH (1)	Turbidity (NTU) (2)	Conductivity (µS/cm) (3)	DO (mg/l) (3)
L1 (Ningthoukhong)	6.59	1.73	100.5	10.4
L2 (Moirang)	7.10	31.1	81.2	10.45
L3 (Sendra)	7.35	0.77	78.9	10.40
L4 (Thanga Machin)	7.06	0.90	114.0	10.41
L5 (Thanga Maton)	6.99	0.82	105.0	10.39
L6 (Karang Machin)	6.91	0.83	115.9	10.36
L7 (Karang Maton)	6.86	1.50	116.6	10.33
L8 (Mayang Imphal A)	6.73	0.58	115.3	10.01
L9 (Mayang Imphal B)	6.80	1.44	134.8	9.61
L10 (Keibul Lamjao)	6.67	0.62	152.0	7.9
Mean	6.90 ± 0.22	4.02 ± 9.52	111.42 ± 22.09	10.03 ± 0.79
Nambul river mean value	6.73 ± 0.10	19.58 ± 4.15	281.08 ± 123.04	8.54 ± 0.094
WHO Limits	6.5 - 9.2	25 NTU	300 mS/cm	4 - 6 mg/L

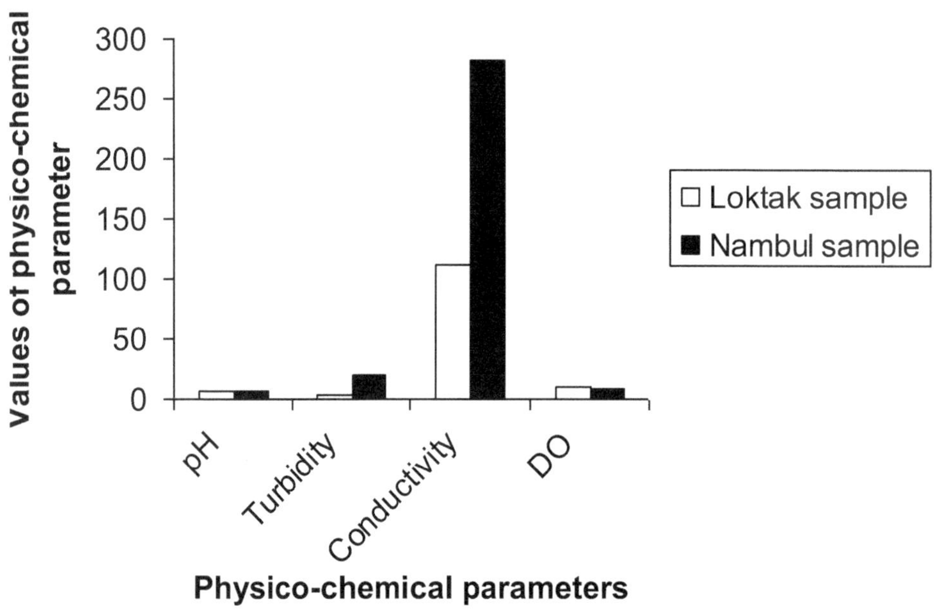

Figure 2. Bar chart of different mean values of physico-chemical parameters of Loktak lake and Nambul river.

consisting of Shale, sandyshale, siltstone, ferruginous sandstone, massive to false bedded ferruginous sandstone, alternation of sandstone and shale with more horizons in the middle and minor conglomerate, transitional characters from flysch to molasses sediments; (e) Metamorphic complex consisting of low to medium grade metamorphic rocks of various composition, phyllitic schist, quartzite, micaceous quartzite, quartz-chlorite-mica achiest, marble etc. Chromium salts impart colour to water. The salts of trivalent chromium are physiologically not harmful but hexavalent chromium salts are known carcinogen and it causes corrosive effects in the intestinal tract and to nephritis when large doses are used. This hexavalent chromium is mutagenic in nature and is highly mobile in soil and aquatic system and also a strong oxidant capable of adsorbed by the skin (Singh et al., 2002). Hexavalent chromium readily crosses cell membranes and is reduced to trivalent chromium in the cellular environment. It is thought that the biological effects of chromium are related to its reduction to trivalent from hexavalent state and formation of complexes with macromolecules present in the biological system. Trace amounts of trivalent chromium are essential in

Table 2. Elemental concentration of Loktak lake water and Nambul river water.

Sample	Element (ppb)								
	V	Cr	Mn	Fe	Co	Ni	Cu	Zn	Hg
L1	0.00	1.37	101.96	105.70	0.00	0.21	1.29	16.44	0.00
L2	5.20	1.96	151.96	421.93	2.49	1.39	2.28	38.94	0.00
L3	0.00	1.19	42.43	53.41	0.15	0.26	1.38	34.69	0.00
L4	1.73	1.80	226.50	153.85	0.16	0.29	1.17	18.83	0.80
L5	1.59	2.71	121.41	339.83	0.36	0.18	1.47	44.95	0.30
L6	1.01	1.60	245.30	194.60	0.60	0.16	1.18	8.27	0.00
L7	0.29	2.49	331.33	360.69	0.40	0.00	1.72	36.35	0.20
L8	0.00	2.51	147.78	168.73	0.75	0.06	1.52	13.78	0.00
L9	1.73	3.73	214.62	646.17	2.68	0.01	1.25	13.73	0.00
L10	2.89	3.58	468.41	615.23	2.73	0.00	2.25	22.93	0.20
Mean	1.44	2.29	205.17	306.01	1.03	0.25	1.55	24.89	0.15
S1	8.39	2.64	1290.88	330.32	0.48	0.98	2.53	244.84	0.80
S2	17.36	0.78	778.08	939.21	1.07	0.32	1.92	48.77	0.00
S3	4.48	3.65	1039.70	1143.04	0.55	1.99	2.16	25.91	0.00
S4	4.63	0.93	1707.86	1486.51	1.72	0.00	2.38	69.53	1.00
S5	4.05	1.78	985.78	558.89	0.66	0.23	1.59	23.55	0.50
S6	2.31	2.76	907.97	169.62	0.29	1.10	1.84	19.80	0.50
S7	6.50	5.46	1713.47	563.04	0.00	0.23	1.94	24.93	0.00
S8	0.00	2.85	1550.41	113.99	0.24	1.21	2.38	43.18	2.90
S9	0.00	3.25	1900.68	290.34	0.78	1.07	2.22	61.81	0.80
S10	2.60	1.05	1966.47	301.81	0.80	0.24	1.72	87.68	0.80
Mean	5.03	2.51	1384.13	589.67	0.65	0.73	2.06	65.00	0.73
WHO limits			500	100		20	2000	5000	
Indian limits			50	300			50	5000	

* Hg concentration was determined from AAS.

carbohydrate metabolism in mammals while this element is a co-factor for the action of insulin. The detection of this element is attributed to the washing of chromite rock from surrounding hills and dumping of industrial wastes like ceramic items in the river stream draining into this lake. The concentration of it was found in the range 1.19 to 3.73 ppb which is well below 10 ppb WHO limit. Chromium is essential functioning as a glucose tolerance factor. It is also used for insulin signaling for biological role and thus to sugar metabolism and diabetes (Chen et al., 1993). The relatively insoluble compounds in the soil containing essential iron, manganese, copper and zinc are solubilized by chelating agents as the citrate anion. For manganese the microorganism is bacteria to dissolve the compounds of it in the soil and rocks by the anaerobic action. Its concentration ranged 42.43 to 468.41 ppb within the permissible WHO limit (500 ppb). In most soils, several of these effective chelating agents are produced naturally from soil organic matter by soil microorganisms (Mack, 1964). Highest concentration was observed in the sample (L10) collected from Keibul Lamjao, the only home of Sangai (*Cervus eldi eldi*). This may be due to the coverage of the entire fresh water by the thick floating weeds (40 km^2) leading to the anaerobic action of the bacteria with the soil at the bottom of water. It is essential for the nutrition of both plants and animals, but it is undesirable in domestic water supplies as it causes unpleasant tastes, deposits on food during cooking, stains and discolours laundry and plumbing fixtures.

Iron in natural water usually exists in ferrous; however it is readily oxidized to ferric form and ferric salts are precipitated as rust colour deposits. It imparts a bitter sweet astringent taste to water. Both iron and manganese tend to precipitate as hydroxide. The iron content is attributed to the solubilization of different compounds in the rocks draining of such water in addition to the industrial effluents of different rivers into the lake. The concentration of this element is slightly higher than the WHO limit (100 ppb). So, proper treatment is required for this element. Metal plating waters when discharged into water courses contribute nickel. Increased used of stainless steel and other nickel containing alloys in water

Table 3. Correlation among the elements

Element	V	Cr	Mn	Fe	Co	Ni	Cu	Zn	Hg
V	1	- 0.011	0.211	0.526	0.159	0.096	0.359	0.343	- 0.177
Cr		1	0.181	0.026	0.052	0.225	0.226	- 0.089	- 0.044
Mn			1	0.315	- 0.127	0.279	0.618	0.413	0.530
Fe				1	0.436	0.047	0.416	0.041	- 0.124
Co					1	- 0.097	0.252	- 0.066	- 0.182
Ni						1	0.541	0.206	0.259
Cu							1	0.529	0.422
Zn								1	0.266
Hg									1

system also contribute nickel (Manivasakam, 1996). Water soluble nickel salts, such as the chloride are quite toxic to higher plants. They are more active against some fugal parasites, particularly wheat rust. Copper is attributable to the corrosive action of the water on copper and brass tubing compounds to industrial effluents or frequently to the use of copper compounds to control plant diseases. But its concentration is much less than the WHO limit (2000 ppb). So it will not lead to any health hazard with this element for the time being. Further copper can be deposited in metallic form in the distribution pipes with the dissolution of iron. Most of the heavy metals are highly toxic, especially to micro organisms. One of the most commonly used metal is copper in inorganic form. Bordeaux mixture is prepared by mixing a slurry of lime in water into a solution of copper sulphate in water (fungicide). Zinc is commonly found in small quantities in domestic water supplies and industrial water due to corrosion of galvanized iron and brass in condensing and distributing systems. The presence of zinc in the water of this lake may be due to effluents of industrial effluents and solubilization of zinc containing soils by the microorganism or zinc compounds used for plant diseases. The concentration of it ranged 8.27 to 44.95 ppb which is in permissible levels of potable water as WHO limit is 15000 ppb. The highest concentration was observed in the samples collected from Moirang (L5). The presence of zinc and manganese may be also because of the constant use of the fungicides like Dhanuka M-45, Dithaine M-45, ABIC M-45, commonly known as Mancozeb, consisting of zinc ion and manganese ethylene. Vanadium occurs in greatest amount in shales and intermediate igneous rock and cobalt is rich in the soils carrying healthy stocks. The detection of these elements may be due to dissolution of their compounds. The toxic element, like Hg was also detected from the river and the lake but in specific locations. However, this will not make any harmful effect as its concentration is less than 2 ppm of USEPA. The availability of it may be due to draining of industrial effluents into the river, dumping of unused compact fluorescence lamp in surrounding water and use of insecticides, pesticides, mercurials fungicides such as Ceresan dry Phennyl Mercuratare-dust 1%, Agrosan G.N., Hexason (Phenylmercuric accutate plus ethyl mercury-dust %), Agallol Tillex (Mercury chloride 36%) by the farmers living at the periphery of the lake. Another factor may be partly due to the inorganic and arylmercury compounds generated by the action of micro organisms in the sediments of the water body. Summarily, the observed elemental concentrations of the river were greater than that of the lake. The detection of elements like Cr, Mn, Fe, Co, Cu and Zn in the river water were also reported by earlier workers (John et al., 1999). The elemental concentrations of Fe and Mn of the river were too high for drinking purpose. Again, the concentration of phosphate ranges from 0.153 to 1.686 mg/L and that of nitrate ranges from 0.013 to 0.585 mg/l for Nambul river (Sharat et al., 2010). So the water in the lower course of the river from Imphal city can be used for the irrigation purpose if the physico-chemical characteristics of the soils are maintained properly such that it may help to provide the micronutrients for the paddy plants leading to the increase of yield.

The correlation among the detected elements in the two series of water samples are shown in Table 3. As evident from the table, the correlation co-efficient between Cu and Mn was found to be highest followed by that of Cu and Zn leading to the conclusion that if Cu concentration was higher, then there would be higher concentration of Mn and Zn. Similarly, for the toxic element viz. Hg, it was found to be highly correlated to Cu but highly correlated negatively to V and Fe. Again, since the correlation between V and Fe was high, the vanadium present might be attributed from iron containing vanadates. Lastly, the element like S and Ca were also detected in the samples, but their concentrations were unable to be calculated as we used certain energy range only.

Conclusion

The trace elements in water samples of Loktak lake and Nambul River were determined by using EDXRF technique as it is one of the quickest, reliable multi-elemental analytical tool. The physico- chemical parameters and trace element concentrations of Nambul river were almost greater than that of the lake leading the conclusion that it might take some significant role for pollution and eutrofication of the lake. However, the pollution in terms of physico-chemical parameters and trace elements were in the permissible range except for iron. This is due to the dilution of fresh water from different rivers (namely: Thonjaorok River, Imphal River, Moirang River, Khuga River) draining into the lake. Further, pretreatment of iron is required before the water is distributed for public use.

ACKNOWLEDGEMENTS

The authors are thankful to the UGC-DAE Consortium for Scientific Research, Kolkata Centre for extending the facility for the analysis work and would like to acknowledge the technical assistance of Mr. Ajay Rathore for carrying out this work. Further, two of the authors namely Ch. Bino Devi and N. Rajmuhon Singh would like to thank Ministry of Environment and Forests GoI, for the financial assistance in the research work.

REFERENCES

Chen KS, Tseng CL, Lin TH (1993). Trace elements in natural drugs determined by INAA. J. Radioanalytical Nuclear Chem. Articles, 170(1):265-280.

Ekbatani A, Pazoki A (2009). Neutron activation analysis of some water sources around Khorram Abad. Asian J. Chem. 21(9):6625-6628.

Heiden ES, Gore DB, Stark SC (2010). Transportable EDXRF analysis of environmental water samples using amberlite IRC748 ion exchange preconcentration. X-ray Spectrom. 39(1):176-183.

John KV, Augusthy A, Varier KM, Magudapathy P, Baskaran V, Panchapakesan S, Nair KGM (1999). PIXE facility for trace element analysis at IGCAR, Kalpakkam, India. Indian J. Phys. 73A(2):159-168.

Joshi GC, Agrawal HM, Mohanta B, Sudarshan M, Sinha AK (2006). Elemental study of Nainital lake water by EDXRF. Nuclear Instrum. Meth. Phys. Res. B., 251(1):223-226.

Mack D (1964). Soil Chemistry and Plant Nutrition. In Firman Bear (Eds.), Chem. Soil Oxford & IBH Publishing Co., pp. 432.

Madan L, Choudury RK (1991). Studies of trace elements in biological systems by energy dispersive X-ray fluorescence (EDXRF) and proton induced X-ray emission (PIXE) methods. Indian J. Phys. 65B(1):30-43.

Manivasakam N (1996). Physico-chemical Examination of Water, Sewage and Industrial Effluents. Pragati Prakashan: Meerut, p. 162.

Devi OG, Kushwaha RAS, Okendro M (2006). Trace Element Contamination in Loktak Lake, Manipur (pp. 313-324). In P.S. Saklani, Himalaya (Geological Aspects) 4th Vol. Satish Serial Publishing House, New Delhi.

Jong-Seok OH, Jun-Gyo B, Hae-iLL B (1990). A study on PIXE spectrum analysis for the determination of elemental contents. J. Kor. Nucl. Soc. 22(2):101-107.

Raychaudhuri S, Mishra M, Salodkar S, Sudarshan M, Thakur AR (2008). Traditional aquaculture practice at east Calcutta wetland: The safety assessment. Am. J. Environ. Sci. 4(2):173-177.

Sharat SNK, Rajmuhon SN (2010). Phosphate and nitrate content of Nambul river in the Imphal city. Ecoscan 4(2&3):175-176.

Singh IB, Singh DR (2002). Cr (VI) removal in acidic aqueous solution using iron bearing industrial solid wastes and their stabilization with cement. Environ. Technol. 23(1):85-95.

Sudarshan M, Dutta RK, Vijain V, Chintalapudi SN (2000). PIXE measurements of drinking waterof Salt lake, Calcutta, India. Nuclear Instrum. Meth. Phys. B., 168:553-558.

24

Geospatial study on morphometric characterization of Umtrew River basin of Meghalaya, India

Pranjit Kumar Sarma[1], Kalyanjit Sarmah[2], Prem Kr.Chetri[3] and Ankita Sarkar[4]

[1]Department of Geography, Mangaldoi College, Assam, India.
[2]North Eastern Hill University, Shillong.
[3]North Orissa University, Orissa.
[4]Gauhati University, Assam, India.

Geospatial technology that is, remote sensing (RS), geographical information system (GIS) and global positioning system (GPS) have proved to be an efficient tool in the field of modern day geographical and geological studies. GIS and Remote sensing techniques have been adopted for the identification of different morphological features and to analyze their existence in a hilly river basin of Meghalaya plateau, which is located in the two North-East Indian states of Meghalaya and Assam. Morphometric analysis of any drainage basin can be very much effective in determining the landform characteristics and the processes and components behind its development. The morphometric parameters of the study area have been discussed with respect to linear, areal and relief aspects. The geological formation of the basin area varies from Shillong group of rocks in the undulated hilly areas of Meghalaya plateau to unconsolidated river alluvium in the plain parts of Brahmaputra valley. It is a 7th order drainage basin and the entire drainage network can be defined as "Trellis" in nature. It has been observed that the drainage density value is in between medium to low which indicates that the basin has highly permeable sub-soil material and thick vegetative cover in the most of its areas. The moderate values of bifurcation ratio in the basin area indicates the possibilities of structural control over the development of the entire drainage network. The circularity ratio value reveals that the basin is elongated in shape and a predominance of highly permeable homogenous geologic material. The aim of the present study is to find out the various sets of inherent factors which were behind the formation of this river basin and its hydrologic nature.

Key words: Umtrew River basin, drainage, morphometric, hydrologic, geographical information system (GIS)

INTRODUCTION

Morphometry is the measurement and mathematical analysis of the configuration of the earth's surface, shape and dimension of its landforms (Agarwal, 1998; Reddy et al., 2002). Morphmetric studies in the field of hydrology were first initiated by Horton (1940) and Strahler (1950). The morphometric analysis of the drainage basin and channel network play a vital role for understanding the geo-hydrological behavior of drainage basin and expresses the prevailing climate, geology, geomorphology, structural control, etc. Besides, the primary aim of studying drainage basin is to understand the hydrologic nature and its morphometric expression of the basin area.

Recently, many researchers have used remote sensing data and analyzed them on Geographical Information System (GIS) platform to understand the inherent morphometric ingredients of the catchment (Nag, 1998). The rapid development of geospatial technology has recently become an effective tool to overcome many problems associated with identification and interpretation of basin morphometric parameters and to design an

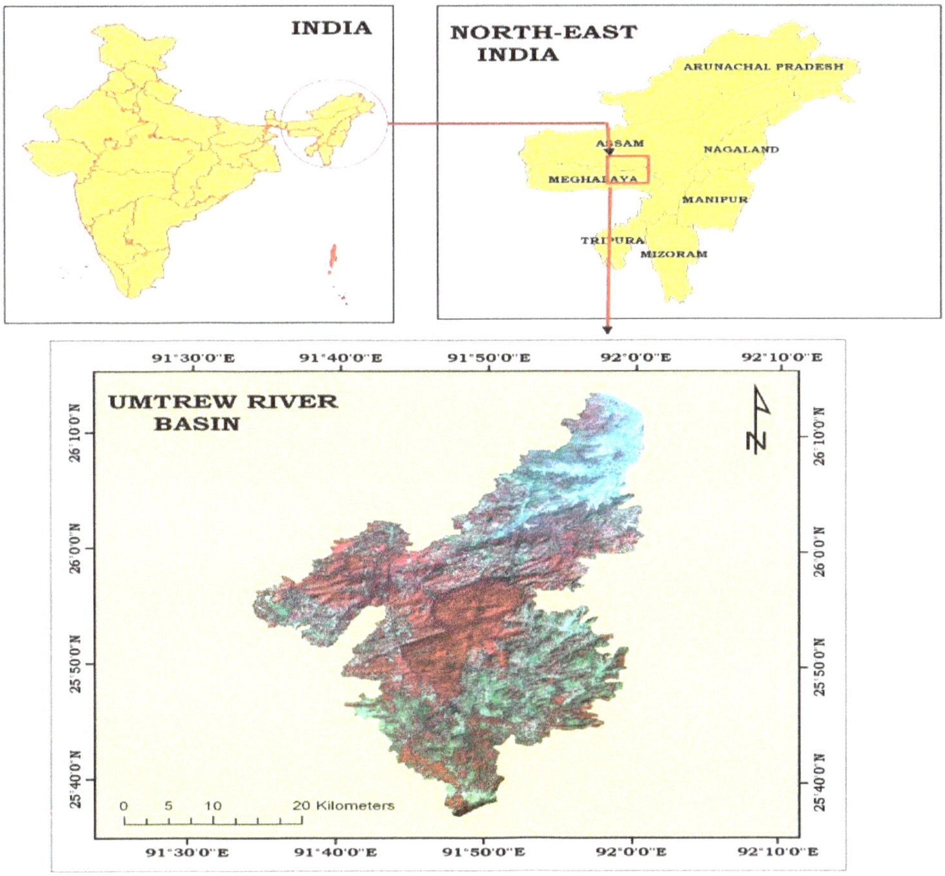

Figure 1. Location map of Umtrew River basin.

effective land and water resource plan for the basin area. Many researchers and scholars including Srinivasa (2004), Nag and Chakraborty (2003), and Srivastava (1997) have studied the morphometric characteristics of different parts of India by using geospatial technology. The objective of the present study is to analyze the morphometric attributes of Umtrew River basin which is a major drainage system in the state of Meghalaya and so far not any systematic work on the morphometry of the basin has been carried out because the optimal and sustainable utilization of the resource is very much essential to avoid future problems regarding its qualitative and quantative availability. This particular study will give a detail picture on the water bearing properties of the basin which could be beneficial in the process of preparing a sustainable river basin or watershed management plan.

The extent of the location of the present study area is very important from geographical and geological perspective. It is located between the confrontation zone of Shillong plateau in its south to plain Brahmaputra valley in the northern part. The findings of this manuscript try to highlight the physical and morphometric nature of this environmentally and geographically important river basin of North-East India. The different morphometric attributes which have been studied in this particular manuscript were observed to get a clear picture of geomorphological, geological and hydrological setting of the study area.

Study area

The investigated area lies between 25°40′00″ N to 26°10′00″N latitude and 91°30′00″ E to 92°00′00″E longitude covering an area of 1369.6 km^2, falling in Survey of India toposheet (SOI) no. 78 N/16, 78 O/9, 78 O/13 and 78 O/14 on 1:50000 scale (Figure 1). Geologically, the study area is occupied by Shillong group of rocks comprising of Phyllites, Quartz schists, Quartzite and Intra formational conglomerates in the Meghalaya part of the basin to unconsolidated alluvium fills in the plain land areas of Assam (Chottopadhyay et al., 1984). The area enjoys sub-tropical climate throughout the year. The topography is undulating in nature having elevation of 171 m in the plains of Assam and about 1790 m at the highest peak of the basin (Figure 2). Physiographically, Umtrew Umtrew River basin

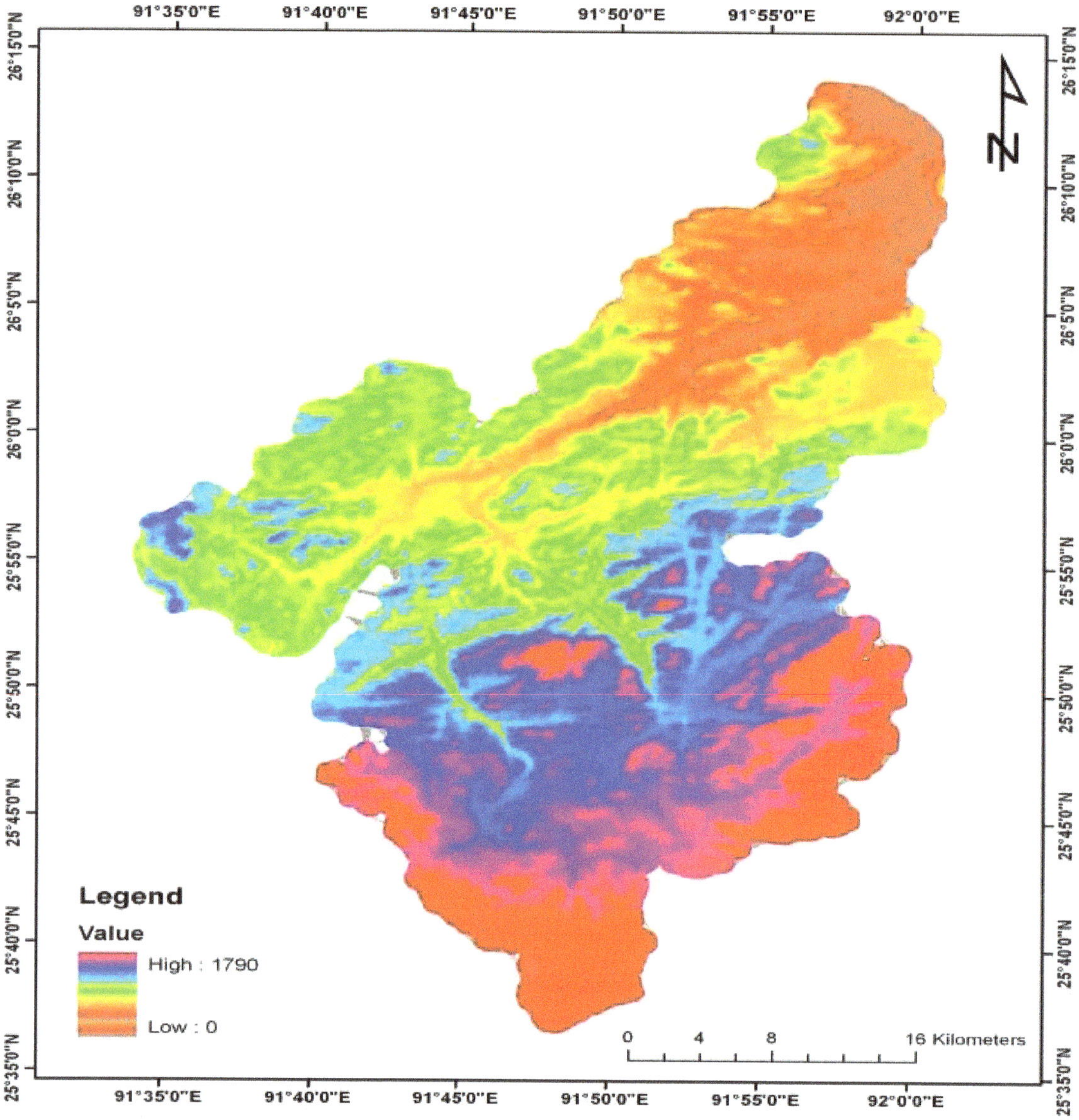

Figure 2. Relief map of Umtrew River basin.

River basin can be grouped into two divisions (i) High elevated southern part and (ii) Low lying Northern River plains. Umtrew River basin comprises of undulated hilly terrain of different altitudes along with inter-mountain deep gorges, small stretch of plains and marshy areas. Soil texture of the basin area varies from clay loam to sandy clay loam in the upper part where as fine loam and coarse loam soil is predominant in the lower territory.

MATERIALS AND METHODS

The drainage map of the study area has been prepared by using IRS P6 LISS-III geo-coded satellite data and verified with Survey of India (SOI) toposheets no. 78 N/16, 78 O/9, 78 O/13 and 78 O/14 on 1:50000 scale with selected ground truths. The SOI toposheets and satellite data were geometrically rectified and georeferenced to World space coordinate system using ERDAS IMAGINE 9.2 software package. The assigned projection system was Universal Transverse Mercator (UTM), WGS 84 datum and spheroid system, the root mean square error (RMS) was accounted as less than 0.25 value. In any Georeferencing process the root mean square error should be below 0.25.For the proper interpretation and identification of different geological and geographical features, satellite imageries were rectified by the techniques like Image enhancement, edege detection and high pass filtering etc. The morphometric parameters of the entire basin have been computed by using Arc GIS 9.3 and ERDAS Imagine 9.1 software packages. The "Spatial analysis" tool of Arc GIS 9.3 software was used immensely throughout the study to carve out different topographic and morphometric components of the study area. The Area (A) and Perimeter (P) of the study site have been calculated by using "measure tool" option of ERDAS Imagine software. Digital Elevation Model (DEM) has been prepared to find out different sets of relevant morphometric data.

Table 1. Linear aspects of Umtrew River basin.

Stream order u	Number of streams (Nu)	Total length of streams Lu (km)	Log Nu	Log Lu
1st	4754	2581.85	3.677	3.411
2nd	1021	763.94	3.009	2.883
3rd	259	366.76	2.413	2.564
4th	74	199.12	1.869	2.299
5th	15	98.23	1.176	1.992
6th	5	33.10	0.698	1.519
7th	1	68.37	3.677	1.834

Table 2. Bifurcation ratio pattern of Umtrew River basin.

1st order/2nd order	2nd order/3rd order	3rd order/4th order	4th order/5th order	5th order/6th order	6th order/7th order	Mean Bifurcation ratio
4.65	3.94	3.5	4.93	3	5	4.17

After the completion of drainage network analysis the stream order to each of the stream was assigned by following Strahler (1964) method. The methodology and mathematical equations adopted for the quantitative analysis of different morphometric parameters has been mentioned in Table 3 and the results summarized have been mentioned in Table 2. To understand the topographic expression of the study area parameters like slope, relief pattern and DEM were analyzed thoroughly. Similarly for the formulation of sustainable basin management strategy factors like length of the overland flow and channel maintenance factors were studied based on the equations developed by Chorley (1969).

RESULTS AND DISCUSSION

The study of morphometric characteristics of any basin area helps in understanding the transmission pattern of hydrological regime and sediment load carried through the entire basin area. In the present study the morphometric analysis is carried out with respect to parameters like stream order Table 1, stream length, bifurcation ratio, stream length ratio, basin length, drainage density, stream frequency, elongation ratio, circularity ratio, form factor, relief ratio, etc. using mathematical equations as given in Table 3. The properties of the stream networks are very important to study the landform development process (Strahler and Strahler, 2002). Morphometric parameters such as relief, shape and length also influence basin discharge pattern strongly through their varying effects on lag time (Gregory and Walling, 1968).

Linear aspects

In the present study the linear aspects of Umtrew River basin include the parameters like stream order, stream length, mean stream length, stream length ratio and bifurcation ratio (Table 1).

Stream Order (Nu)

The designation of stream order is the first step in drainage basin analysis and expresses the hierarchical relationship between stream segments, their connectivity and the discharge coming from contributing watersheds or sub-watersheds. In the present study, the channel segment of the basin area has been ranked according to Strahler's stream ordering method. According to Strahler (1964), the smallest fingertip tributaries are designated as order I. Where two first order channels join, a channel segment of order II is formed; Where two of order II joins, a segment of order III is formed, etc. The trunk stream through which all discharge of water and sediment passes is therefore the stream segment of the highest order. The present study area is a VII order drainage basin in nature. The total number of 6129 streams were identified of which 4754 are Ist order streams, 1021 are IInd order, 259 are IIIrd order, 74 in IVth order, 15 in Vth order, 5 in VIth order and 1 is indicating as VIIth order stream. The overall drainage pattern of the basin area highlights a Trellis pattern because the entire drainage network has been constituted by a network of tributaries and the master consequent stream which is flowing in the regional slope direction and are well adjusted to the geological structures.

Stream Number (u)

The total order of wise stream segments is known as stream number. Horton (1945) states that the numbers of stream segments of each order form an inverse geometric sequence with order number. It has been mentioned earlier that the total number of streams in the study area is 6129. This shows that Umtrew River basin is spreading in a vast stretch of land throughout both hill and plain areas of Meghalaya and Assam State.

Table 3. Different morphometric parameters of Umtrew River basin.

Morphometric Parameters	Formula	Value
Area (km^2)	A Where, A is the total area of the basin	1369.6
Perimeter (km)	P	339.8
Length of Basin (km)	Lb	69.45
Drainage Density (km/km^2)	Dd = Lu/A	3.00
Stream Frequency (Fs)	Fs = Nu/A	4.47
Drainage Texture (Rt)	Rt = Nu/P	18.03
Infiltration Number (If)	If = Rt * Fs	80.59
Length of overland flow (Lg)	Lg = 1/D x 2	0.16
Form Factor (Rf)	$Rf = A/ Lb^2$	0.28
Circulatory Ratio (Rc)	$Rc = 4\pi A/P^2$	0.148
Basin Relief (H)	H = Hr - Hd	1640
Relief Ratio (Rh)	Rh = H/Lh	23.59
Relative Relief (Rr)	Rr = H / P*100	482.63
Ruggedness Number (Rn)	Rn = H x Dd	4920
Channel maintenance factor (Cm)	Cm = 1/Dd	0.33

Stream Length (Lu)

The total stream network of Umtrew River basin has been categorized into several orders, which have computed with the help of SOI topographical sheets and “Measure” tool option of Arc GIS software. Horton's law of stream lengths supports the theory that geometrical similarity is preserved generally in the basin of increasing order (Strahler, 1964). The stream length based on the law proposed by Horton (1945) has been computed.

Bifurcation Ratio (Rb)

The bifurcation ratio is the ratio of the number of stream segments of given order to the number of segments of next higher order. Bifurcation ratio is an index of relief and dissection (Horton, 1945). It is well demonstrated that bifurcation ratio shows a small range of variation for different regions or for different environment except where geological control dominates. It has been found that the bifurcation ratio characteristically ranges between 3.0 and 5.0 for the basin in which geology is reasonably homogeneous and with no structural disturbances. The lower values of Rb indicate that the study area has suffered less structural disturbances and the drainage pattern is not distorted. Furthermore, the low Rb values signify a high drainage density, low permeability of the terrain and indicate areas with uniform surficial materials where geology is reasonably homogeneous.

Mean Bifurcation Ratio

To arrive at a more representative bifurcation number Strahler (1952) used a mean bifurcation ratio obtained by multiplying the bifurcation ratio for each successive pair of orders by the total numbers of streams involved in the ratio and taking the mean of the sum of these values. Mean bifurcation ratios vary from about 2.0 for flat or rolling basins to > 3.0 for mountainous hilly dissected basins (Horton, 1945). As the mean bifurcation ratio of our present study area is showing a value greater than 3, it denotes that the area is very hilly and dissected in nature.

Length of the Main Stream

This is the length along the longest watercourse from the outflow point of designated basin to the upper limit to the basin boundary. During the study period, main channel length was computed using ArcGIS-9.3 software. The analysis shows that the length of the main stream is 69.45 km.

Areal aspects of Umtrew River Basin

The areal aspects include drainage density, drainage texture, stream frequency, form factor, circularity ratio, elongation ratio and length of overland flow. Basin area is hydrologically important because it directly affects the size of the storm hydrograph and the magnitudes of peak and mean runoff. It is interesting that the maximum flood discharge per unit area is inversely related to size of the basin (Chorley, 1957).

Drainage density (Dd)

The drainage density is an important indicator of the linear scale of landform element in stream eroded topography and defined as the total length of stream of all orders/drainage area and may be an expression of the closeness of spacing of channels (Horton, 1932). The significance of drainage density is recognized as a factor determining the time travel by water (Schumm, 1956). The low drainage density is favored in regions of highly permeable subsoil material, under dense vegetative cover where relief is low, while high drainage density is favored in regions of weak or impermeable sub-surface materials, sparse vegetation and high mountain relief. The low drainage density is also indicative of relatively long overland flow of surface water. The drainage density of the study area is 3.0/ km^2; this highlights a medium category of drainage density. The drainage density analysis shows that the region is a hilly area with sparse vegetation and loose sub-surface material.

Drainage Texture (Dt)

An important geomorphic concept is drainage texture which means the relative spacing of drainage lines (Smith, 1950). Similarly, Horton (1945) defined drainage texture on the basis of stream frequency. Drainage texture less than 2 indicates very coarse, between 2 and 4 as coarse, between 4 and 6 as moderate, between 6 and 8 as fine and greater than 8 as very fine drainage texture (Smith, 1950). In the present study the drainage texture value of Umtrew basin stands at 18.03, it indicates that the texture pattern of Umtrew basin is very fine drainage texture.

Stream Frequency (Fs)

The stream frequency is defined as the total number of stream segment of all order per unit area (Horton, 1932). The greater the drainage density and stream frequency in a basin, the faster the runoff, and therefore, flooding is more likely in basins with a high drainage and stream frequency (Kale and Gupta, 2001). The stream frequency of Umtrew River basin stands at 4.47. Thus, the stream frequency of the study area falls in the moderate frequency class. The stream frequency of the study area represents a positive correlation with the drainage density, which indicates that the stream population increases with the increase of drainage density.

Infiltration Number (If)

Infiltration number of a drainage basin is the product of drainage density and stream frequency of a basin. It is the number by virtue of which an idea regarding the infiltration characteristics of the basin is obtained. The higher value indicates low infiltration and high runoff. The study area has the infiltration value of 80.59 which shows a comparatively higher nature of infiltration number.

Form Factor (Rf)

The form factor is the ratio of basin area to square of the basin length and it is a quantitative expression of drainage basin outline (Horton, 1932). It is a dimensionless ratio of basin area to the square of basin length. Basin shape may be indexed by simple dimensionless ratios of the basic measurements of area, perimeter and length (Singh, 1998). The form factor value of Umtrew River basin is 0.28 which indicate lower value of form factor and thus represents an elongated shape. The elongated basin with low form factor indicates that the basin will have a flatter peak of flow for longer duration.

Circulatory Ratio (Rc)

Circularity ratio is defined as the ratio of basin area to the area of a circle having the same perimeter as the basin and it is pretentious by the lithological character of the basin. Miller (1953) has described the basin of the circularity ratios range between 0.4 to 0.5, which indicates strongly elongated and highly permeable homogenous geologic materials. Circulatory ratio also indicates the tendency of small drainage basin in homogeneous geologic materials to preserve geometrical similarity (Suresh, 2000). The circulatory ratio value of 0.148 corroborates with Miller's range, which is indicating that the basin is elongated in shape, low discharge of runoff and high permeability of subsoil material.

Length of Overland Flow (Lg)

Length of the overland flow, considered as a dominant hydrologic and morphometric factor, is the -mean horizontal length of the flow-path from the water divide to the stream in a first order basin and is a measure of stream spacing and degree of dissection and approximately one half the reciprocal of the drainage density (Chorley, 1969). Length of the overland flow is one of the most important morphometric variables, which affects the hydrological and topographic development of the basin. The length of overland flow depends primarily on the degree of relief fragmentation, and hence on the drainage density. Overland flow is significantly affected by infiltration (exfiltration) and percolation through the soil, both varying in time and space (Schmid, 1997). In this study, the computed value of length of overland flow is 0.16 which shows low surface runoff of the study area.

Relief aspects of Umtrew River Basin

The relief aspects determined include relief ratio, relative relief and ruggedness number. The results of the relief aspects are given in Table 3.

Relief Ratio (Rh)

Difference in the elevation between the highest point of a basin and the lowest point on the valley floor is known as the total relief of the river basin. The relief ratio may be defined as the ratio between the total relief of a basin and the longest dimension of the basin parallel to the main drainage line. The possibility of a close correlation between relief ratio and hydrologic characteristics of a basin suggested by Schumm who found that sediments loose per unit area is closely correlated with relief ratios. In the study area the value of relief ratio is 23.59 (Table 3). It has been observed that the area is hilly in nature, steeply sloping and the intensity of erosion is high, besides the presence of waterfalls here and there in the basin area increases the intensity of erosion process due to the presence of fault/share zones and the structural controls.

Relative Relief (Rr)

The relative relief value of Umtrew River basin have been determined and given in Table 3 which has been computed by the equation developed by Melton (1958). Furthermore, the visual study of the ASTER G-DEM (Figure 2) indicates that the average elevation of the basin area varies from 171 to 1790 m.

Ruggedness Number (Rn)

It is the product of maximum basin relief (H) and drainage density (D), where both parameters are in the same unit. An extreme high value of ruggedness number occurs when both variables are large and slope is not only steep but long as well (Strahler, 1956). In the present study, the value of ruggedness number (Table 3) is low which indicates moderate to steep slope of the basin area. The value of ruggedness number stands at 4920.

Slope analysis

Slope is the most important and specific feature of the earth's surface form. Maximum slope is well marked in the direction of a channel reaching downwards on the ground surface. There are many contributions to slope-geomorphology and various methods of representing the slope, but the contributions made by Wentworth (1930), Smith (1938-1939), Calef (1950), Calef and Newcomb (1953), Strahler (1956) and Miller (1953) are very important. Slope can evaluated as a quantitative parameter. Slope map (Figure 4) in the present study has been prepared by using "3D Analysis" tool of Arc GIS 9.2 software package.

The slope analysis of Umtrew River basin shows that a substantial portion of area falls under the slope category of more than 15° (Figure 5). Though 0 to 5° slope category is dominant in low lying areas and in some parts of southern flatlands but more than 20° slopes are still spreading everywhere in the study area. Low slope (0 to 5°) category is mostly visible in the plain patches of land which fall in the plain stretches of Assam. Highly steep slope areas (more than 25°) are located in the central-mid land to entire eastern and western part of Umtrew River basin. Moreover, the slope map of the study area enables to highlight the impact of underlying geological structures on the morphometric expression of the region. So, from environmental point of view gradually accelerating rate of deforestation in these patches of land may lead to the excessive rate of soil loss in the entire basin area.

Conclusion

The quantitative analysis of morphometric parameters is found to be of immense utility in river basin evaluation, watershed prioritization for soil and water conservation, and natural resources management at a micro level. The morphometric analysis carried out in the Umtrew River basin shows that the basin is having a mixture of landform comprising of high hilly region to low lying plains. Umtrew River basin holds a very important position in the entire physiographic setting of Meghalaya plateau. Drainage network of the study area is trellis in nature, due to the entire basin area elongated tributaries flowing parallel or sub-parallel to the main stream; and also multiplicity in drainage texture and more significance of structural control over the formation of drainage network. In some parts of the basin, the dipping and jointing of the topography reveals parallel and radial pattern. The analysis of drainage pattern shows it that in the central upland part of the basin the drainage has taken a more or less dendritic shape, which has contributed immensely in the overall spread of the network in all the directions. Low value of length of the overland flow also signifies that most part of the basin has been developed in the undulating hilly topography, where different geological structures has played an important role in the development of a true mountainous drainage network. The study reveals that remotely sensed datasets merged with GIS based tools in evaluation of drainage morphometric parameters and their influence on landforms, soils and eroded land characteristics at river basin level is more appropriate than the conventional methods. The bifurcation ratio in the basin indicates that the area has received significant structural disturbances and the drainage pattern is not distorted. Furthermore,

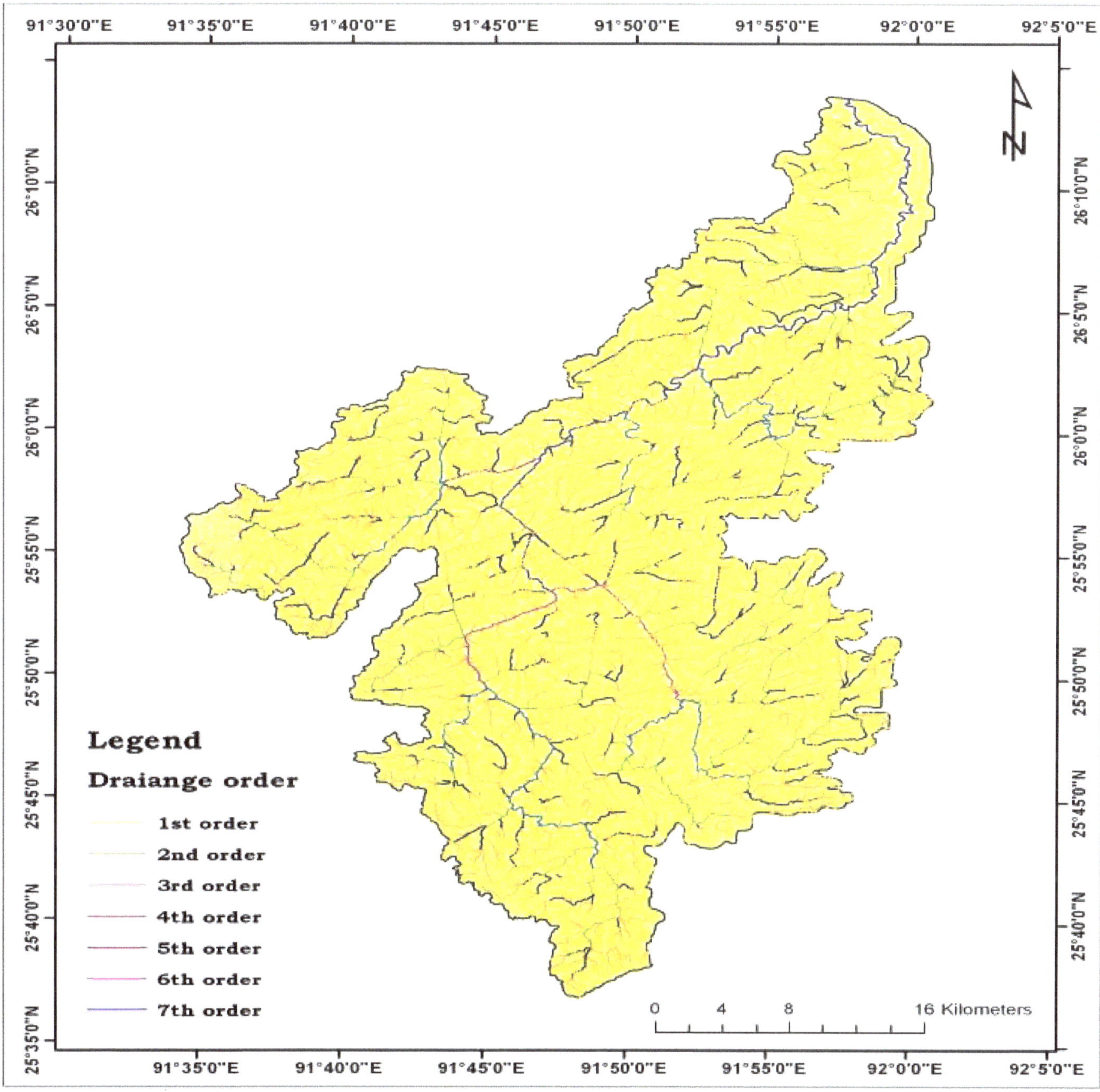

Figure 3. Drainage map of Umtrew River basin.

the low Bifurcation ratio values signify a high drainage density, low permeability of the terrain and indicate areas with uniform surficial materials where geology is reasonably homogeneous. The stream frequency of the study area shows a positive correlation with the drainage density, which indicates that the stream population increases with the increase of drainage density. GIS based approach facilitates analysis of different morphometric parameters and to explore the relationship between the drainage morphometry and properties of landforms, soils and eroded lands. GIS techniques characterized by very high accuracy of mapping and measurement prove to be a competent tool in morphometric analysis. The morphometric analyses in the present study were carried out through measurement of linear, areal and relief aspects of the basin area with different sets of morphometric parameters. The variation in stream characteristics might be due to change in slope

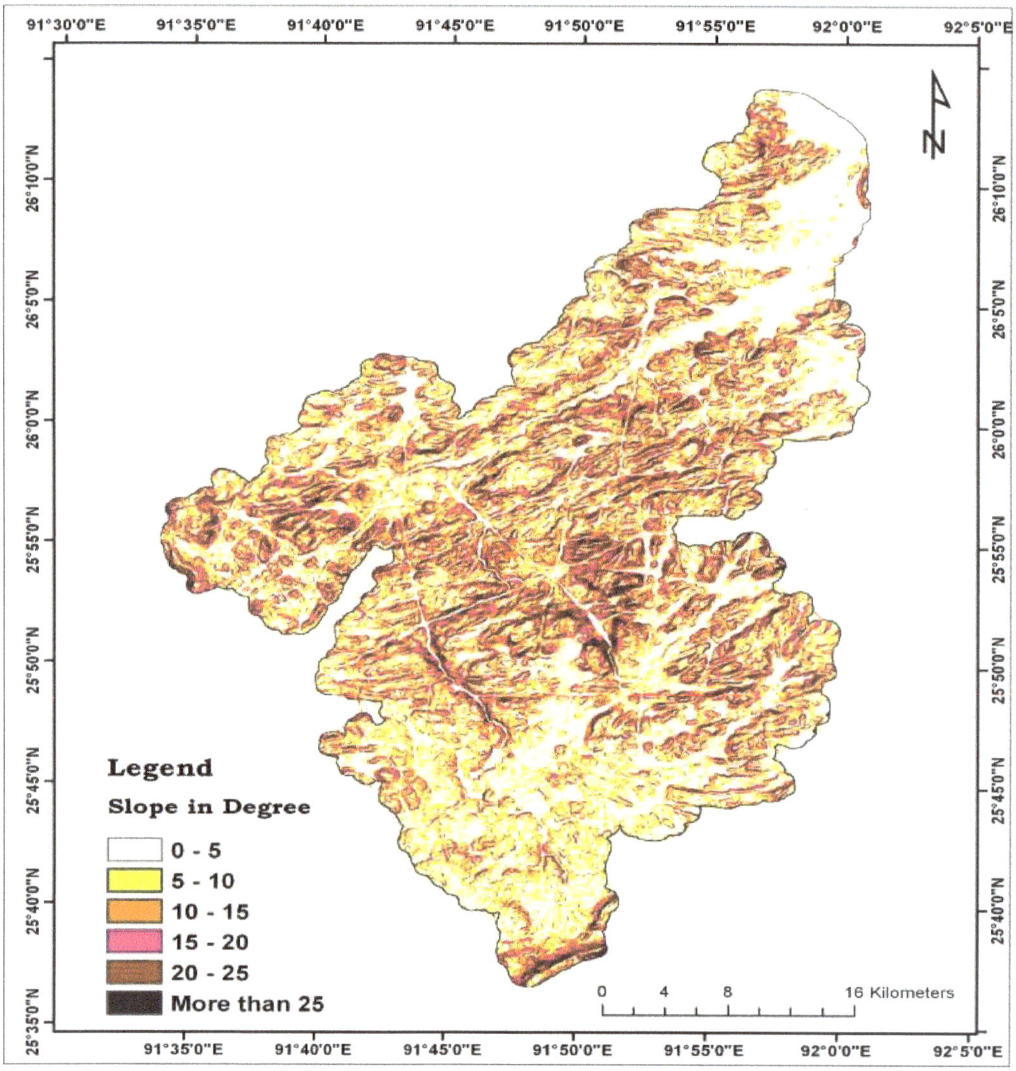

Figure 4. Slope map of Umtrew River basin.

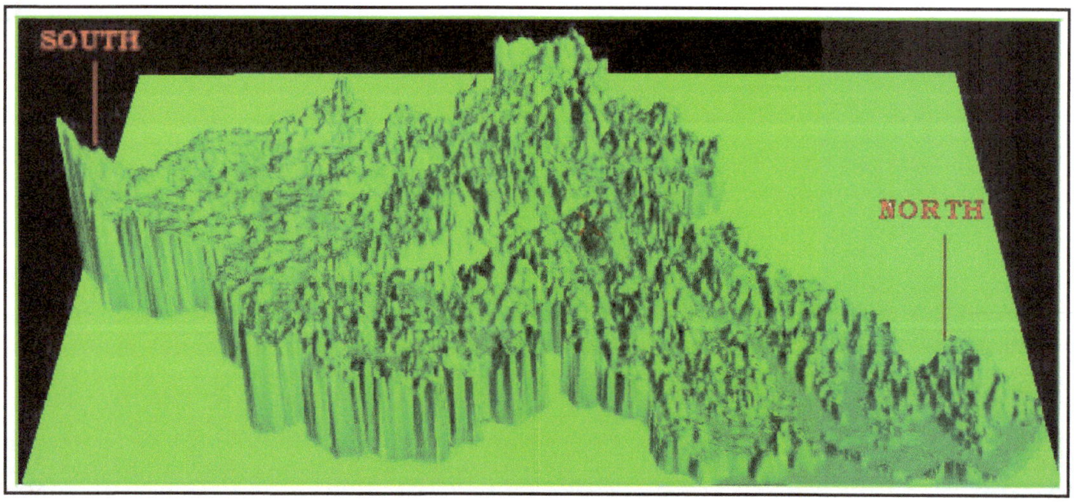

Figure 5. Digital Elevation Model (DEM) of Umtrew River basin.

and topography.

REFERENCES

Agarwal CS (1998). "Study of drainage pattern through aerial data in Naugarh area of Varanasi district, U.P.". J. Indian Soc. Rem. Sens. 26:169-175.

Calef WC (1950). "Form and Process." Cambridge University Press, London, p. 473.

Calef WC, Newcomb R (1953). "An average slope map of Illlnois". Annals Association of American Geographers, 43:305-316.

Chottopadhyay N, Hasmi S (1984). "The Sung Valley Alkaline Ultramafic Carbonalite Complex, East Khasi Hills and Jaintia Hills district, Meghalaya." G.S.I. Record. 113(4):24-33.

Chorley RJ, Donald EGM, Pogorzelski HA (1957). "A new standard for estimating drainage basin shape." Am. J. Sci. 255:138-141.

Gregory KJ, Walling DE (1968). "The Variation of Drainage Density within a Catchment." Int. Assoc. Sci. Hydrol. – Bull. 13:61-68.

Horton RE (1932). "Drainage basin characteristics." Trans. Am. Geophys. Union 13:350-361.

Horton RE (1940). "An approach toward a physical interpretation of infiltration capacity." Proc. Soil Sci. Soc. Am. 5:399-417.

Horton RE (1945). "Erosional development of streams and their drainage basins: Hydrophysical approach to quantitative morphology." Bull. Geol. Soc. Am. 5:275-370.

Kale VS, Gupta A (2001). "Introduction to geomorphology." pp. 84-86.

Miller VC (1953). "A quantitative geomorphic study of drainage basin characteristics in the Clinch Mountain area, Varginia and Tennessee", Project NR 389042,Tech. Rept. 3.,Columbia University, Department of Geology, ONR, Geography Branch, New York.

Melton MA (1958). "Geometric properties of mature drainage basin systems and their representation in their E4 phase space." J. Geol. 66:35-56.

Nag SK (1998). "Morphometric analysis using remote sensing techniques in the Chaka sub basin Purulia district, West Bengal." J. Indian Soc. Rem. Sens. 26:69-76.

Nag SK, Chakraborty S (2003). "Influence of Rock types and Structures in the Development of Drainage Network in Hard rock Area." J. Indian Soc. Rem. Sens. 31(1):25-35.

Reddy O, Maji GE, Gajbhiye KS (2002). "GIS for morphometric analysis of drainage basins." GIS India 11(4):9-14.

Schumn SA (1956). "Evaluation of drainage systems and slopes in badlands at Perth Amboy, New Jersy." Bull. Geol. Soc. Am. 67:597-646.

Singh S (1998). Physical Geography, Prayag Pustak Bhawan, Allahabad, India.

Smith KG (1950). "Standards for grading texture of erosional topography." Am. J. Sci. 248:655-668.

Srivastava BK (1997). "Study of Drainage Pattern of Jharia coalfield (Bihar), India through Remote sensing Technology." J. Indian Soc. Rem. Sens. 25(1):41-46.

Srinivasa VS (2004). "Morphometric analysis of sub-watersheds of in the Pavagada area of Tumkur, district, South India using remote sensing and GIS techniques." J. Indian Soc. Rem. Sens. 32(4):351-362.

Strahler AN (1950). "Equilibrium theory of erosional slopes approached by frequency distribution analyses." Am. J. Sci. 248:673-696, 800-814.

Strahler AN (1952). "Hypsometric (area-altitude) analysis of erosional topography." Bull. Geol. Am. 63:1117-1142.

Strahler AN (1964). "Quantitative geomorphology of drainage basins and channel networks." Handbook of Applied Hydrology, McGraw Hill Book Company, New York, Section 4II.

Strahler AN, Strahler AH (2002). "A Text Book of Physical Geography." John Wiley & Sons, New York.

Suresh K (2000). Soil and Water Conservation Engineering 3rd Ed.24. Watershed-Concept, Management, pp. 785-813.

Wentworth CK (1930). "A Simplified Method of Determining the Average Slope of Land Surfaces." Am. J. Sci. 21:184-194.

25

Integrated water resources management in Iran: Environmental, socio-economic and political review of drought in Lake Urmia

N. Zarrineh and M. Azari Najaf Abad

Department of Hydrology and Hydrological Engineering, Vrije Universiteit Brussel (VUB), Brussels Belgium.

In recent years Lake Urmia, the largest saline lake in the Middle East located in northwestern Iran has undergone severe environmental changes. As a result of drought and anthropogenic impacts, the area of the Lake has been shrunk and the water level has been dropped. In this article the environmental, political and socio-economic impacts of drought in Lake Urmia basin has been reviewed and the obstacles regarding institutional water frameworks in national and regional levels has been studied and assessed. Furthermore, lack of sufficient mitigation and adaptation policies and inadequate attention to the environmental impact assessment during megaprojects has been discussed. The most crucial impacts have been realized as ecological and environmental consequences of lake drying up on all over the Lake Urmia catchment. In associated with these consequences, several scarce species of flora and fauna are exposed to the danger of extinction, and polluted air ensued by the salt storms affect the daily life of people in the region. In addition, this event has a direct influence on the economy of the region. Therefore, the implementation of a holistic institutional-based remediation program to accomplish lake restoration seems to be inevitable.

Key words: Lake Urmia management, Lake drought, human activities, climate change, environmental impact assessment.

INTRODUCTION

Lake Urmia (formerly called Rezaiyeh) is a shallow landlocked hypersaline lake located in northwestern corner of Iran close to the border with Turkey. It is known as one of the largest continental salt lake in the world and the largest one in the Middle East (Figure 1) (Hassanzadeh et al., 2012). Recently, area and volume of the lake have been shrinking seriously. The lake was declared a Wetland of International Importance by Ramsar Conversation in 1971 and designated a UNESCO Biosphere Reserve in 1976 (Chander, 2012).

Lake Urmia is one of the hyper saline lakes throughout the world. In addition, this basin is an endorheic basin that retains water and does not allow any outflow to other external bodies of water. According to the information issued by The Ramsar Convention on Wetlands, the primary outflow of Lake Urmia is evaporation and the primary inflows are precipitation, freshwater discharge from the rivers and getting fed by springs. Increase in evaporation and decrease in rainfall, freshwater inflow and springs inflow has led to salinization and water level decline.

The surface area of Lake Urmia has been estimated about 5200 km^2 and the area of the basin is estimated 51,876 km^2. Lake Urmia possesses more than 102 rocky

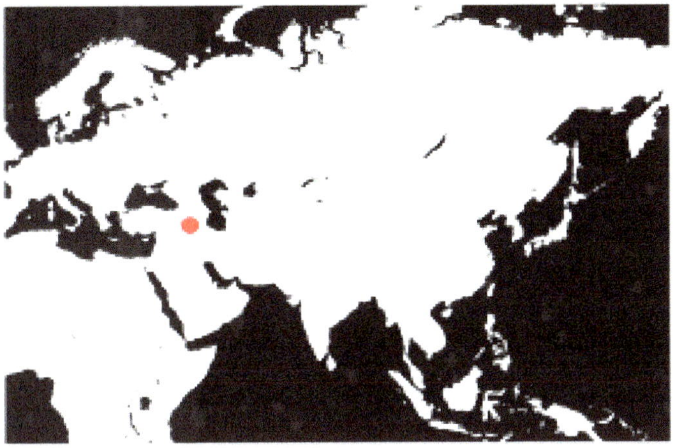

Figure 1. Location of the Lake Urmia.

saline islands and its ecosystem is the habitat for more than 212 species of birds, 41 reptiles, 7 amphibians, and 27 species of mammals which most of them are endemic species (Rezvantalab et al., 2011).

According to the U.S. Geological Survey anticipation, the Lake Urmia is sharply drying, and it may completely vanish in just two years (USGS, 2012). If this scenario happens just a vast salt desert will remain from this lake which is an undeniable treat to the local ecosystem. Therefore two main problems have been recognized and defined lately based on water quantity and water quality.

In terms of water quantity, according to the official statistics, more than 60% of Lake Urmia is disappeared since 1960s (Karmi, 2011). The water level decline is another drought indicator. Based on current investigations, in a recent drought of the Urmia basin, started in 1999, the water level of Lake Urmia dropped from 1277.80 masl to 1273.35 masl (Tabari et al., 2012). During August 1998 to August 2001, water level of Lake Urmia declined up to three meters and caused serious decrease in the area of the lake, approximately from 6000 to 5200 km^2.

In terms of water quality, the lake has been decayed and sudden change in salinity regime has been occurred. Currently the amount of the salt, dissolved in the lake water, increased from 160 to 170 gL^{-1} up to 400 gL^{-1} (Zarghami, 2011). The mentioned disaster has affected the biota of the lake. In these conditions, water with so high salinity is no longer a suitable and sustainable environment for fauna and flora and has caused to destroy the ecology of the lake.

Figure 2 represents shrinking in the area of the Lake and severe drought has been observed (left picture) and the trend to this drought is likely to be continued (right picture). As a plausible consequence, the eastern and southern areas of the lake would be completely dried up until 2019 and this can be associated with irreparable environmental impacts in the ecosystem of the Lake Urmia catchment.

Objective

This article aims to develop a holistic review on ecological and environmental impacts and political and socio-economic implications of Lake Urmia basin drought in response to anthropogenic drivers such as increased water demand due to population growth (Statistical Center of Iran, 2012) and environmental pressures (e.g. climate variability and climate change) within the context of current institutional set-up. Political and institutional obstacles will be presented and comprehensively discussed in this review. Thus the final conclusion of this review reflects the previous and the current river basin's characteristics and the impacts of human activity and climate change on the status of water bodies in the basin and the possible measures. For this aim, this study will also address how the water quantity and quality and ecological states of Lake Urmia Basin has been changed over the past decades. The ultimate objective of this review paper is to create and develop a recommended list of measures that might improve the management of the Lake Urmia Basin.

LITERATURE REVIEW

Main drivers and pressures

Two major aspects of drivers which are responsible for this environmental crisis are anthropogenic driving forces and natural driving forces. In this occurrence, anthropogenic drivers are more effective than environmental drivers. Sharp population growth during 1980 to 2011 and intensive immigration towards big cities raised the food and water demand. Regarding to the amount of water which is consumed instead of being discharged in the Lake Urmia, these participations can be obtained that agriculture sector is responsible for 91% of the water consumption, whereas industry and domestic consumption are responsible for 6 and 3% of the water consumption respectively (Faramarzi, 2012). Raise of demand for water forced the government and farmers to provide numerous damming and water diversion projects and groundwater pumping stations to supply water demand.

The main freshwater resource is Zarrineh River, located in the south of the lake, supplying more than 50% of the annual inflow. Simineh River and Ghadar River are providing 35% of the inflow (Mohaggeg, 2002). Some of the freshwater from these sources flows unimpeded to Lake Urmia, but much of it is impounded within reservoirs or water diversion dams for the sake of irrigation, drinking water, power generation (hydro-power plant) and industrial usage of water. For instance, Bukan dam was constructed on Zarrineh River in order to launch hydropower plant and supply water demand. The location of the Bukan Dam is specified in Figure 3. In addition, several damming and water projects are currently under

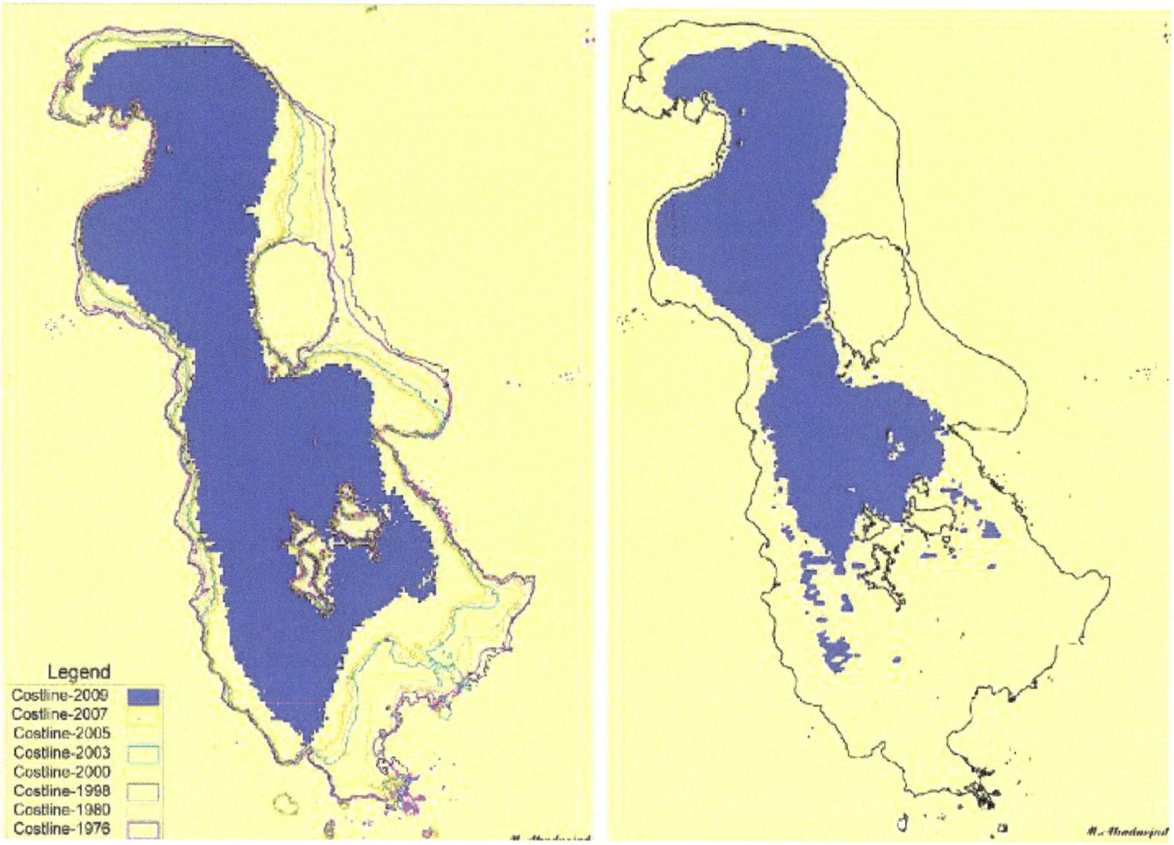

Figure 2. Lake Uremia water level changes in period 1976-2009 (left) and predicted change map by 2019 (right) (Ahadnejad Reveshty, 2010).

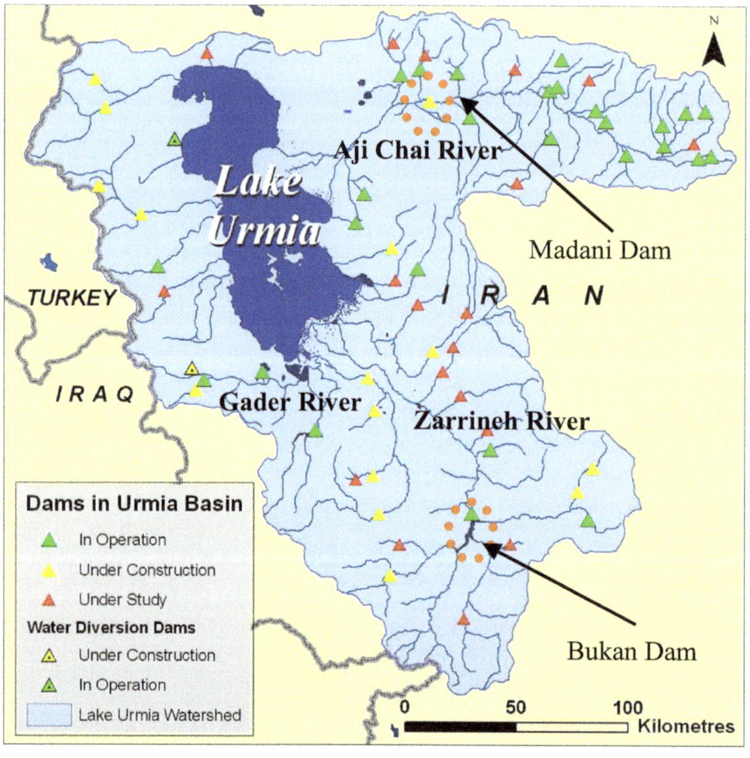

Figure 3. Dams location in Urmia Basin.

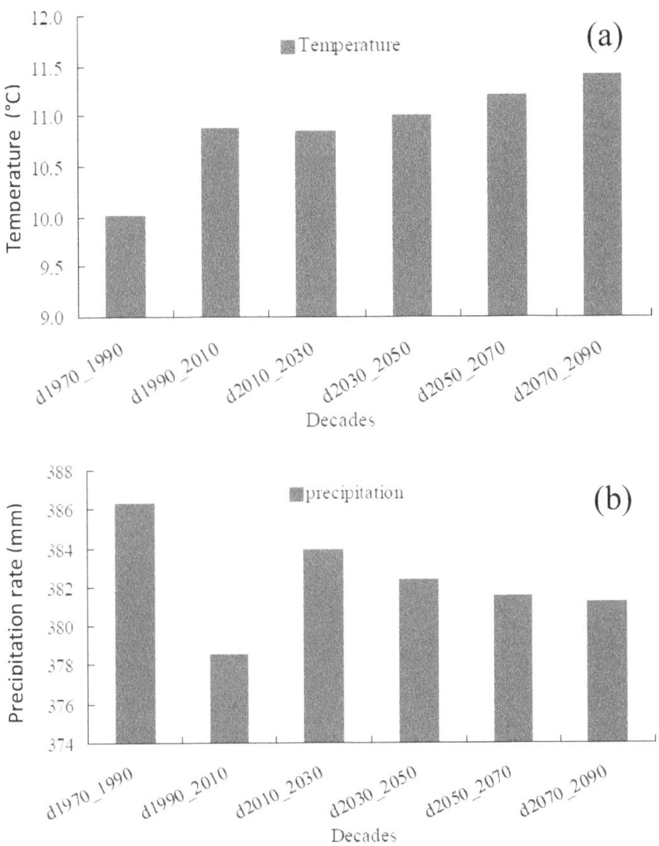

Figure 4. (a) Ten-year average (1970-2100) of trend of changes in temperature components of the Lake; (b) the ten-year average (1970-2100) of trend of changes in precipitation components of the Lake.

study and under construction. Since large number of dams and water diversion projects are already in operation on the rivers feeding the lake, Lake Urmia is shrinking, getting shallower and extremely drying, and a salt desert has been remained which can lead to several environmental consequences. Furthermore, groundwater excessive unsupervised overuse caused groundwater depletion which decreased the amount of water that is fed into the lake by the springs. This event may lead to the salt intrusion in the catchment of Lake Urmia.

Moreover, the construction of a causeway, from 1999 until 2007, in order to decrease the distance between Tabriz city, capital of East Azerbaijan province, and Urmia city, capital of West Azerbaijan Province, was believed to pressurize the flow and salinity regime of the lake. Various studies has shown the influence of the causeway construction on the lake's hydrodynamics and water quality by using two-dimensional and three-dimensional numerical models (Zeinoddini et al., 2009). The necessity to build a new road is the direct effect of population and trade growth as the main drivers and consequently increasing transportation demand is the main pressure (Khosravifard, 2010). This bridge divides the lake into two parts. The severe drying of the lake was simultaneously coincided with the project construction duration and has been intensified after running the project. The flow and salinity regimes are affected by the presence of this new causeway (Zeinoddini et al., 2009).

Furthermore, due to climatic variability and climate change the quantity of water flowing into the lake varies annually. Higher temperature and less precipitation have led to higher evaporation from the lake and less water yield into it. As evaporation is the primary output and precipitation is the input, salinization will increase dramatically due to less dilution.

Figure 4 shows the trend of the average temperature and precipitation changes which two time scales of the past and future are considered. To characterize the changes in climate components for time scale of the past and future the longitudinal data set between 1968 and 2011 and between 2012 and 2100 considering the climate change effect have been used respectively. The changes of annual average of precipitation display a decreasing trend every decade from 1968 to 2011 which is equal to -14.8 mm/decade. The trend of changes in average temperature between 1968 and 2011 displays a significant rise which is equal to 0.59°C/decade. Also the increasing trend of annual temperature considering climate change scenario, indicates an increase in temperature of +0.88°C by the year 2100 in the Lake Urmia region. Furthermore, it is expected that the average of precipitation is projected equal to -0.37 mm/decade. Totally it will decrease as 3.3 mm by the year 2100 (Tisseuil et al., 2013).

These results of latter study are fully compatible with the map given in Figure 5. It shows that areas around Lake Urmia are being threatened by drought.

The decreasing factor in precipitation rate is estimated between 50% (severe drought) to 80% (weak drought). The map is based on the percentage of rainfall from September 2010 to January 2011 in comparison with the long-term average.

Therefore it is concluded that the precipitation (primary inflow of Lake Urmia) will decrease and the evapotranspiration will increase (primary outflow the Lake) by increase in the temperature.

States

All of these drivers and pressures have led to the current lake situation. As the Lake Urmia basin is an internal drainage system, the climatic factors have driven the lake to the more evaporation and less precipitation and the anthropogenic factors have led to less freshwater input. Less freshwater inflow and higher evaporation have led to higher salinity and water level decline.

Figure 6 represents the gradual change in surface area of the lake. After fluctuations between 1963 and 1998, sharp drop of surface area has been observed.

As shown in Figure 7, in terms of water quantity, although there are fluctuations in water level over the past

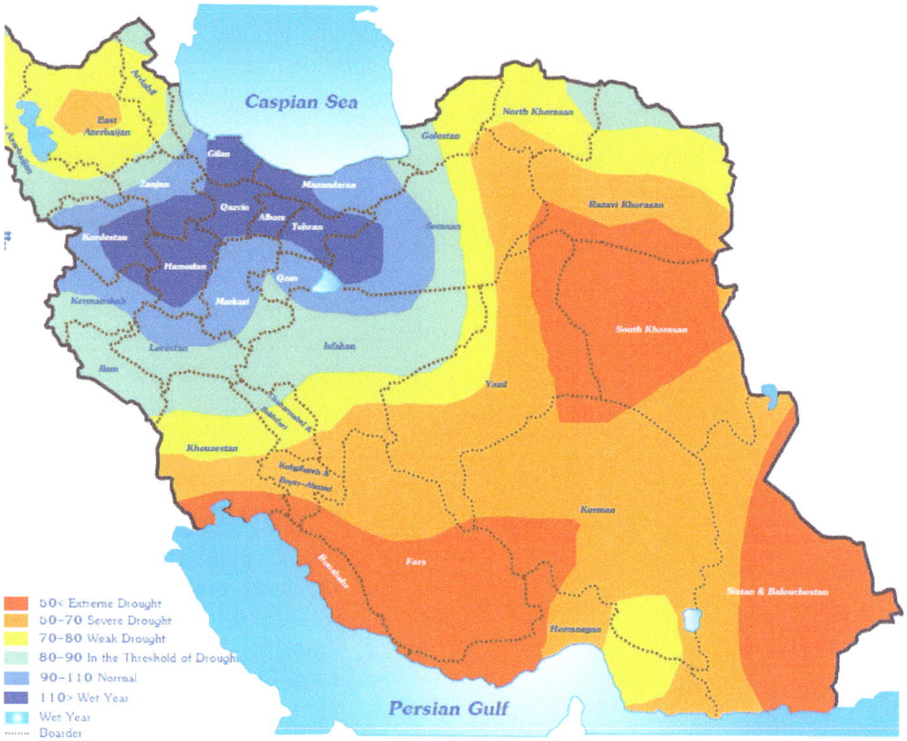

Figure 5. Drought map of Iran.

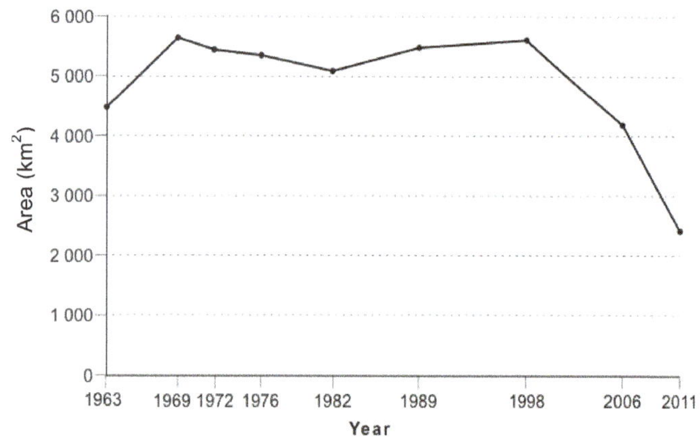

Figure 6. Lake Urmia surface area, 1963-2011 (Chander, 2012).

100 years, the trend line slope is negative which implies that the lake situation is getting worse.

In terms of water quality, recent investigations denotes that in early 2003 salinity of the lake has been raised up to 250 g/L in the southern part, whereas salinity was as high as 280 g/L in the northern part. More recent studies illustrate the amount of salt in the lake water increased from 160 to 170 g/L up to 400 g/L (Zarghami, 2011).

Environmental and ecological Impacts

Environmental campaigners claim that the construction of dams on rivers and the recently built causeway across the narrowest part of the lake have reduced the water level, affected the water circulation, increased the water salinity and jeopardised animals as well as aquatic organism. The most important flora and fauna which are presently endangered and may extinct in the future are aquatic organisms, regional vegetation, flamingos and Iranian yellow deer. Lake Urmia possesses more than 102 rocky islands. The islands and the wetlands around the lake are major staging areas for migrating flamingos, however their numbers have recently declined (Zafarnejad, 2010) (Figure 8).

Significant alterations in the biota of the lake endangered aquatic organisms such as Artemia, a kind of brine shrimp which is a unique cosmopolitan anostracan living in hypersaline and saline lakes. Artemia serve as food source for the migratory birds such as flamingos and is also used in aquaculture sector. Because of high salinity, Artemia was not able to fully recover in the lake's water (Abbaspour and Nazaridoust, 2007). The other consequence on the biota of the lake is fish mortality. Due to the high level of salinity, the Lake Urmia is no longer able to sustain any fish species.

The hydrological pressures on the lake have profound impacts not only on the lake itself, but also on the land cover and the water quality and the ecosystem of surrounding wetlands especially those located in estuary parts. Water quality conditions differed among the wetlands and ranged from mostly freshwater, nutrient-rich (eutrophic) conditions to more saline, nutrient-poor

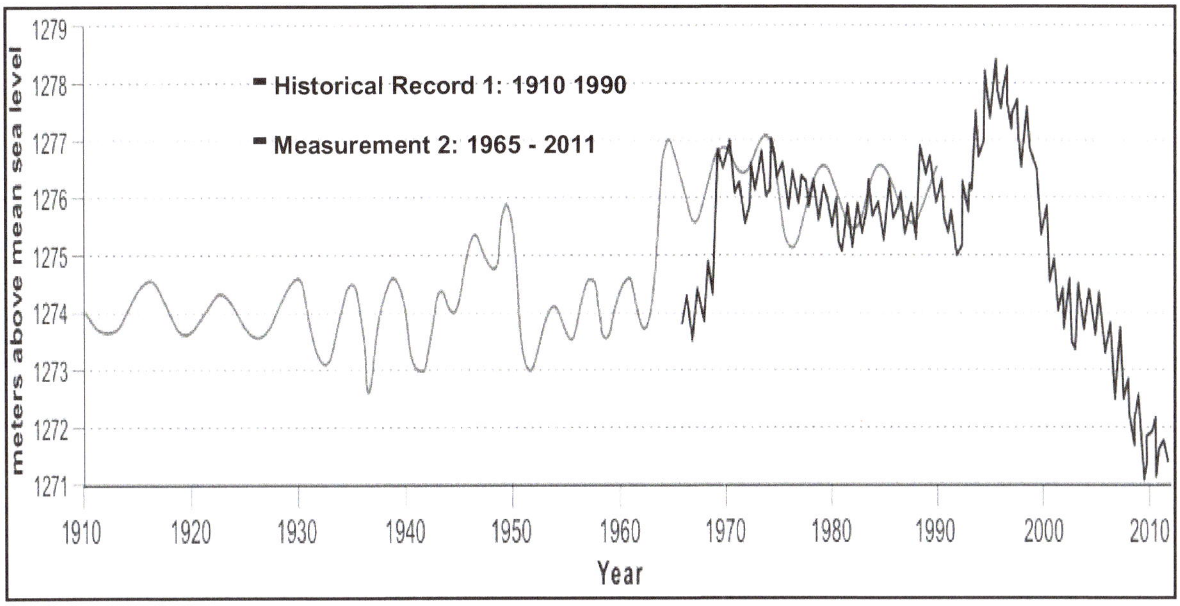

Figure 7. Lake Urmia water level change and its fluctuations, 1910-2011 (Chander, 2012).

Figure 8. Environmental Impact of Lake Urima Drying on biota of the lake basin: (Left: flamingos, right, and a dead flamingo).

(oligotrophic) conditions. One principal example is the aquatic macroinvertebrates living in Zarrineh River which are currently endangered upon poor nutrient conditions and highly saline water (Ahmadi et al., 2011).

Socio-economic impacts

The use of Artemia in fish and shrimp aquaculture is commercially important and recently has been restricted. Due to fish kill in Lake Urmia no fishing benefits can be expected which has led to reduce economic benefits.

Increase in the number of salt storm events is noticed over the last few years. Destructive effects of chemical pollutants of salt storms on bridge constructions and other infrastructures such as harbor and port facilities, cargo warehouses and passenger boats have been observed lately. In addition, when the toxins and minerals from the salt flats are blown into the air by these storms and inhaled, the toxins and minerals may cause throat and lung cancer, infant mortality, decreasing the life expectancy and increase in birth defects. Vice president of the Iranian Department of Environmental Protection believes that the consequences may lead to exacerbate air pollution in adjacent regions (Fazel, 2012; Interview).

The environmental specialists call these events "Salt Tsunami" (Khabbaz, 2012). The other impact on the social life is related to the mud bathing properties of Lake Urmia salt marshes. This is a medicinal bath in heated mud to treat some diseases such as Rheumatism.

Recently due to severe drought and unsatisfactory environmental conditions, this region may not be as a target for patients.

Political and Institutional implications

Integrated Management Plan for Lake Urmia Basin was signed by Ministry of Agriculture, Ministry of Energy, Chief of Iran's Department of Environment Protection and the governors of provinces responsible for the Lake Urmia basin. This IWRM plan consists of two main programs which are currently under implementations by Ministry of Energy, the main client for projects related to supplying water demand, and Ministry of Agriculture, the main responsible for irrigation sector.

The first and main suggested plan is to reduce agricultural water use to provide 3100 MCM Lake Urmia's water demand by expanding pressurized irrigation and changing cropping pattern (Morid, 2012). The proposed duration is 3 years and Ministry of Agriculture is the main supervisor of this plan.

As the second plan, constructing dams in the Lake Urmia basin will be suspended, and the possibility of transfer water from out of the basin for instance from Aras River and Caspian Sea will be explored. Water transfer from Aras River has started from February 2012, through this way annually 600 MCM water can be transferred into the basin, but the ecological consequences have seriously concerned ecologists (Chit Chian, 2013; Interview).

Prospective barriers of current Lake regenerative programs are insufficient collaboration between corresponding responsible ministries and organizations, mismanagement in responsible regional institutions and inconsistency in other administrative sectors which are leading to a notable delay in ongoing projects.

Furthermore, most studies on Lake Urmia are based on a few limited samples, so investment on research by universities and research institutes are crucial. Lack of government supervision and inadequate attention to environmental issues and human activities impacts is undeniable.

In terms of obstacles caused by publics due to overconsumption and lack of sufficient information, an educational program for public awareness about how the potential volunteers can cooperate to lake restoration programs is inevitable. Volunteers can be involved directly in Non-Governmental Organizations (NGO), charity foundations and public companies to maintain a public-private partnership structure (PPP) in order to save Lake Urmia.

RECOMMENDATIONS

The authors recommend the current restoration plans for saving Lake Urmia could include some other measures. We have integrated our new proposals and categorized the program of measures into three main procedures.

Procedure A is adjusting water allocation in the basin to allow an adequate flow for sustainable management of the Lake Urmia basin. Full descriptions on the proposed measures for water allocation adjustment (Procedure A) are given in Table 1. It has been presumed that the combination of all measures given in this procedure would be the best ecological option to save the Lake Urmia.

Procedure B could be carried out by importing water from outside the basin to supply water demand and let more water flow to the lake (Kordavani, 2012, Reiisi, 2011). Although procedure B as transferring the water has been proposed by many specialists as a time-efficient way to rescue the lake, the head of Iran's Environmental Protection Organization, noted that the corresponding Organization does not consent with the idea of transferring water from Caspian Sea or Aras River to Lake Urmia. Due to this prospective view, the following plans are detrimental to the ecosystem of the basin and may bring severe environmental degredation. Regarding to the remarkable economic costs for the government and the unbeneficial impact on the agronomy of the neighboring catchments, the procedure B cannot be justified. Therefore, Instead of procedure B, Iran's Environmental Protection Organization is intended in strategies more compatible with region's ecology (Ebtekar, 2013; interview).

Procedure C is proposing investment on research, designing and implementing novel techniques e.g. cloud seeding. That would be possible by providing the artificial rains on the Urmia Lake basin. Although the limited cloud seeding projects are under study, but the operation of these projects are still controversial.

CONCLUSION

Lake Urmia located in northwestern of Iran has experienced the severe drought over the past years, where over 70% of the lake water has dried up. Studies show the water level has declined since 1960s. Climate change and climatic variability in climatic patterns, increase in sowing areas, construction of dams and occurrence of excessive water wells around the lake are the main natural and anthropocentric reasons which caused the lake to diminish. These driving forces result in major changes in the area's ecosystem and significant changes in economy of the region.

On August 2013, the Iranian president established a working group to tackle the issue of saving the Lake Urmia. The work group was commissioned to use the background of the already conducted studies and technical research works to resolve this environmental problem and to present their proposals for saving the Lake Urmia to the government. However, in a long-term vision, more studies have to be accomplished on establishing an innovative and holistic IWRM framework

Table 1. Full description on the proposed measures for water allocation adjustment (Procedure A).

S/N	Proposition	Remarks
1	Stopping under construction dams and all ongoing surveys	Pros: Providing sustainable ecosystem for future More budget will be saved Cons: Unemployment at consultancy sector Farmers dissatisfaction
2	Opening remaining dams permanently or temporarily during wet periods	Pros: Providing water for restoring the Lake periodically Cons: Farmers dissatisfaction
3	Reducing agricultural water demand: 3.1) Deficit irrigation, 3.2) Reducing cropped area 3.3) Elimination of some of the annual crops 3.4) Changing the cropping patterns 3.5) Expanding pressurized irrigation 3.6) wastewater reuse 3.7) Enhancing agricultural drainage system	Pros: Accomplish more than 3000 MCM/year to the lake Cons: Social tension Unemployment Endangering food security Consuming huge budget Stakeholder management and engagement is needed Subsidies for supporting frames are required
4	Reducing municipal and industrial water demand: 4.1) Studying modern approaches regarding to drinking water supply and distribution systems, urban drainage and sewage collection and disposal systems 4.2) Public participation programs 4.3) Water and wastewater reuse	Pros: Can accomplish more than 600 MCM annually Cons: Lack of education High budget is required No studies are implemented so far
5	Reducing available Lake Urmia basin water quota for different sectors in each province only in drought spells	Pros: Sustainable allocation during drought spell Cons: Stakeholder participation More collaboration between regional sectors
6	Enhancing the water quality of the lake towards a proper ecological status: 6.1) Enhancing water and wastewater treatment techniques to lower salt loads into the water bodies (e.g. membrane filtration) 6.2) Salt and minerals extraction from accumulated salts around the lake	Pros: Sustainable management of ecosystem of basin Minerals extraction from the lake is profitable Cons: High budget Insufficient studies

to study the casual relations between the different drivers and responses of this crisis within the catchment. Once these frameworks are established, a work plan of remediation consisting of structural and non-structural measures can be conceptualized and implemented for sustainable management of Lake Urmia basin.

REFERENCES

Abbaspour M, Nazaridoust A (2007). Determination of environmental water requirements of Lake Urmia, Iran: an ecological approach. Int. J. Environ. Stud. 64(2):161-169.

Ahmadi R, Mohebbi F, Hagigi P, Esmailly L, Salmanzadeh R (2011). Macro-invertebrates in the Wetlands of the Zarrineh estuary at the south of Urmia Lake (Iran). Int. J. Environ. Res. 5(4):1047-1052.

Chit Chian A (2013). Official Interview. Accesed: 12/10/2013. Available form: http://www.azernews.az/region/60438.html

Ebtekar M (2013). Official Interview. Accessed: 12/10/2013. Available form: http://isna.ir/fa/news/92071308466

Faramarzi N (2012). Agricultural Water Use in Lake Urmia Basin, Iran: An Approach to Adaptive Policies and Transition to Sustainable Irrigation Water Use. Master Thesis, Upsala University, Department of Earth Science, p. 44.

Fazel A (2012). Official Interview. Accessed by: 12/10/2013. Available from: http://aftabnews.ir/fa/news/171331

Hassanzadeh E, Zarghami M, Hassanzadeh Y (2012). Determining the Main Factors in Declining the Urmia Lake Level by Using System Dynamics Modeling. Water Resour. Manage. 26:129-145.

Karmi N (2011). Iran's largest lake turning to salt. Accessed: 16/12/2012. Available from: http://news.yahoo.com/s/ap/20110525/ap_on_re_mi_ea/ml_iran_environmental_disaster/print

Khabbaz M (2012). Salt Tsunami. Jame Jam News Paper. Accessed: 16/12/2012. Available from: http://jamejamonline.ir/newstext.aspx?newsnum=100818506546

Khosravifard S (2010). Campaigners Fear Lake Urmia Drying Up. Institute for War & Peace Reporting (IWPR). Accessed: 19/12/2012. Available from: http://iwpr.net/report-news/campaigners-fear-lake-urmia-drying.

Kordavani P (2012). Little Zab is the way to rescue the Lake Urmia. Young Journalists Agency. Accessed: 16/12/2012 Available from: http://www.yjc.ir/fa/news/40802114
Mohaggeg MH (2002). Reducing on the water level of Urmia lake, future view and recomondations. Urmia lake and its potentials to development. Urmia University pp. 1:69
Morid S (2012). Introducing Lake Urmia Basin (Iran) Drought Risk Management Plan. High Level Meeting on National Drought Policy, Geneva, Switzerland.
Chander A (2012). The drying of Iran's Lake Urmia and its environmental consequences. Environ. Dev. 2(2):128-137. http://dx.doi.org/10.1016/j.envdev.2012.03.011.
Reiisi B (2011). Official speech, Mehr News Agency. Accessed by: 16/12/2012. Available from: http://www.mehrnews.com/fa/news detail.aspx?NewsID=1322.
Rezvantalab S, Amrollahi, MH (2011). Investigation of Recent Changes in Urmia Salt Lake. Int. J. Chem. Environ. Eng. 2(3):168–171.
Statistical Center of Iran (2012). Result of Census. Accessed by 12/16/2012, Available from: http://www.amar.org.ir/Portals/0/Files/abstract/1390/sarshomari90_nahaii.pdf.
Tabari J, Nikbakht J, Talaee PH (2012). Hydrological Drought Assessment in NorthwesternIran Based on Streamflow Drought Index (SDI). Water Resources Management
Tisseuil C, Roshan G, Nasrabadi T, Asadpour GA (2013). Statistical modeling of future lake level under climatic conditions: Case study of Urmia Lake (Iran). Int. J. Environ. Res. 7(1):69-80.
U.S. Geological Survey (2012). Changes Over Time: Lake Oroumeih (Urmia), Iran. Accessed by: 16/12/2012. Available from: http://remotesensing.usgs.gov/feature/description.php?id=223
Zafarnejad F (2010). The contribution of dams to Iran's desertification. Int. J. Environ. Stud. 66(3):327-341.
Zarghami M (2011). Effective watershed management: Case study of Urmia Lake, Iran. Lake Reservoir Manage. 27:87-94.
Zeinoddini M, Tofighi MA, Vafaee F (2009). Evaluation of dike-type causeway impacts on the flow and salinity regimes in Lake Urmia, Iran. J. Great Lakes Res. 35(1):13-2.

Assessment of water quality of Betwa River, Madhya Pradesh, India

Santosh Vishwakarma[1], Alok Varma[2] and Geeta Saxena[1]

[1]Department of Zoology, Government Science and Commerce College, Benazeer, Bhopal, India.
[2]Department of Zoology, Government Raja Bhoj College, Mandideep Raisen, India.

The present paper deals with the monthly interval variation of physico-chemical characteristics of River Betwa during October 2011 to August 2012. Five major sampling station (10 sites) were selected: station- I Jhirri (1 to 2 sites), station- II Nayapura (3 to 4 sites), station- III Mandideep (5 to 6 sites), station- IV Bhojpur (7 to 8 sites) and station- V Vidisha (9 to 10 sites). Various physico-chemical parameters like temperature, pH, hardness, dissolved oxygen (DO), biological oxygen demand (BOD) and chemical oxygen demand (COD) and correlation co-efficient were measured, and standard methods were applied in the present work. Studies revealed high values of hardness, BOD and COD at station II Nayapura and station III Mandideep, and low values of DO and pH at the station - station I Jhirri, IV Bhojpur, and Vidisha low values of hardness, BOD and COD. High value of hardness, BOD and COD indicate the river water polluted at station II Nayapura and station III Mandideep.

Key words: Betwa River, Mandideep, Nayapura, physico-chemical parameters.

INTRODUCTION

India is a land of many rivers. The total length of all our major rivers together with their tributaries is 27,359 km. From ancient time in India, the rivers are not only considered as a mass of flowing fresh water. According to Hindu mythology the water is believed to be one of the five essential ingredients of all the living beings, or the "Panch-Mahabhutas" ever since the Vedic times. Rivers have always been the most important fresh water resources and most developmental activities are still dependent upon them. Rivers play a major role in assimilating or carrying industrial and municipal waste water, manure discharge and runoff water from agricultural field, road ways and streets which are responsible for river pollution (Ward and Elliot, 1995). The River Betwa is a river in Northern India, and tributary of the Yamuna originating in the Kumra (Jhirri) village in Raisen dirtrict of Madhya Pradesh, India. The River Betwa plays a significant role in the human life of the villages villages located in Mandideep, Nayapura and Bhojpur areas. It has become polluted at some places of Mandideep due to industrial activities and the confluence of sewage, domestic wastes and industrial effluents of many big and small enterprises with various types of organic compounds and heavy metals deterioted to human health and aquatic organisms. Urban areas, farms, factories and individual households – all contribute to the contamination of this river. The water quality in the stretch of the River Betwa extending from its origin near Mandideep industrial area up to Bhojpur remains poor because of the regular inflow of domestic waste of the Bhopal city through the Kaliyasot River and industrial/domestic waters from Mandideep (Kori et al., 2006). Study of different physico-chemical parameter revealed that the intensity of pollution increased as the river was subjected to sewage and industrial waste. In order to manage the pollution load of River Betwa that pass nearby Mandideep and Nayapura, it is recommended that various methods of sewage industrial

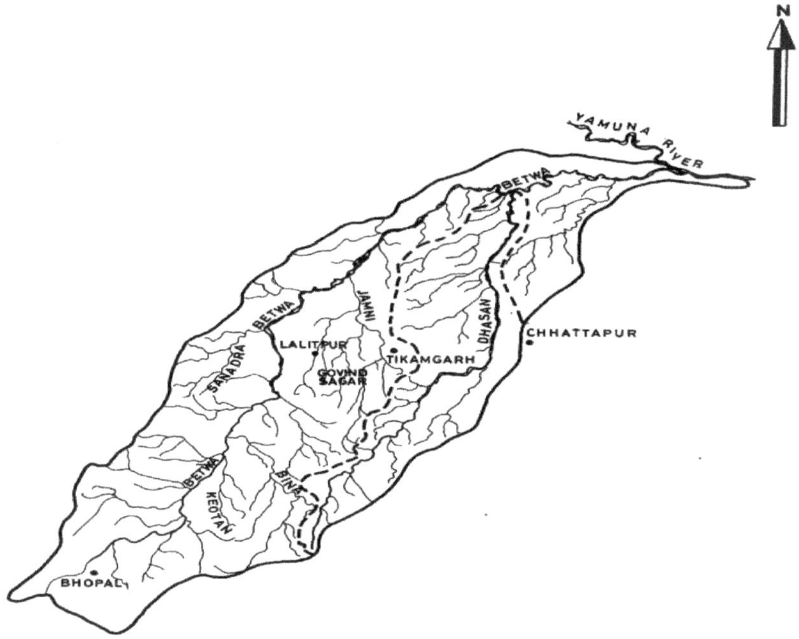

Figure 1. Betwa River Besin.

waste treatment should be used before disposal of effluents.This is a study to aware the people how they have contaminated the source of life. In view of increased demand of water due to expanding needs it seems evitable to study surface water bodies particularly river for proper utilization and conservation. For generation to come it becomes prime necessity to study River Betwa, of Madhya Pradesh which is facing many changes due to the development of urbanization and industrial evolution. The problem of resettlement of submergence, rehabilitation and change in ecosystem is creating major environmental impact as riverine ecosystem is being disturbed. The present work is aimed at understanding the following points:

1. Study of industrial pollution in River Betwa at Mandideep, and approximately 200 km of river stretch at 20 stations or sites.
2. To observe the impact of pollution on different sites of River Betwa, both biologically as well as chemically.
3. The analysis of water quality in terms of various selected physico-chemical parameters.

MATERIALS AND METHODS

Study area (sampling sites)

The river basin lies between the latitudes of 22* 54' N and 26*00, N and the longitudes of 77*10' E and 80*20' E. The total catchments area of the basin in 43895 Km^2 out of which 30217 Km^2 lies in Madhya Pradesh and the remaining 13678 Km lies in Uttar Pradesh (KBLP, 1995).

On the basis of the survey conducted and literature available, five stations and ten monitoring sites were selected to sample water.

The sites were selected mostly on the basis of various activities occurring on surrounding area of the river (Figure 1).

Samples were taken at the following points during alternate month

Station I: This station is located near village Jhirri. River Betwa originates from this point. It is fifteen kilometer away from Bhopal on Kolar Dam Road, touching the border of Ratapani sanctuary; home too many wildlife and birds (1 to 2 sites).

Station II: It is a village near Mandideep, the Betwa touches the boundary of this village near Road Bridge where it confluences with the Kaliyasot tributaries (3 to 4 sites).

Station III: Mandideep is the municipality in Goharganj district Raisen in MP. Mandideep is 20 km away from Bhopal and popular as industrial township which came in to existence in late 1970s. It has an industrial area, the major industries that are closer are Hindustan Electro Graphite (HEG), Procter & Gamble, Eicher tracters Ltd, Lupin laboratories, and national and international level companies have their manufacturing units at Mandideep. It is located near River Betwa. River Betwa encircle this industrial area, thus, Mandideep was also selected as a study site in the present research work (5 to 6).

Station IV: Bhojpur is situated on the bank of Betwa River 28 km away from Bhopal, the state capital of MP. The site is located on sandstone ridges, typical of central India, next to a deep gorge through which the Betwa River follows (7 to 8).

Station V: Vidisha is a city in the state of Madhya Pradesh, India located near the state capital Bhopal. The city was also known as Bhilsa during the medieval period. The town is situated east of Betwa River in the fork of the Betwa and Bes River, 10 km from Sanchi (9 to 10).

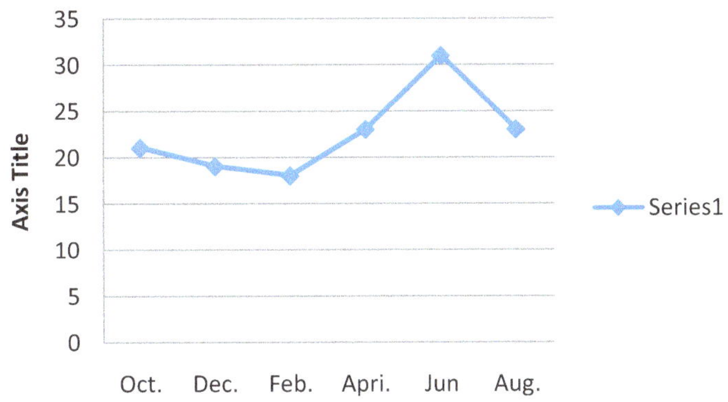

Graph I. Value of Temperature (°C) in different months.

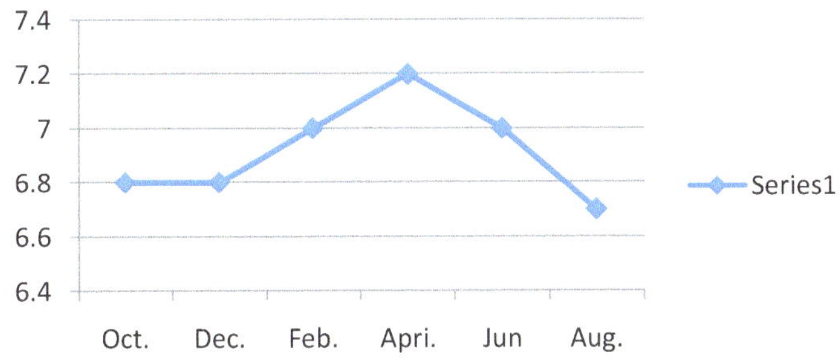

Graph II. Value of pH in different months.

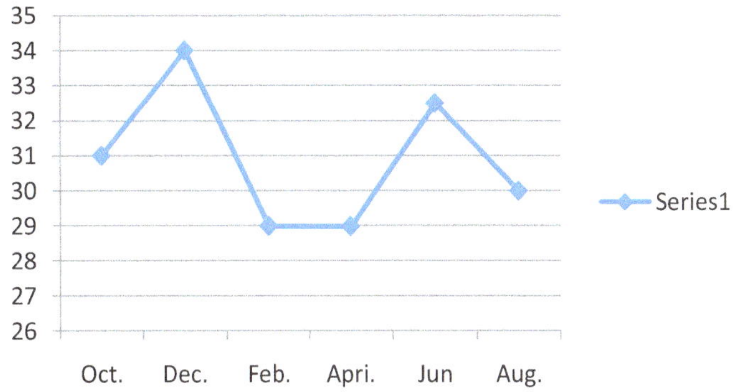

Graph III. Value of hardness (mg/L) in different months.

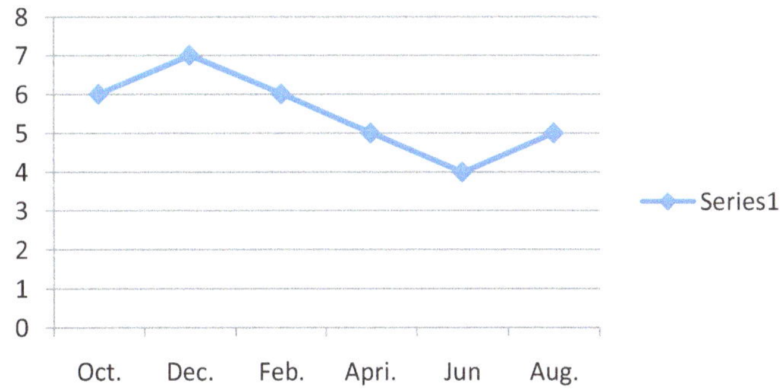

Graph IV. Value of Dissolved oxygen (mg/L) in different months.

Measurement of physico-chemical parameters

The physico-chemical parameters such as temperature, pH, dissolved oxygen, biological oxygen demand and chemical oxygen demand and hardness were measured as according to APHA (1976). The correlation between abiotic (physico-chemical) parameters was done by using coefficient of correlation Karl Pearson`s formula

$$r = \frac{\sum d_x d_y}{\sqrt{\sum d_{x^2} \sum d_{y^2}}}$$

RESULTS

Temperature influences the life of all biological organisms. During the period of study temperature was recorded ranging from 16 to 31°C (Graph I). pH expresses the intensity of acidity or alkalinity of an aquatic environment. The pH ranged from 5 to 9.2, (Graph II) the water was acidic at station II Nayapura and station III Mandideep - the change in pH values of station II and III. It was observed that pH decreased. Hardness is often used to assess the quality of water. Hard water contains large concentrations of alkaline earths dissolved from the drainage of calcium deposits (Wetzel, 1975). In the present study the observed values ranged from 14.6 to 58.6 mg/L (Graph III). Dissolved oxygen is very crucial for the survival of aquatic organism, Yakub and Ugwvmba,(2009). The DO ranged from 2 to 12.3 mg/L. minimum DO was recorded at station II and III Nayapura and Mandideep (Graph IV), and negative relation showed with BOD (r = - 0210). BOD value of river water ranged from 2 to 15 mg/L (Graph V) and high value of BOD was recorded at station II and III; BOD showed a significant positive correlation with COD (r = 0.314). In the present study the value of COD varies from 8 to 81.5 mg/L.

DISCUSSION

The temperature variation is mainly related with the temperature of atmospheric and weather condition (Adebowale et al., 2008). The reduction in the pH of River Betwa could have been due to the discharged industrial effluents. This result agrees with the reports by previous scientist (Edema et al., 2006; Rim-Rukeh et al., 2006; Dhakyanika and Kumara, 2010). The increase in hardness can be attributed to the decrease in water volume in the rate of evaporation at high temperature, high loading organic substances, detergent, chlorides and other pollutants (Rajgopal et al., 2010; Eruola et al., 2011) (Table 1). Minimum dissolved oxygen due to

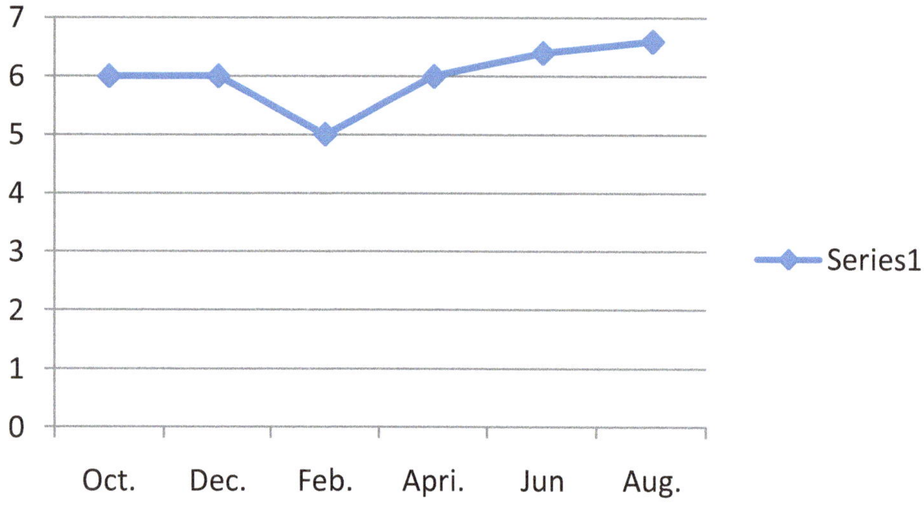

Graph V. Value of BOD (mg/L) in different months.

Table 1. Values of physico-chemical parameter at different sites of river.

Month	Site										
	1	2	3	4	5	6	7	8	9	10	M
						Temperature					
Oct.	21	20	20.5	21	21.4	22	20	21.1	22	22.4	21
Dec.	19	18	19.4	19	17.3	18.3	19.2	20	20.5	17.7	18.6
Feb.	16.3	17.5	17.2	18	16.5	18.5	19.2	17	19	18.5	17.5
Apri.	23	23.5	24.3	22	23	24	21	23	23.2	23.5	23
Jun.	30	30.2	31.3	29.5	32	31	31.6	30.2	32.2	31.2	31
Aug.	24	24.3	23	22	24	23	21	23.2	23	24.3	23.1
						pH					
Oct.	7.3	7.1	6.2	6.4	5.4	5.8	7.5	6.7	8.9	7	6.8
Dec.	7.4	7.8	5.5	6	5.8	6.3	7.4	8.3	6.8	7.5	6.8
Feb.	7.2	7	6.3	6.5	5.7	5.3	8.2	8	7.8	7.1	7
Apri.	8.1	8.4	6.2	6.1	6	5.5	7.6	6.7	7.3	8.2	7.2
Jun.	7	7.2	6.4	6.3	6.1	6	7.7	7.4	8.2	8.5	7
Aug.	8.8	8.4	5.6	5	5.3	5.5	8.1	7.4	6.8	7.2	6.7
						Hardness					
Oct.	16	16.2	40	40.4	52.4	50.5	20.2	20.5	22	22.6	30.8
Dec.	15	15.8	42.4	42.8	50.5	52.2	18.4	18.9	20.2	20.8	34
Feb.	14.2	14.8	38.2	36.8	48.4	48.6	20.4	20.6	24.5	24.4	29
Apri.	15.6	15.9	35.6	35.8	50.5	50.8	22.4	22.8	21.4	21.6	29.2
Jun.	18.4	18.8	36.2	36.5	58.4	58.6	24.4	24.6	24.8	24.9	32.5
Aug.	16	16.8	35.4	35.6	54.8	54.2	22	22.4	21.5	21.8	30
						Dissolved oxygen					
Oct.	8.4	8.2	2.1	2.2	2	1.8	7.6	8	8.2	7.5	5.6
Dec.	10.2	12.3	2.4	2	2.1	2.5	10	10.7	9.8	9.7	7
Feb.	8.8	9.9	2.2	2.	3	3.1	7.2	8.2	8	8.4	5.7
Apri.	7.1	7.2	2.4	2.3	2.4	2.1	7.8	8.3	7.5	7.1	5.4
Jun.	5.8	6.2	1.5	1.3	1.6	1.3	6.2	6.9	6.5	6.1	4.3
Aug.	6.8	6.5	2.2	2.4	3.2	3.4	7.6	7.8	6	6.5	5.2

Table 1. Contd.

					Biological oxygen demand						
Oct.	2.4	2.5	7.4	7.5	12.3	14.6	2.5	2.4	4.2	3.4	6
Dec.	3.4	2.2	8	8.5	10.2	12.3	4.7	4.4	2.8	2.5	6
Feb.	2.5	2.9	8.5	8	9	11.5	2.1	2.3	2.4	2.5	5.1
Apri.	3.2	4.4	10.5	10.2	9.8	10.4	3.5	4.8	2.5	2.8	6.2
Jun.	4.4	4.2	11	10.5	10.6	11.8	2.1	3.8	2.5	3.5	6.4
Aug.	3.5	3.8	9.4	9.2	12.1	14.6	2.5	2.6	4.4	4.2	6.6
					Chemical oxygen demand						
Oct.	10.1	11.4	42.3	42.2	70.5	71.2	12.5	12.2	11.6	11.5	29.5
Dec.	8.4	8.5	38.6	38.2	78.9	72.5	11.3	13.5	12.2	13.8	29.5
Feb.	9.5	9.8	39.2	40	60.6	64.2	10.6	10.5	10.8	10.4	26.5
Apri.	10.5	10.6	42.5	40.3	64.2	68.5	11.4	10.3	11.5	12.2	28.2
Jun.	12.5	12.2	43.3	41.4	82.5	82.2	10	10.4	11.5	12.4	28.8
Aug.	11.5	10.3	42.8	40.6	80.8	81.8	12.5	12.4	13.2	13.6	31.9

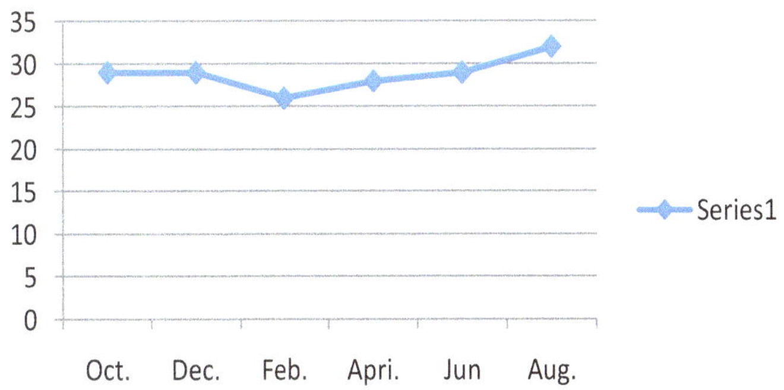

Graph VI. Value of COD (mg/L) in different months.and increase.

effluents discharge, as suggested by Emongor et al. (2005), and Agrawal and Saxena (2011), that the industries were releasing some organic substances that were of high oxygen demanding wastes. Biological oxygen demand is an important parameter which is widely used to determine the pollution load of waste water. The aim of BOD test is to determine the amount of bio-chemically oxidisable carbonaceous matter (Gupta et al., 2003; Azumi and Bichi, 2010). Values of BOD were due to higher rate of decomposition of organic matter at higher temperature, turbidity and less water current (Sanap et al., 2006). Chemical oxygen demand is a test which is used to measure pollution of domestic and industrial waste. This gives valuable information about the pollution potential of industrial effluents and domestic sewage (Gupta et al., 2003; Ijeoma and Achi, 2011). Highest value of COD indicates that most of the pollution in study zone in Betwa River is caused by industrial effluents discharged by industrial units. Similar results were also reported by Pande and Sharma (1998), Murhekar (2011) (Table 1, Graph VI).

ACKNOWLEDGEMENTS

The authors are thankful to Dr. Alok Varma and Dr. Geeta Saxena for their valuable suggestions and facilities provided during this work.

REFERENCES

Adebowale KO, Agunbiade FO, Olu-Owolabi BI (2008). Impacts of natural and anthropogenic multiple sources of pollution on the environmental conditions of Ondo State Costal Water Nigeria. J. Environ. Agric. Food Chem. 7(4):2797-2811.

Agrawal A, Saxena M (2011). Assessment of pollution by physicochemical water parameter using regression analysis: A case study of Gagan River at Moradabad, India. Adv. Appl. Sci. Res. 2(2):185-189.

Azumi DS, Bichi MH (2010). Industrial pollution and heavy metal profile of Challawa River in Kano, Nigeria. J. Appl. Sci. Environ. Sanit. 5(1):2, 3-29.

APHA (1976). American Public Health Association.

Edema MO, Omenu AM, Fapeta OM (2006). Microbiology and physiological analysis of different sources of drinking water in Abeokuta. Nig. J. Microb. 13(1):57-61.

Dhakyanika K, Kumara P (2010). Effect of pollution in river Krishni on hand pump water quality. J. Eng. Sci. Technol. Rev. 3(1):14-22.

Eruola AO, Ufoegbune GC, Awomeso JA, Adeofun CO, Idowu OA, Abhulimen SI (2011). An assessment of the effect of industrial pollution on Ibese River, Lagos, Nigeria. Afr. J. Environ. Sci. Technol. 5(8):608-615.

Gupta S, Bhatnagar M, Jain R (2003). Physico-chemical characteristics and analysis of Fe and Zn in tubewell water and sewage water of Bikaner City. Asian J. Chem. 15:727.

Ijeoma K, Achi OK (2011). Industrial effluents and their impact on water quality of receiving river in Nigeria. J. Appl. Sci. Environ. Sanit. 1:75-86.

Kori R, Shrivastava PK, Upadhyay N, Singh R (2006). Studies on presence of heavy metals and halogenated hydrocarbons in river Betwa (M.P), India. 2/3:147-153.

Murhekar GH (2011). Determination of physicochemical parameter of surface water sample in and around Akot city. Int. J. Res. Chem. Environ. 1(2):183-187.

Pande KS, Sharma SD (1998). Studies of toxic pollutants in Ramganga River at Moradabad India. Environ. Geo. 1(2):93-96.

Rim-Rukeh A, Ikhifa GO, Okokoyo PA (2006). Effects of agricultural activities on the water quality of Orogodo River, Agbor, Nig. J. Appl. Sci. Res. 2(5):256-259.

Rajgopal T, Thangamani A, Sevarkodiyone SP, Sekar M, and Archunan G (2010), Zooplankton diversity and physicochemical conditions in three perennial ponds of Virudhunagar district, Tamilnadu. J. Environ. Biol. 31:265-272.

Sanap RR, Mohite AK, Pingle SD, Gunale VR (2006). Evaluation of water qualities of Godawari River with reference to physicochemical parameters, district Nasik (M.S.) India. Pollut. Res. 25(4):775-778.

Yakub AS, Ugwumba AAA (2009). Study on the macroinvertebrates fauna of cower Ogun river at Ishasi, Ogun State - south west Nigeria. Zoologist 7:65-74.

Wetzel RG (1975). Limnology. WB. Saundrs Co., Philadelphia, p. 980.

Ward AD, Elliot WJ (1995). Environmental Hydrology. Lewis publishers, Boca Raton, FL.P. pp. 404-407.

Permissions

All chapters in this book were first published in IJWREE, by Academic Journals; hereby published with permission under the Creative Commons Attribution License or equivalent. Every chapter published in this book has been scrutinized by our experts. Their significance has been extensively debated. The topics covered herein carry significant findings which will fuel the growth of the discipline. They may even be implemented as practical applications or may be referred to as a beginning point for another development.

The contributors of this book come from diverse backgrounds, making this book a truly international effort. This book will bring forth new frontiers with its revolutionizing research information and detailed analysis of the nascent developments around the world.

We would like to thank all the contributing authors for lending their expertise to make the book truly unique. They have played a crucial role in the development of this book. Without their invaluable contributions this book wouldn't have been possible. They have made vital efforts to compile up to date information on the varied aspects of this subject to make this book a valuable addition to the collection of many professionals and students.

This book was conceptualized with the vision of imparting up-to-date information and advanced data in this field. To ensure the same, a matchless editorial board was set up. Every individual on the board went through rigorous rounds of assessment to prove their worth. After which they invested a large part of their time researching and compiling the most relevant data for our readers.

The editorial board has been involved in producing this book since its inception. They have spent rigorous hours researching and exploring the diverse topics which have resulted in the successful publishing of this book. They have passed on their knowledge of decades through this book. To expedite this challenging task, the publisher supported the team at every step. A small team of assistant editors was also appointed to further simplify the editing procedure and attain best results for the readers.

Apart from the editorial board, the designing team has also invested a significant amount of their time in understanding the subject and creating the most relevant covers. They scrutinized every image to scout for the most suitable representation of the subject and create an appropriate cover for the book.

The publishing team has been an ardent support to the editorial, designing and production team. Their endless efforts to recruit the best for this project, has resulted in the accomplishment of this book. They are a veteran in the field of academics and their pool of knowledge is as vast as their experience in printing. Their expertise and guidance has proved useful at every step. Their uncompromising quality standards have made this book an exceptional effort. Their encouragement from time to time has been an inspiration for everyone.

The publisher and the editorial board hope that this book will prove to be a valuable piece of knowledge for researchers, students, practitioners and scholars across the globe.

List of Contributors

N. B. Abdelmageed
Faculty of Engineering, Shobra, Benha University, Egypt

K AMU-MENSAH Frederick
CSIR Water Research Institute, Ghana

YAMAMOTO Tahei
Arid Land Research Center, Tottori University, Japan

INOUE Mitsuhiro
Arid Land Research Center, Tottori University, Japan

Sudevi Basu
Department of Biotechnology, Sir MVisvesvaraya Institute of Technology, Bangalore - 562157, Karnataka

K S. Lokesh
Department of Environmental Engineering, Sri Jayachamarajendra College of Engineering, Mysore - 570006, Karnataka

YAMAMOTO Tahei
Arid Land Research Center, Tottori University, JAPAN

INOUE Mitsuhiro
CSIR Water Research Institute, Ghana

A. C. C. Ezeabasili
Heriot Watt University, England, UK

O. L. Anike
Nnamdi Azikiwe University, Awka, Anambra State, Nigeria

A.U. Okonkwo
Anambra State University, Uli, Anambra State, Nigeria

Mohammed H. I. Dore
Brock University, St. Catharines, Ontario, Canada

Rajiv G. Singh
Brock University, St. Catharines, Ontario, Canada
Bloomsburg University of Pennsylvania, United States

Arian Khaleghi-Moghadam
Brock University, St. Catharines, Ontario, Canada

Gopal Achari
University of Calgary, Alberta, Canada

S. Shah
Department of Civil Engineering, Birla Vishvakarma Mahavidyalaya Engineering College (BVM, Vallabhviyanagar),Gujarat India

H. J. Dalwadi
Department of Civil Engineering, Birla Vishvakarma Mahavidyalaya Engineering College (BVM, Vallabhviyanagar),Gujarat India

Girish Chopra
Department of Zoology, Kurukshetra University Kurukshetra-136119 India

Anita Bhatnagar
Department of Zoology, Kurukshetra University Kurukshetra-136119 India

Priyanka Malhotra
Department of Zoology, Kurukshetra University Kurukshetra-136119 India

Mohammad Hafizul Islam
Public Works Department, Sub-Divisional Engineer, Bangladesh

Md. Mafizur Rahman
Department of Civil Engineering, BUET, Bangladesh

Fahmidah Ummul Ashraf
Water Resources Engineering, BUET, Bangladesh

Bloodless Dzwairo
Tshwane University of Technology, Faculty of Engineering and the Built Environment, Civil Engineering Department, Private Bag X680, Pretoria, 0001, South Africa

Fred A. O. Otieno
Durban University of Technology, Deputy Vice Chancellor for Technology, Innovation and Partnerships,P. O. Box 1334, 4000, South Africa

George M. Ochieng
Tshwane University of Technology, Faculty of Engineering and the Built Environment, Civil Engineering Department, Private Bag X680, Pretoria, 0001, South Africa

Sati M. Al-Rawi
Mosul University Technical College/Kirkuke Iraq

Kahtan A. Al-Najjar
Mosul University Technical College/Kirkuke Iraq

Sule Magaji
Department of Economics, Faculty of Social Sciences, University of Abuja, Abuja Nigeria

Eke, Chukwuemeka Ifegwu
Department of Economics, Faculty of Social Sciences, University of Abuja, Abuja Nigeria

A. A Koinyan
Department of Geography and Environmental Management, University of Port Harcourt, Nigeria

H. O. Nwankwoala
Department of Geology, College of Natural and Applied Sciences, University of Port Harcourt, Nigeria

O. S. Eludoyin
Department of Geography and Environmental Management, University of Port Harcourt, Nigeria

B. V. N. P. Kambhammettu
New Mexico Water Resources Research Institute, MSC 3167, P. O. Box. 30001, Las Cruces,NM-88003, U.S.A

James P. King
Civil Engineering Department, New Mexico State University, Las Cruces, NM-88003, U.S.A

Praveena Allena
Office of the State Engineer, 301 S Tin St., Deming, NM-88030, U.S.A

Lokesh Kumar
120 MGD Water Treatment Plant, Water Works Wazirabad, Delhi Jal Board, Municipal Water Supply Department, Government of NCT Delhi, Delhi 110054, India

S. Olu Adesogan
Civil Engineering Department, Federal University, Oye-Ekiti, Ekiti State, Nigeria

Aloyce W. Mayo
Department of Water Resources Engineering, University of Dar es Salaam, Tanzania

Marwa Muraza
Department of Water Resources Engineering, University of Dar es Salaam, Tanzania

Joel Norbert
Department of Water Resources Engineering, University of Dar es Salaam, Tanzania

S Mariraj Mohan
Civil Engineering Department, Alagappa Chettiar College of Engineering and Technology, Karaikudi, Tamilnadu, India

P Vanalakshmi
Civil Engineering Department, Kalasalingam University, Srivilliputhur, Virudhunagar, India

Abdeen Mustafa Omer
Energy Research Institute (ERI), UK

O. D. Mireku
Department of Geography and Regional Planning, University of Cape Coast, Cape Coast, Ghana

A.K. Mensah
Department of Geography, Kenyatta University, Nairobi, Kenya

Binyam Alemu Yosef
Department of Soil and Water Resources Management, College of Agriculture, Wollo University, Ethiopia

Desale Kidane Asmamaw
Department of Geography, College of Social Science, Samara University, Ethiopia

I. M Sujaul
School of Environment and Natural Resource Sciences, Faculty of Science and Technology, Universiti Kebangsaan Malaysia, 43600 UKM Bangi, Selangor, Malaysia

B. S Ismail
School of Environment and Natural Resource Sciences, Faculty of Science and Technology, Universiti Kebangsaan Malaysia, 43600 UKM Bangi, Selangor, Malaysia

G Muhammad Barzani
School of Environment and Natural Resource Sciences, Faculty of Science and Technology, Universiti Kebangsaan Malaysia, 43600 UKM Bangi, Selangor, Malaysia

A. R Sahibin
School of Environment and Natural Resource Sciences, Faculty of Science and Technology, Universiti Kebangsaan Malaysia, 43600 UKM Bangi, Selangor, Malaysia

T. Mohd Ekhwan
School of Social Development and Environmental Studies, Faculty of Social Science and Humanities, Universiti Kebangsaan Malaysia, 43600 UKM Bangi, Selangor, Malaysia

N. K. Sharat Singh
Department of Physics, Manipur University, Canchipur - 795 003, Manipur, India

Ch. Bino Devi
Department of Chemistry, Manipur University, Canchipur - 795 003, Manipur, India

M. Sudarshan
UGC-DAE Consortium for Scientific Research, Kolkata Centre, Bidhan Nagar, Kolkata - 700 098, India

N. Sanamacha Meetei
Ecology and Enviroment Wing, Government of Manipur, Porompat - 795 005, India

T. Brajakumar Singh
Ecology and Enviroment Wing, Government of Manipur, Porompat - 795 005, India

N. Rajmuhon Singh
Department of Chemistry, Manipur University, Canchipur - 795 003, Manipur, India

Pranjit Kumar Sarma
Department of Geography, Mangaldoi College, Assam, India

Kalyanjit Sarmah
North Eastern Hill University, Shillong

Prem Kr.Chetri
North Orissa University, Orissa

Ankita Sarkar
Gauhati University, Assam, India

N. Zarrineh
Department of Hydrology and Hydrological Engineering, Vrije Universiteit Brussel (VUB), Brussels Belgium

M. Azari Najaf Abad
Department of Hydrology and Hydrological Engineering, Vrije Universiteit Brussel (VUB), Brussels Belgium

Santosh Vishwakarma
Department of Zoology, Government Science and Commerce College, Benazeer, Bhopal, India

Alok Varma
Department of Zoology, Government Raja Bhoj College, Mandideep Raisen, India

Geeta Saxena
Department of Zoology, Government Science and Commerce College, Benazeer, Bhopal, India